T0362711

Technical Drawing

THIRD EDITION

An Australian Course in Graphics

A. W. Boundy
AssDipMechEng, MPhil,
MIIE Aust., MIIE USA

I. L. Hass
Subject Master (Retired)
Manual Arts

NELSON
CENGAGE Learning™

Australia • Brazil • Japan • Korea • Mexico • Singapore • Spain • United Kingdom • United States

NELSON
CENGAGE Learning

Technical Drawing
An Australian Course in Graphics
3rd Edition
A. W. Boundy
I. L. Hass

Sponsoring editor: Matthew Sandblom
Production editor: Christine Eslick
Design: George Sirett
Cartoonist: Greg Gaul
Reprint: Jess Lovell

Any URLs contained in this publication were checked for currency during the production process. Note, however, that the publisher cannot vouch for the ongoing currency of URLs.

First published 1974.
Second edition 1981.
Reprinted 1982 (twice), 1983, 1984, 1985, 1986, 1987, 1988.
Third edition 1990 by McGraw Hill Australia.
Reprinted 1990, 1991, 1993, 1994, 1995, 1996, 1998, 1999, 2002, 2003, 2004, 2006, 2007, 2008 by McGraw Hill Australia.
This edition published in 2010 by Cengage Learning Australia.

© 2010 Cengage Learning Australia Pty Limited

Copyright Notice
This Work is copyright. No part of this Work may be reproduced, stored in a retrieval system, or transmitted in any form or by any means without prior written permission of the Publisher. Except as permitted under the Copyright Act 1968, for example any fair dealing for the purposes of private study, research, criticism or review, subject to certain limitations. These limitations include: Restricting the copying to a maximum of one chapter or 10% of this book, whichever is greater; providing an appropriate notice and warning with the copies of the Work disseminated; taking all reasonable steps to limit access to these copies to people authorised to receive these copies; ensuring you hold the appropriate Licences issued by the Copyright Agency Limited ("CAL"), supply a remuneration notice to CAL and pay any required fees. For details of CAL licences and remuneration notices please contact CAL at Level 15, 233 Castlereagh Street, Sydney NSW 2000, Tel: (02) 9394 7600, Fax: (02) 9394 7601
Email: info@copyright.com.au
Website: www.copyright.com.au

For product information and technology assistance,
in Australia call **1300 790 853**;
in New Zealand call **0800 449 725**

For permission to use material from this text or product, please email **aust.permissions@cengage.com**

National Library of Australia Cataloguing-in-Publication Data
Boundy, A. W. (Albert William).
Technical drawing.

3rd ed.
9780170218092

1. Mechanical drawing. I. Hass, I. L. (Ian Lloyd). II. Title.

604.2'4

Cengage Learning Australia
Level 7, 80 Dorcas Street
South Melbourne, Victoria Australia 3205

Cengage Learning New Zealand
Unit 4B Rosedale Office Park
331 Rosedale Road, Albany, North Shore 0632, NZ

For learning solutions, visit **cengage.com.au**

Printed in Australia by Ligare Pty Limited.
2 3 4 5 6 7 15 14 13 12

Contents

Preface

This book has been written for students of technical drawing. It has been designed to give sound educational training in the important fundamentals of technical drawing without any specified bias towards one particular vocation.

The course has been divided into twelve sections: Introductory and Standards Information; Geometrical Constructions, Charts and Diagrams; Descriptive Geometry; Orthogonal Projection; Drawing Analysis; Primary Auxiliary Views; Pictorial Drawing; Working Drawings; Surveying and Setting Out; Cabinet Drawing; Architectural Drawing and Light Construction; and Developments and Intersections of Solids. Each section of the book has been given thorough coverage, with a large number of exercises for each section. Practice gained from solving these exercises should make the students better drafters, and broaden their knowledge and understanding of technical drawing.

We have tried at all times to conform with the latest Australian Drawing Standards.

Introductory and standards information

Engineering drawing is the main method of communication between all persons concerned with the design and manufacture of components; the building and construction of works; and engineering projects required by management or professional engineering staff.

The practice of drawing is in many ways so repetitive that, in the interests of efficient communication, it is necessary to standardise methods to ensure the desired interpretation.

Standards Australia has recommended standards for drawing practice in all fields of engineering, and these are set out in their publications Australian Standards (AS) 1100 (Technical Drawing) Parts 101 (General principles) and 201 (Mechanical drawing).

This section presents the standards that are relevant to mechanical drawing and provides other introductory information that is often required by drafters and students.

Standard abbreviations

The abbreviations in Table 1.1 have been selected from the more comprehensive list found in *AS 1100* Part 101 and are those that are commonly used on mechanical engineering drawings.

Table 1.1 *Standard abbreviations*

Term	Abbreviation	Term	Abbreviation
A		dimension	DIM
abbreviation	ABBR	distance	DIST
absolute	ABS	distance point	DP
across flats	AF	drawing	DRG
addendum	ADD	**E**	
approximate	APPROX	elevation	ELEV
arrangement	ARRGT	equivalent	EQUIV
assembly	ASSY	external	EXT
assumed datum	ASSD	eye	E
automatic	AUTO	**F**	
auxiliary	AUX	figure	FIG
average	AVG	fillister head	FILL HD
B		flange	FLG
bearing	BRG	flat	FL
bottom	BOT	**G**	
bracket	BRKT	galvanise	GALV
brass	BRS	galvanised iron	GI
building	BLDG	galvanised-iron pipe	GIP
C		general arrangement	GA
capacity	CAP	general-purpose outlet	GPO
cast iron	CI	geometric reference frame	GRF
cast-iron pipe	CIP	grade	GR
cast steel	CS	grid	GD
centre line	CL	ground line	GL
centre of gravity	CG	ground plane	GP
centre of vision	CV	**H**	
centre-to-centre, centres	CRS	head	HD
chamfer	CHAM	height	HT
channel	CHNL	hexagon	HEX
cheese head	CH HD	hexagon head	HEX HD
chrome plated	CP	hexagon-socket head	HEX SOC HD
circle	CIRC	high strength	HS
circular hollow section	CHS	high-tensile steel	HTS
circumference	CIRC	horizontal	HORIZ
coefficient	COEF		
cold-rolled steel	CRS	**I**	
computer-aided design and drafting	CAD	inside diameter	ID
computer-aided manufacture	CAM	internal	INT
concentric	CONC	**J**	
contour	CTR	joint	JT
corner	CNR	junction	JUNC
counterbore	CBORE		
countersink	CSK	**L**	
countersunk head	CSK HD	least material condition	LMC
cross-recess head	C REC HD	left hand	LH
cup head	CUP HD	length	LG
cylinder	CYL	longitudinal	LONG
D		**M**	
dedendum	DED	machine	M/C
detail	DET	malleable iron	MI
diagonal	DIAG	material	MATL
diagram	DIAG	maximum	MAX
diameter	DIA	maximum material condition	MMC
diametral pitch	DP	measuring point	MP
diamond pyramid hardness number (vickers)	HV	mechanical	MECH

Term	Abbreviation	Term	Abbreviation
mild steel	MS	Rockwell hardness C	HRC
minimum	MIN	rolled-hollow section	RHS
modification	MOD	rolled-steel angle	RSA
modulus of elasticity	E	rolled-steel channel	RSC
modulus of section	Z	rolled-steel joist	RSJ
moment of inertia	I	roughness value	R_a
mounting	MTG	round	RD
mushroom head	MUSH HD	round head	RD HD
N		**S**	
negative	NEG	schedule	SCHED
nominal	NOM	section	SECT
nominal size	NS	sheet	SH
not to scale	NTS	sketch	SK
number	NO	spherical	SPHER
		spigot	SPT
O		spotface	SF
octagon	OCT	spring steel	SPR STL
outside diameter	OD	square	SQ
		square hollow section	SHS
P		stainless steel (corrosion-resistant steel)	CRES
parallel	PAR	standard	STD
part	PT	Standards Association of Australia	SAA
pattern	PATT	steel	ST
picture plane	PP	switch	SW
pipe	P		
pipeline	PL	**T**	
pitch-circle diameter	PCD	tangent point	TP
phosphor bronze	PH BRZ	temperature	TEMP
position	POSN	thread	THD
positive	POS	tolerance	TOL
prefabricated	PREFAB	true position	TP
pressure	PRESS	true profile	TP
pressure angle	PA		
		U	
Q		undercut	UCUT
quantity	QTY	universal beam	UB
		universal column	UC
R			
radius	RAD	**V**	
raised countersunk head	RSD CSK HD	vanishing point	VP
rectangular	RECT	vertical	VERT
rectangular hollow section	RHS	volume	VOL
reference	REF		
regardless of feature size	RFS	**W**	
required	REQD	wrought iron	WI
right hand	RH		
Rockwell hardness A	HRA	**Y**	
Rockwell hardness B	HRB	yield point	YP

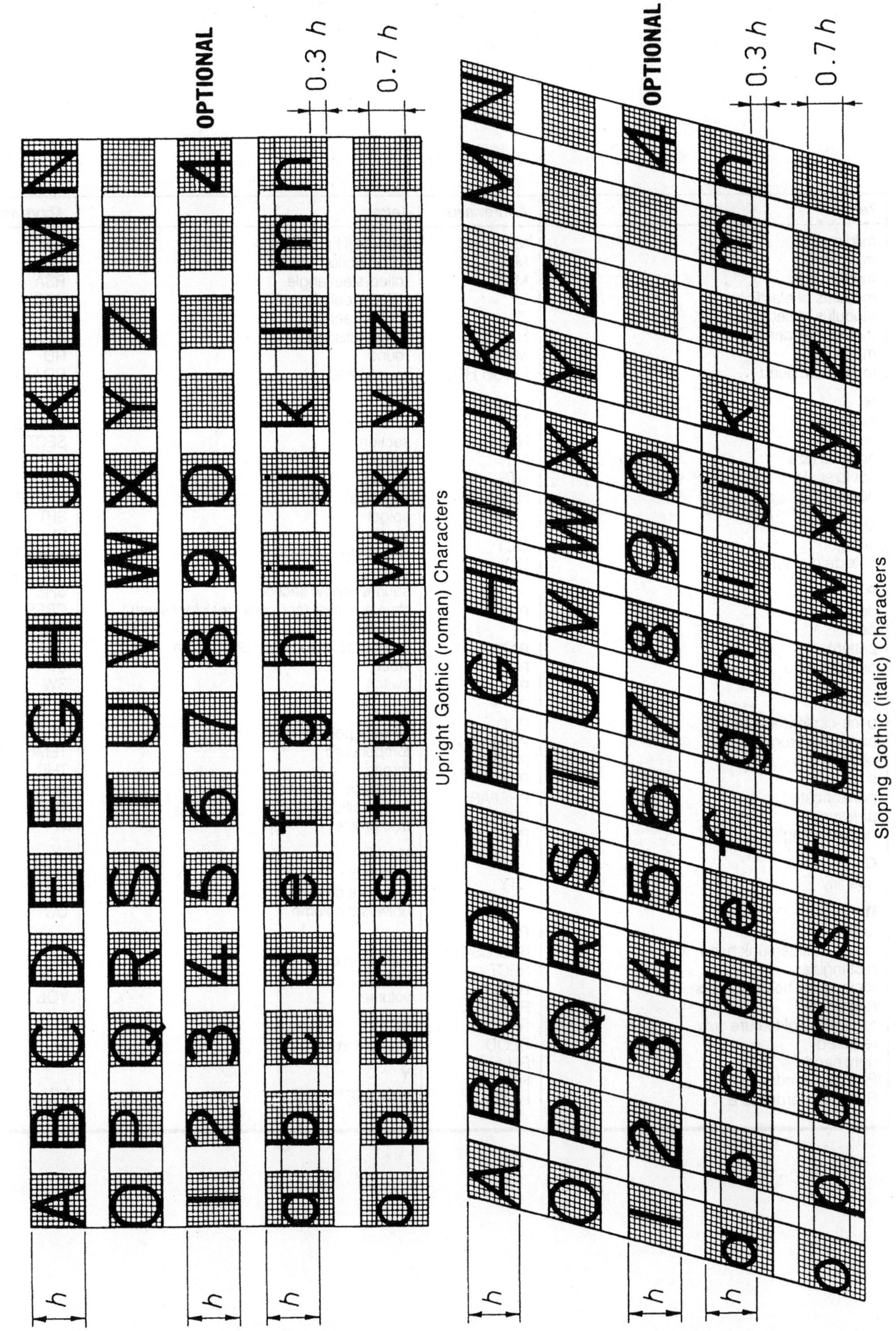

ABCDEFGHIJKLMN
OPQRSTUVWXYZ
1234567890
abcdefghijklmn
opqrstuvwxyz

OPTIONAL
4

0.3 h
0.7 h

h
h
h

Upright Gothic (roman) Characters

ABCDEFGHIJKLMN
OPQRSTUVWXYZ
1234567890
abcdefghijklmn
opqrstuvwxyz

OPTIONAL
4

0.3 h
0.7 h

h
h
h

Sloping Gothic (italic) Characters

Fig. 1.1 *Lettering and numerals suggested by Standards Australia*   COPYRIGHT

Letters and numerals

Standards Australia has established the following standards for lettering and numerals.

1. Letters and especially numbers, which do not often occur in self-identifying patterns, should be drawn clearly and their style, spacing and size should be consistent (Fig. 1.2).
2. The letters and numerals in Figure 1.1 are suggested for use. Capital letters should be used.
3. Letters and numbers used for dimensions and notes should not be less than 3.5 mm high and titles and drawing numbers are usually larger. The following heights (H) are suggested for use: 2.5, 3.5, 5, 7, 10, 14 and 20 mm.
4. Notes and captions should be positioned to be read in the same direction as the title block.
5. To draw attention to a note or caption, use larger letters. Do not underline.

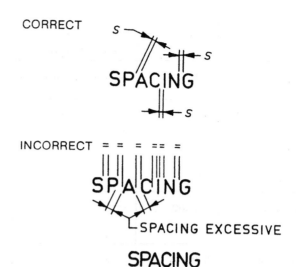

Fig. 1.2 *Spacing of characters*

Types of line

The types of line commonly used in engineering drawings are illustrated in Table 1.2.

Figure 1.3 includes examples of nine types of lines, lettered to correspond with those in Table 1.2 (with the exception of type F).

1. The *visible outline* of the bracket, type A, is heavy and dark enough to make it stand out clearly on the drawing sheet. This line should be of even thickness and darkness.
2. The *dimension, projection, cross-hatching* and *leader line*, type B, is illustrated. Leader lines are of

two types, one that terminates with an arrowhead at an outline and the other that terminates in a dot (4) within the outline of the part to which it refers. Leaders should be nearly at right angles to any line or surface. Further uses of type B lines are to partly outline the adjacent part to which the bracket is bolted and to represent fictitious outlines such as minor diameters of male threads and major diameters of female threads (the latter are not illustrated).

3. The *short break line*, type C, is drawn freehand to terminate part views and sections as shown. It is also

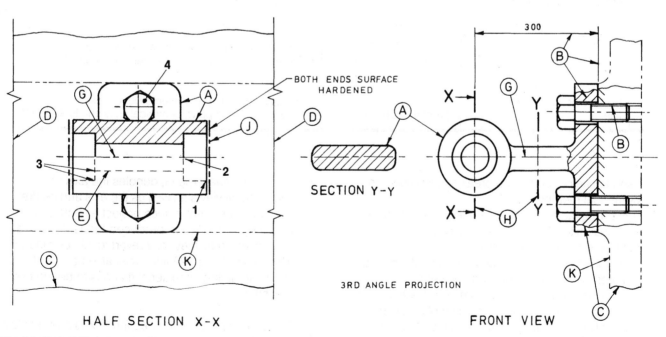

Fig. 1.3 *Use of different types of line*

Table 1.2 *Types of lines*

Type	Description	Drawing example	Usage
A	continuous thick line	——————————	to indicate visible outlines
B	continuous thin line	————————	for fictitious outlines, dimensions, projection, hatching and leader lines; also for the imaginary intersection of surfaces, revolved sections, adjacent parts, fold and tangent bend lines, short centre lines, and for indicating repeated detail
C	continuous thin freehand line	∼∼∼∼∼∼	on part sectional boundary lines or to terminate a part view, and for short break lines
D	continuous thin ruled line with intermittent zigzag	—∧—∧—	to show a break on an adjacent member to which a component is attached; also to indicate a break in a long continuous series of lines on architectural or structural drawings
E	thin dashed line	⊢s⊣ ⊢q⊣ – – – – – – s = 1 mm minimum q = 2s to 4s	to show outlines of hidden features: • for complete hidden features, the line should begin and end with a dash • dashes should meet at corners • where a hidden line is a continuation of a visible outline, it should commence with a space
F	medium dashed line	— — — — — — — (proportions as for E)	in electrotechnology drawing only; for assemblies, boxes and other containers
G	thin chain line	s⊢ ⊣q⊢ ⊣ p ⊢ —·—·—·— s = 1 mm minimum q = 2s to 4s p = 3q to 10q	to indicate centre lines, pitch lines, path movement, developed views, material for removal and features in front of a cutting plane
H	chain line, thick at the ends and at change of direction but thin elsewhere	— · — · — · — (proportions as for G)	to indicate a cutting plane for sectional views
J	thick chain line	— · — · — · — (proportions as for G)	to indicate surfaces that must comply with certain requirements such as heat treatment or surface finish
K	thin double-dashed chain line	— · · — · · — (proportions as for G)	to indicate adjacent parts, alternative and extreme positions of moving parts, centroidal lines and tooling profiles

used to sketch the curved break section used on cylindrical members.

4. A *ruled zigzag line*, type D, is used for long break lines that extend a short distance beyond the outlines on which they terminate.

5. The *hidden outline line*, type E, represents internal features which cannot normally be seen.

   A hidden outline should commence with a dash (1) except where it is a continuation of a visible outline (2), where there is a space first. Corners and junctions (3) should be formed by dashes.

6. The *centre line*, type G, denotes the axis of symmetrical views as well as the axis and centre lines of holes. Centre lines project a short distance past the outline. When produced further for use as dimension lines, they may revert to thin continuous (type B) lines. Type G lines may also be used to show the outline of material that has to be removed (not shown).

7. The *cutting plane of the section* X–X is represented by the type H line. Arrows are located at right angles

to the thick ends of the line, and point to the direction in which the sectional view is being taken.

In the case of the removed section Y–Y which merely shows the cross-sectional shape of the member, it is immaterial which direction the view is taken from, and the arrows may be left off the cutting plane.

8. *Surfaces requiring special treatment* such as heat treatment or surface finish may be indicated with a type J line drawn parallel to the profile of the surface in question.

9. Where it is necessary to show the relationship of a component to an adjacent part, the latter is outlined using a type K line. This type of line is also used to indicate extreme positions of moveable parts, and to outline tooling profiles in relation to work set up in machine tools.

Scales

The scales recommended for use with the metric system are:
1. full size   1:1
2. enlargement   2:1, 5:1, 10:1
3. reduction   1:2, 1:2.5, 1:5, 1:10

Use of scales

Technical drawings may be prepared full size, enlarged or reduced in size. Whatever scale is used, it is important that it be noted in or near the title block.

Indication of scales

When several scales are used, they should be shown close to the view(s) to which they refer and a note in the title block should read, 'scales as shown'.

If a drawing has predominantly one scale, the main scale should be shown in the title block, together with the notation 'or as shown' to indicate the use of other scales elsewhere on the drawing.

Sometimes it is necessary to use different scales on the one view, for example on a structural steel truss where the cross-sections of members are drawn to a larger scale than the overall dimensions of the truss. Such variations are indicated on the drawing, for example:

> Scales
> Member cross-sections   1:10
> Truss dimensions   1:100

If a particular scale has to be used on a drawing, it may be shown by one of the following methods:
1. a scale shown on the drawing, for example:

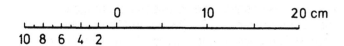

2. the word 'scale' followed by the appropriate ratio, for example 'scale 1:10';
3. the words 'scale:none' in or near the title block, for example on pictorial drawings.

Line thicknesses

Thicknesses for the various types of line are divided into specific groups according to the size of drawing sheet being used. Table 1.3 shows the metric sheet size, the line type and thickness applicable in each case.

Sizes of drawing paper
Preferred sheet series

Standards Australia has recommended that paper sizes be based on the A series of the International Organization for Standardization (ISO), and these sizes are specified in *AS 1100* Part 101. This series is particularly suitable for reduction onto 35 mm microfilm because the ratio of $1:\sqrt{2}$ is constant for the sides of the paper (Fig. 1.4(a)) and this ratio is also used for the microfilm frame.

Paper sizes are based on the A0 size, which has an area of 1 m². This allows paper weights to be expressed in grams per square metre.

Table 1.3 *Line thicknesses for various sheet sizes*

Sheet size	Line type and thickness (mm)									
	A	B	C	D	E	F	G	H	J	K
A0	0.7	0.35	0.35	0.35	0.35	0.5	0.35	0.35 0.7	0.7	0.35
A1	0.5	0.25	0.25	0.25	0.25	0.35	0.25	0.25 0.5	0.5	0.25
A2, A3, A4	0.35	0.18	0.18	0.18	0.18	0.25	0.18	0.18 0.35	0.35	0.18

The relationship between the various paper sizes is illustrated in Figure 1.4(a) and (b), where the application of the $1:\sqrt{2}$ side ratio can be seen. An A0 size sheet can be divided up evenly into the various other sizes simply by halving the sheet on the long side in each case. This is shown in Figure 1.4(c). The dimensions of metric sheets from size A0 to size A4 are given in Table 1.4, together with appropriate border widths for each sheet size.

Non-preferred sheet series

The B series of sheet sizes provides for a range of sheets designated B1, B2, B3, B4 and so on, which are intermediate between the A sizes. The relationship of the B and A sizes is shown in Figure 1.4(b); B sizes are in broken outline.

Rolls

The standard widths of rolls are 860 mm and 610 mm. Drawing sheets can be cut off the roll to suit individual drawings.

Layouts of drawing sheets

Standard layouts for drawing sheets of all sizes are given in *AS 1100* Part 101. Figures 1.5 and 1.6 show typical layouts of A1 and A2 sheets, illustrating the paper size, drawing frame with microfilm camera alignment marks and zoning or grid referencing details. Figure 1.6 also includes a parts list and a revisions table. The layout of Figure 1.5 is suitable for detail drawings, and that of Figure 1.6 is suitable for multidetail and assembly drawings.

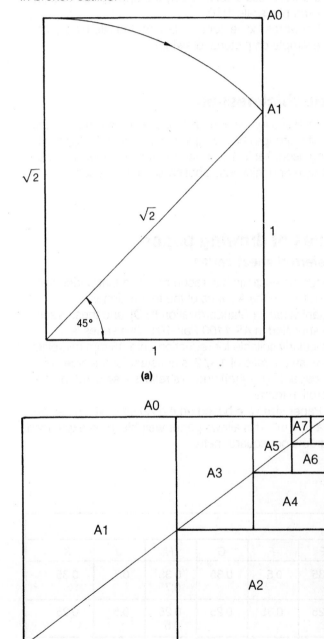

(a)

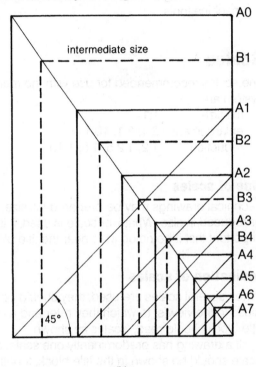

(b)

(c)

(d)

Fig. 1.4 *Paper sizes*

Table 1.4 *Drawing frames without a filing margin (see Figure 1.4(d))*

| Paper size | Border width (mm) | | Dimensions of drawing frame (mm) |
| | both sizes | top and bottom | |
	a	b	A × B
A0	28	20	1133 × 801
A1	20	14	801 × 566
A2	14	10	566 × 400
A3	10	7	400 × 283
A4	7	5	283 × 200

Sheet frames (borderlines)

It is usual for each sheet to be provided with a drawing frame a short distance in from the edge of the paper. Drawing frames are standardised for the various sizes of paper, and Figure 1.4(d) details this information for frames without a filing margin.

Title block

The title block represents the general information source for a drawing. It is normally placed in the bottom right-hand corner of the drawing frame. Figure 1.7 illustrates title block dimensions for various sheet sizes, together with the type of information that should be contained in the title block and its location.

Material or parts list

When a drawing includes a number of parts on the one sheet, or when an assembly drawing of a number of parts is required, a tabulated list of the parts is attached to the top of the title block and against the right-hand drawing frame, as shown in Figure 1.6.

The list should give the following information:
1. the item or part number
2. the part name
3. the quantity required
4. the material and its specification
5. the drawing number of each individual part
6. other information considered appropriate

A separate standard drawing sheet may be used to set out a parts list alone when it is desirable or when the list is very large. Such a list should be provided with a standard title block and a revisions table. For further details see *AS 1100* Part 101.

Revisions table

A table of revisions is located normally in the upper right-hand corner of the drawing frame, as shown in Figure 1.6.

The ability to effect revisions or modifications to existing drawings is an important requirement in all drawing and design offices. In many instances when only minor modifications are required, it is much easier to revise an existing drawing than to create a new one. However, such modifications must be tabulated to record existing details of the feature as well as the modification.

Each change should be identified by a symbol, such as a letter or number placed close to the revision on the drawing. The letter or number may (but need not) be encircled on the drawing. Reference is made to the symbol in the tabulated details of the change (see Figs 8.1 and 8.2). Drawings so modified should be given a new issue number or letter, which should be situated in the title block adjacent to the drawing number.

If a particular modification affects the interchangeability of a part, the modified part should be allocated a new drawing number.

Zoning

Drawings may be divided into zones by a grid reference system based on numbers and letters as shown in Figures 1.5 and 1.6. Zoning is located inside the drawing frame.

The purpose of a grid reference system is to assist location of detail. It is particularly useful on large drawings.

Horizontal zones are designated by capital letters, starting with A, reading from top to bottom. Vertical zones are designated by numbers reading from left to right.

The number of zones and widths of zone margins to be used on various sheet sizes are detailed in Table 1.5. Further use of zoning is shown in Figure 8.2, where a revision in the table designates a change of thread form (Whitworth to metric), and the reference C2 is a grid reference indicating the position on the sectional view of the thread in question, that is symbol Ⓐ.

Table 1.5 *Details of grid references*

| Detail | Size of drawing | | | | |
	A0, B1	A1, B2	A2, B3	A3, B4	A4
Number of vertical zones designated (1, 2, etc.)	16	12	8	6	4
Number of horizontal zones designated (A, B, etc.)	12	8	6	4	4
Width of margins for grid reference (mm)	10	7	7	5	5

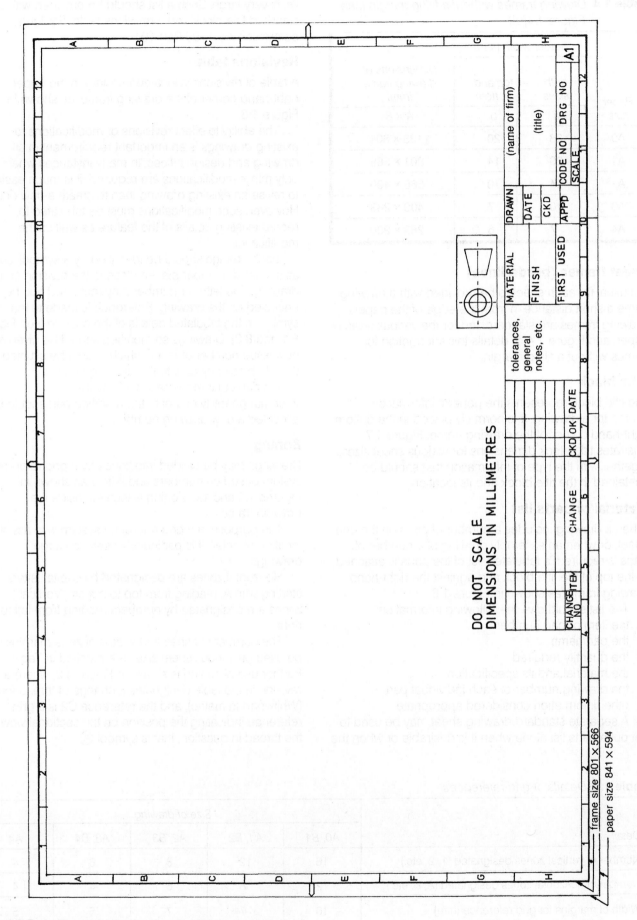

Fig. 1.5 *Typical layout of a drawing sheet (suitable for detail drawings)*

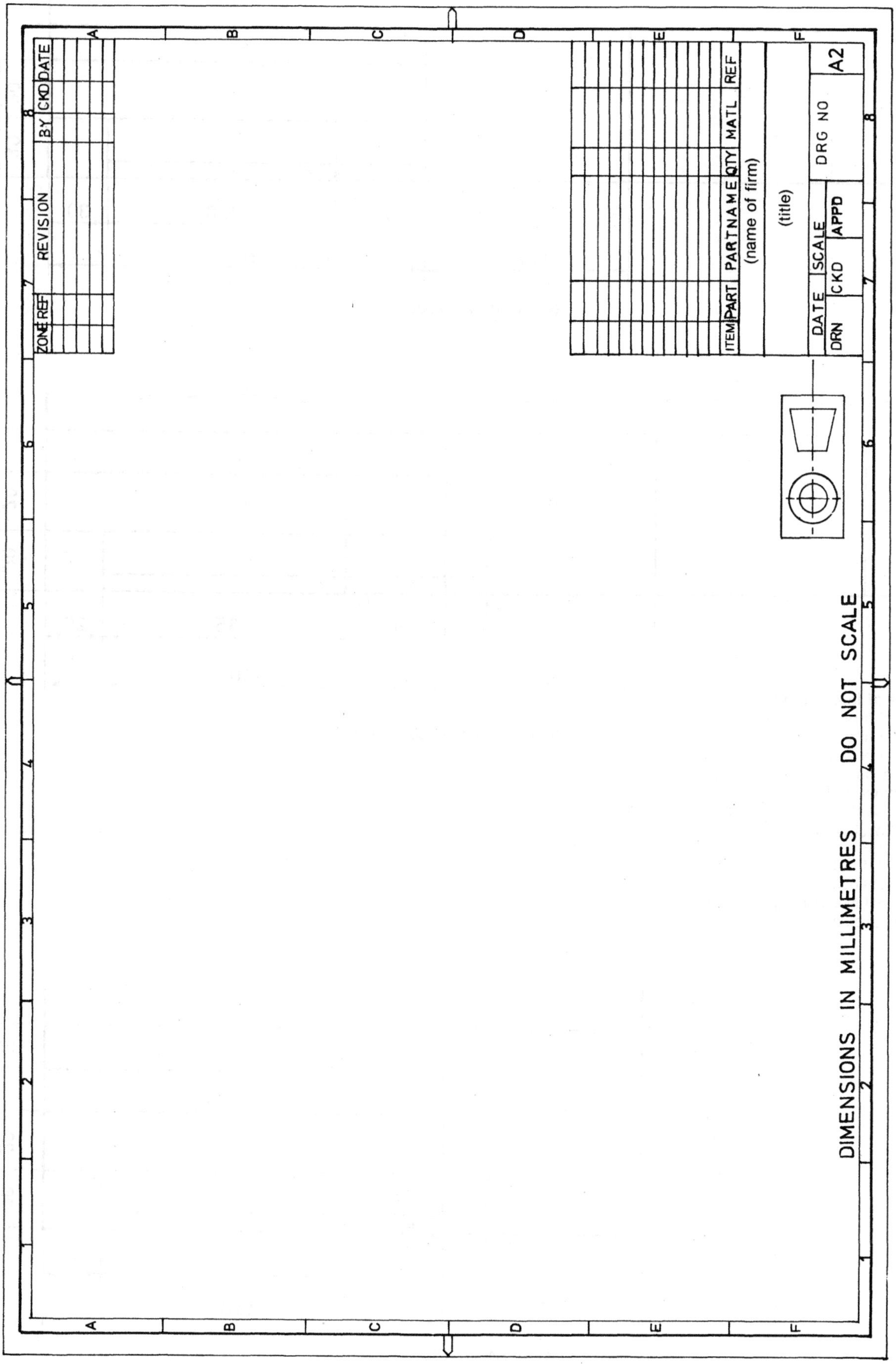
ZONE
REF
REVISION
BY
CKD
DATE
ITEM
PART
PARTNAME
QTY
MATL
REF
(name of firm)
(title)
DATE
SCALE
DRN
CKD
APPD
DRG NO
A2
DIMENSIONS IN MILLIMETRES
DO NOT SCALE

Fig. 1.6 *Typical layout of a drawing sheet (suitable for assembly and multidetail drawings)*

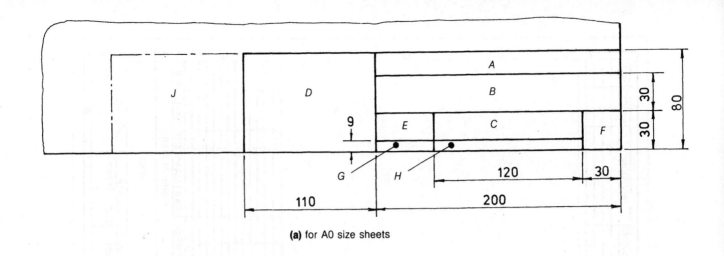

(a) for A0 size sheets

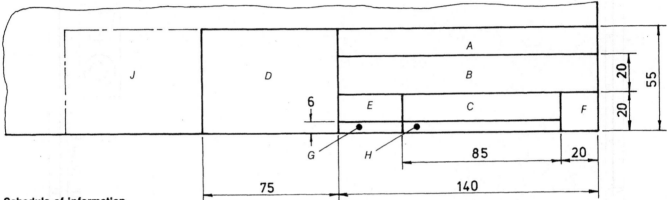

(b) for A1, A2, A3 size sheets

Schedule of information
A name of firm
B drawing title
C drawing number
D information regarding drawing preparation, e.g.
   signatures of drafter, checkers, etc.
E code identification number of the design
   authority (if required)
F sheet size
G scale of drawing
H miscellaneous information
J additional blocks for general information such
   as tolerancing, material, finish, etc.

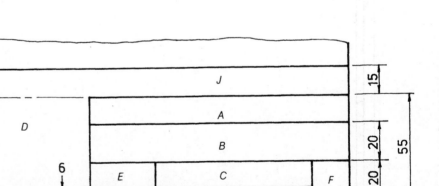

(c) for A4 size sheets

Fig. 1.7 *Typical title blocks for various size sheets*

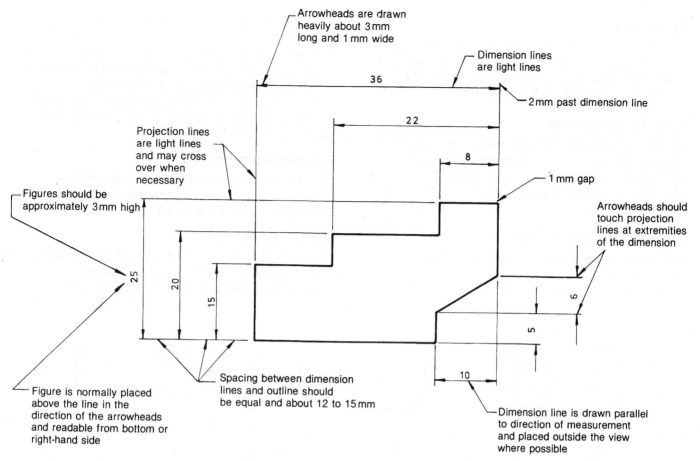

Fig. 1.8 *Use of projection and dimensioning lines*

Dimensioning
Dimension and projection lines

These lines are thin, light, continuous type B lines drawn outside the outline wherever possible.

Projection lines are used as follows:

1. to project from one view to another in order to transfer detail

2. to allow dimensions to be inserted—projection lines indicate the extremities of a dimension

Dimension lines are necessary to indicate the extent of a measurement.

Figure 1.8 shows the use of projection and dimension lines with appropriate measurements indicating spacing and so on.

Figure 1.9 illustrates correct and incorrect methods of employing centre lines and projection lines for dimensioning purposes.

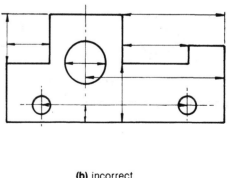

(a) correct   **(b)** incorrect

Fig. 1.9 *Use of centre and projection lines in dimensioning*

13

Linear dimensions

These should preferably be expressed in millimetres. It is not necessary to write the symbol 'mm' after every figure. A general note such as 'all dimensions are in millimetres' in the title block is sufficient.

Angular dimensions

Angular dimensions should be stated in degrees, in degrees and minutes, or in degrees, minutes and seconds, for example 36.5°, 36°30′, 36°29′30″. A zero should be used to indicate an angle of less than 1°, for example 0°30′0.5″.

Methods of dimensioning

Two methods of indicating measurements are in common use:

1. *unidirectional*, where the dimensions are drawn parallel to the bottom of the drawing, that is horizontal
2. *aligned*, where the dimensions are drawn parallel to the related dimension line and are readable from the bottom or right-hand side of the drawing

Dimensions and notes indicated by leaders should use the unidirectional method. The two methods are illustrated in Figure 1.10.

Staggered dimensions

Where a number of parallel dimensions are close together they should be staggered to ensure clear reading, as shown in Figure 1.11.

Overall dimensions

When a length consists of a number of dimensions, an overall dimension may be shown outside the dimensions concerned (see Fig. 1.12). The end

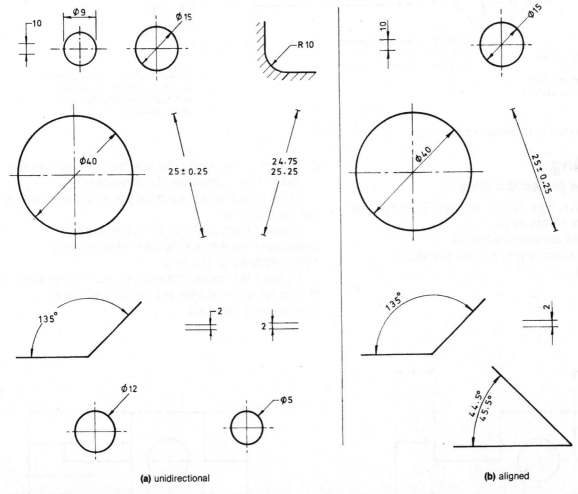

(a) unidirectional

(b) aligned

Fig. 1.10 *Methods of dimensioning*

projection lines are extended to allow this. When an overall dimension is shown, however, one or more of the dimensions that make up the overall length is omitted. This is done to allow for variations in sizes that may occur during production. The omitted dimension is always a non-functional dimension, that is, one that does not affect the function of the product. Functional dimensions are those that are necessary for the operation of the product; these dimensions are essential.

Dimensions not complete

Where a dimension is defining a feature that cannot be completely inserted on a drawing, for example a large distance or diameter, the free end is terminated in a double arrowhead pointing in the direction the dimension would take if it could be completed:

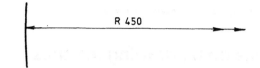

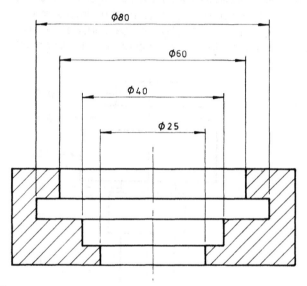

Fig. 1.11 *Use of staggered dimensions*

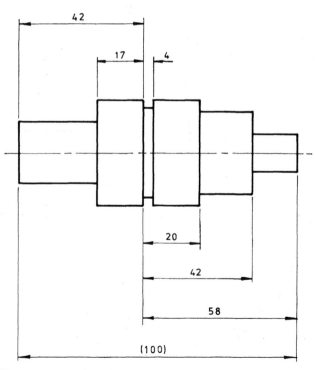

Fig. 1.12 *Use of overall dimensions*

Auxiliary dimensions

When all the dimensions that add up to an overall length are given, the overall dimension may be added as an auxiliary dimension. This is indicated by enclosing the dimension in brackets.

Auxiliary dimensions are never toleranced and are in no way binding as far as machining operations are concerned. Figure 1.13 illustrates the use of an auxiliary dimension, namely (100).

If the overall length dimension is important, then one of the intermediate dimensions is redundant, for example the width of the narrow groove in the centre. This dimension may be inserted as an auxiliary.

Dimensions not to scale

When it is desirable to indicate that a dimension is not drawn to scale, the dimension is underlined with a full, heavy, type A line, for example:

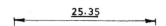

Fig. 1.13 *Use of auxiliary dimensions*

15

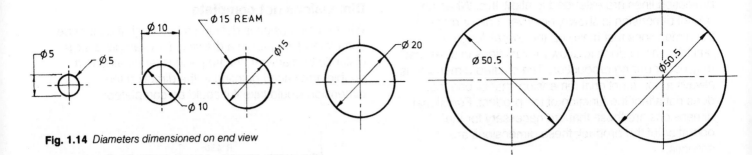

Fig. 1.14 *Diameters dimensioned on end view*

Dimensioning drawing features
Diameters
End view

The symbol ∅ may be used to precede the figure indicating a hole or cylinder. See Figure 1.14 for methods used on circles of various diameters.

Side view

This may be indicated, as shown in Figure 1.15(a), by the use of the symbol ∅ preceding the dimension or, as shown in Figure 1.15(b), by the use of leaders at right angles to the outline in conjunction with the symbol ∅.

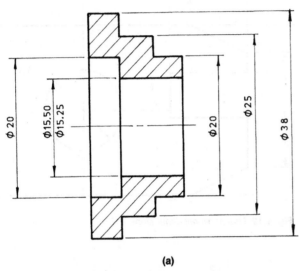

(a)

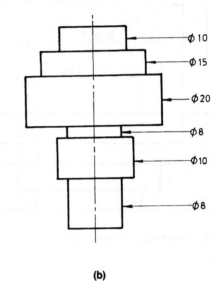

(b)

Fig. 1.15 *Diameters dimensioned on side view*

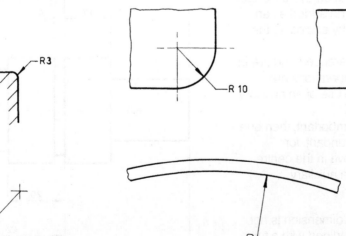

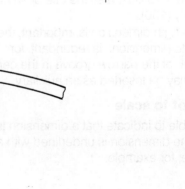

Fig. 1.16 *Methods of dimensioning radii and small spaces*

Radii and small spaces

Figure 1.16 illustrates methods of dimensioning these features. A radius is preceded by the letter R. Leaders should pass through or be in line with the centres of arcs to which they refer.

Spherical surfaces

These are dimensioned as shown in Figure 1.17. Note the distinction made between spherical diameters and spherical radii.

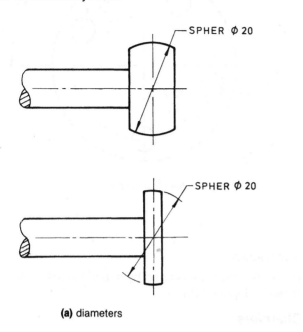

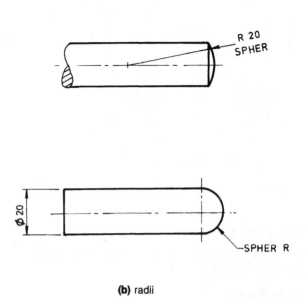

(a) diameters

(b) radii

Fig. 1.17 *Methods of dimensioning spherical surfaces*

Squares

The symbol □ is used to indicate a square section, as shown in Figure 1.18.

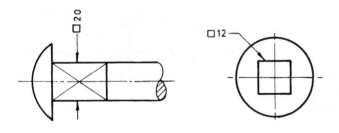

Fig. 1.18 *Methods of dimensioning squares*

Holes

Holes go either right through a material or to a certain depth, and this must be specified as well as the diameter. If no indication is given, a hole is taken as going right through. Figure 1.19 illustrates methods of dimensioning holes using both end and top views.

Flanges

Bolt holes on flanges may be positioned round the PCD (pitch-circle diameter) by either of the methods shown in Figure 1.20.

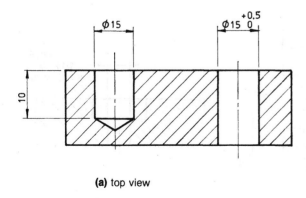

(a) top view

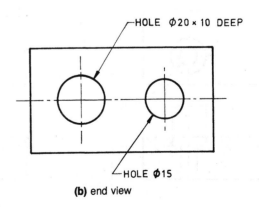

(b) end view

Fig. 1.19 *Methods of dimensioning holes*

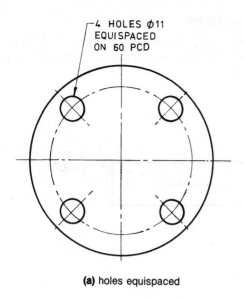

(a) holes equispaced

(b) holes not equispaced

Fig. 1.20 *Methods of dimensioning flanges*

Countersinks

These may be dimensioned by one of the methods shown in Figure 1.21.

Counterbores

These may be dimensioned by one of the methods shown in Figure 1.22.

Spotfaces

These may be dimensioned by one of the methods shown in Figure 1.23.

Chamfers

These may be dimensioned by one of the methods shown in Figure 1.24.

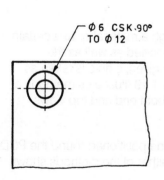
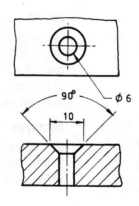
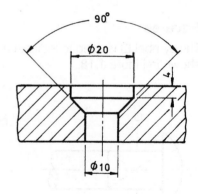

Fig. 1.21 *Methods of dimensioning countersinks*

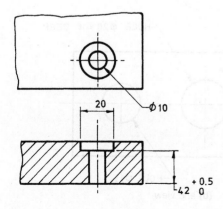
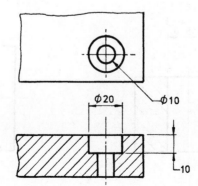
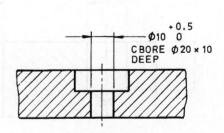

Fig. 1.22 *Methods of dimensioning counterbores*

18

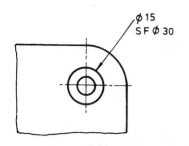
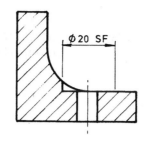
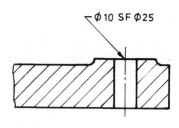

Fig. 1.23 *Methods of dimensioning spotfaces*

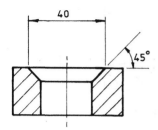
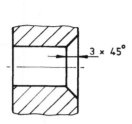
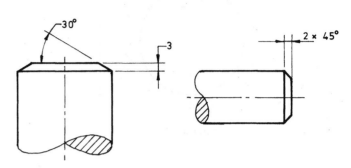

Fig. 1.24 *Methods of dimensioning chamfers*

Keys—square and rectangular

Methods of dimensioning keyways in shafts and hubs, both parallel and tapered, are shown in Figure 1.25,

together with suitable proportions for drawing rectangular keys.

For design purposes, correct keyway proportions should be obtained from *BS 4235* Part 1 (1977).

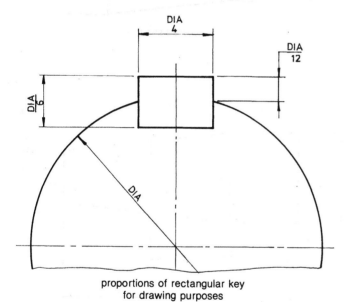

proportions of rectangular key
for drawing purposes

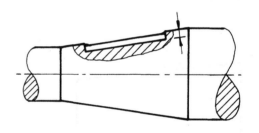

parallel keyway in a tapered
shaft

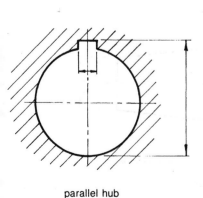

parallel hub

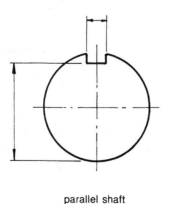

parallel shaft

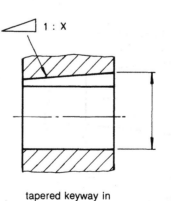

tapered keyway in
a parallel hub

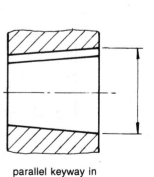

parallel keyway in
a tapered hub

Fig. 1.25 *Methods of dimensioning keys and keyways*

19

Woodruff keys

Methods of dimensioning Woodruff keyways in shafts and hubs, both parallel and tapered, are shown in Figure 1.26.

Tapers

Tapers are dimensioned by one of the four methods shown in Figure 1.27.

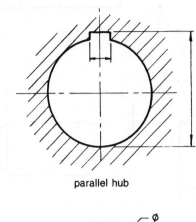

parallel hub

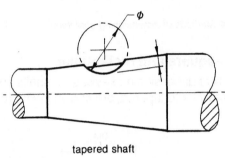

tapered hub

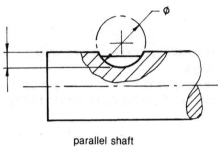

parallel shaft

tapered shaft

Fig. 1.26 *Methods of dimensioning Woodruff keys*

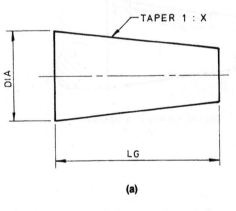

(a)

(b)

(c)

(d)

Fig. 1.27 *Methods of dimensioning tapers*

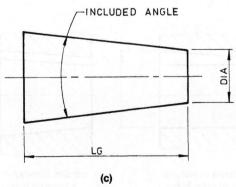

20

Screw threads

General representation

The methods shown in Figure 1.28 are recommended for right-hand or left-hand representation of screw threads. The diameter (ϕ DIA) of a thread is the nominal size of the thread, for example for a 10 mm thread (M10, see p. 23), DIA = 10 mm.

Threads on assembly and special threads

Figure 1.29(a) illustrates the method of representing two threads in assembly. Figure 1.29(b) shows the assembly of two members by a stud mounted in one of them. Special threads are usually represented by a scrap sectional view illustrating the form of the thread, as shown in Figure 1.29(c).

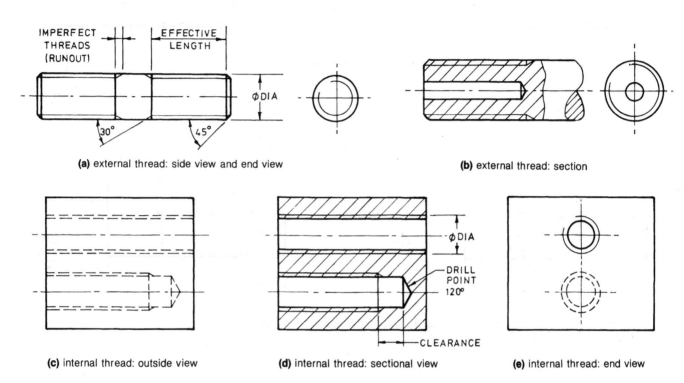

(a) external thread: side view and end view

(b) external thread: section

(c) internal thread: outside view

(d) internal thread: sectional view

(e) internal thread: end view

Fig. 1.28 *Methods of representing screw threads*

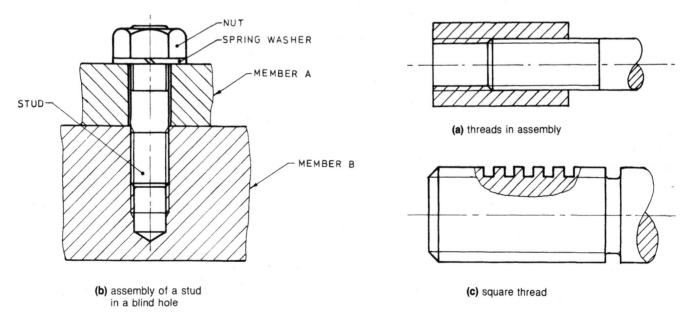

(b) assembly of a stud in a blind hole

(a) threads in assembly

(c) square thread

Fig. 1.29 *Methods of representing assembled and special threads*

Designation of threaded members

When full and runout threads have to be distinguished, the methods of designation shown in Figure 1.30 are recommended. Where there is no possibility of misreading, the runout threads need not be dimensioned.

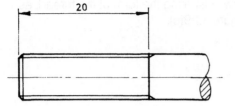

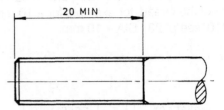

(a) dimensioning length of full threads

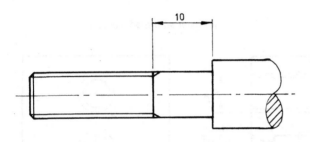

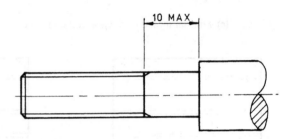

(b) dimensioning to end of full threads

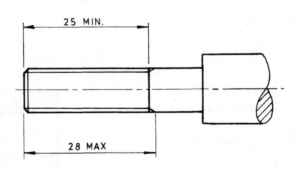

(c) dimensioning length of full threads and runout threads

Fig. 1.30 *Methods of designating threaded members*

Dimensioning full and runout threads in holes

Figure 1.31 shows various methods used to dimension threaded holes. The diameter of the thread is always preceded by the capital letter M, which indicates metric threads.

The coarse thread series is designated simply by the letter M followed by a numeral, for example M12. However, fine threads should show the pitch of the thread as well, for example M12 × 1.25.

If it is not important, the runout threads need not be dimensioned. However, in blind holes it is often important to have fully formed threads for a certain depth, and dimensioning must be provided to control this.

The ISO metric thread

Figure 1.32 shows the profile of the ISO metric thread, together with proportions of the various defined parts of the thread.

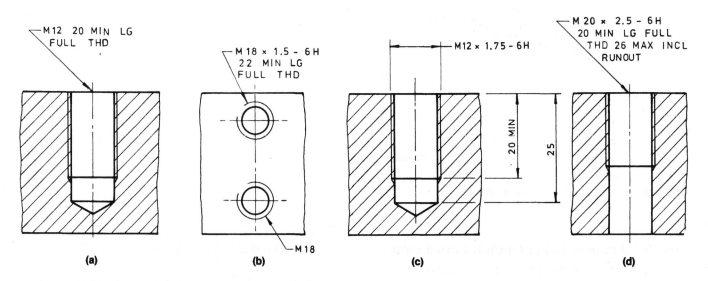

Fig. 1.31 *Methods of dimensioning full and runout threads in holes*

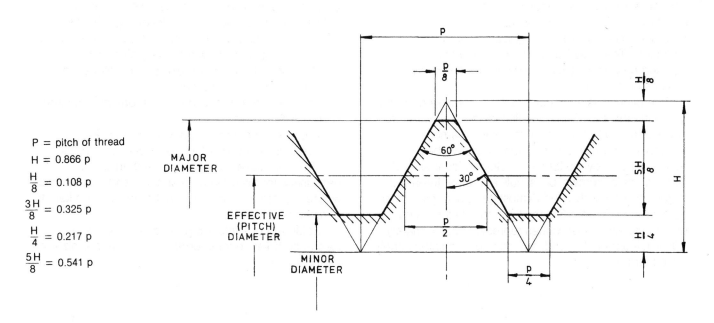

P = pitch of thread

H = 0.866 p

$\frac{H}{8}$ = 0.108 p

$\frac{3H}{8}$ = 0.325 p

$\frac{H}{4}$ = 0.217 p

$\frac{5H}{8}$ = 0.541 p

Fig. 1.32 *Basic profile and proportions of the ISO metric thread*

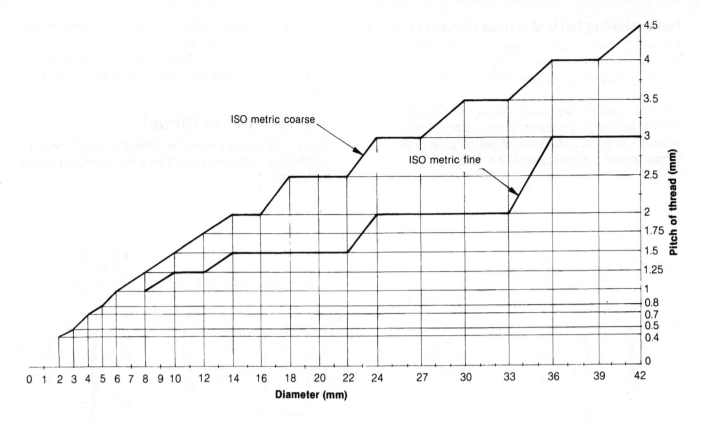

Fig. 1.33 *Graphical comparison of metric threads*

Graphical comparison of metric thread series

ISO metric threads are of two kinds: coarse and fine thread. A graphical comparison of these two series is shown in Figure 1.33.

Tapping size and clearance holes for ISO metric threads

Tapping sizes and clearance holes for metric threads are shown in Table 1.6. In this table column 1 represents first and second choices of thread diameters. The sizes listed under second choice should be used only when it is not possible to use sizes in the first choice column.

The pitches listed in column 2 are compared on the graph in Figure 1.33. These pitches, together with the corresponding first and second choice diameters of column 1, are those combinations that have been recommended by the ISO as a selected 'coarse' and 'fine' series for screws, bolts, nuts and other threaded fasteners commonly used in most general engineering applications. Column 3 is the tapping size for the coarse and fine series. These values represent approximately 100% full depth of thread, and can be calculated simply by the formula:

tapping drill size = outside diameter − pitch
$$3.3 = 4.0 - 0.7$$

Sometimes the drill size has to be rounded off to the next largest stock drill size; this can be obtained from Table 1.7.

Column 4 of Table 1.6 gives tapping sizes for coarse threads in mild steel only; these will give approximately 75% of the full depth of thread. In most general engineering applications this depth of thread is sufficient and desirable for the following reasons:
1. Tapping 100% depth of thread necessitates about three times more power than tapping 75%.
2. The possibility of tap breakage is greater as the depth of thread increases.
3. The 100% thread has only 5% more strength than the 75% thread.
4. The amount of metal removed for a 75% depth thread is only 56% of that removed for a 100%.

There are cases when a full depth thread is necessary, for example on machines and in situations where movement in the mating threads is to be kept to a minimum.

Column 5 of Table 1.6 gives three classes of clearance holes recommended for the various sizes of metric threads.

Table 1.6 *Tapping sizes and clearances (mm) for metric threads*

1		2		3		4	5		
Nominal diameters		Pitch (mm)		Tapping size diameter (mm)		Tapping size (75% full depth coarse threads in mild steel)	Clearance holes		
1st choice	2nd choice	coarse threads	fine threads	coarse threads	fine threads		close	medium	coarse
1.6		0.35		1.25		1.3	1.7	1.8	2
	1.8	0.35		1.45		1.5	1.9	2	2.2
2		0.4		1.6		1.65	2.2	2.4	2.6
	2.2	0.45		1.75		1.8	2.4	2.6	2.8
2.5		0.45		2.05		2.1	2.7	2.9	3.1
3		0.5		2.5		2.55	3.2	3.4	3.6
	3.5	0.6		2.9		2.95	3.7	3.9	4.1
4		0.7		3.3		3.4	4.3	4.5	4.8
	4.5	0.75		3.8		3.8	4.8	5	5.3
5		0.8		4.2		4.3	5.3	5.5	5.8
6		1		5		5.1	6.4	6.6	7
8		1.25	1	6.8	7	6.9	8.4	9	10
10		1.5	1.25	8.5	8.7	8.6	10.5	11	12
12		1.75	1.25	10.2	10.8	10.4	13	14	15
	14	2	1.5	12	12.5	12.2	15	16	17
16		2	1.5	14	14.5	14.25	17	18	19
	18	2.5	1.5	15.5	16.5	15.75	19	20	21
20		2.5	1.5	17.5	18.5	17.75	21	22	24
	22	2.5	1.5	19.5	20.5	19.75	23	24	26
24		3	2	21	22	21.25	25	26	28
	27	3	2	24	25	24.25	28	30	32
30		3.5	2	26.5	28	27	31	33	35
	33	3.5	2	29.5	31	30	34	36	38
36		4	3	32	33	32.5	37	39	42
	39	4	3	35	36	35.5	40	42	45

Table 1.7 *Stock sizes of metric drills (mm)*

0.32	0.68	1.1	1.8	2.5	3.4	4.8	6.2	7.6	9	10.4	11.8	13.2	15.5	19	22.5
0.35	0.7	1.15	1.85	2.55	3.5	4.9	6.3	7.7	9.1	10.5	11.9	13.3	15.75	19.25	22.75
0.38	0.72	1.2	1.9	2.6	3.6	5	6.4	7.8	9.2	10.6	12	13.4	16	19.5	23
0.4	0.75	1.25	1.95	2.65	3.7	5.1	6.5	7.9	9.3	10.7	12.1	13.5	16.25	19.75	23.25
0.42	0.78	1.3	2	2.7	3.8	5.2	6.6	8	9.4	10.8	12.2	13.6	16.5	20	23.5
0.45	0.8	1.35	2.05	2.75	3.9	5.3	6.7	8.1	9.5	10.9	12.3	13.7	16.75	20.25	23.75
0.48	0.82	1.4	2.1	2.8	4	5.4	6.8	8.2	9.6	11	12.4	13.8	17	20.5	24
0.5	0.85	1.45	2.15	2.85	4.1	5.5	6.9	8.3	9.7	11.1	12.5	13.9	17.25	20.75	24.25
0.52	0.88	1.5	2.2	2.9	4.2	5.6	7	8.4	9.8	11.2	12.6	14	17.5	21	24.5
0.55	0.9	1.55	2.25	2.95	4.3	5.7	7.1	8.5	9.9	11.3	12.7	14.25	17.75	21.25	24.75
0.58	0.92	1.6	2.3	3	4.4	5.8	7.2	8.6	10	11.4	12.8	14.5	18	21.5	25
0.6	0.95	1.65	2.35	3.1	4.5	5.9	7.3	8.7	10.1	11.5	12.9	14.75	18.25	21.75	25.25
0.62	1	1.7	2.4	3.2	4.6	6	7.4	8.8	10.2	11.6	13	15	18.5	22	
0.65	1.05	1.75	2.45	3.3	4.7	6.1	7.5	8.9	10.3	11.7	13.1	15.25	18.75	22.25	

Sectioning—symbols and methods

General symbol

A sectional view is one that represents the part of an object that remains after a portion has been removed. It is used to reveal interior detail. Only solid material that has been cut is sectioned. The main types of sectional views used in mechanical drawing are illustrated on pages 29–31. As far as possible the general sectioning symbol (cross-hatching) should be used (Fig. 1.34(a)).

A useful aid for drawing equally spaced sectioning lines is shown in Figure 1.34(b).

Sectioning lines (cross-hatching)

These are light lines (type B) and are normally drawn at 45° to the horizontal, right or left. If the shape of the section would bring the sectioning lines parallel to one or more of the sides, another angle may be used (Fig. 1.35).

Adjacent parts

In section, adjacent parts should have their sectioning lines at right angles (Fig. 1.36(a)). When more than two parts are adjacent, as in Figure 1.36(b), they may be distinguished by varying the spacing or the angle of the hatching lines.

Dimensions

Dimensions may be inserted in sectioned areas by interrupting the sectioning lines, as shown in Figure 1.36(c).

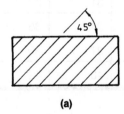

(a)

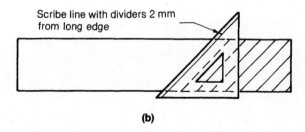

Scribe line with dividers 2 mm from long edge

(b)

Fig. 1.34 *(a) General symbol for hatching (b) aid for drawing section lines*

 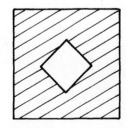

Fig. 1.35 *General application of sectioning lines*

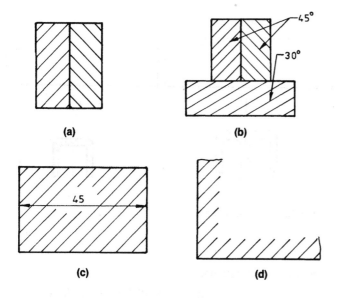

Fig. 1.36 *Special application of sectioning lines*

Large areas

These can be shown sectioned by placing section lines around the edges of the area only, as in Figure 1.36(d).

Section cutting plane: application

Section cutting planes are denoted by a chain line (type H) drawn across the part as shown in the front view of Figure 1.37. Arrowheads indicate the face of the section and the direction of viewing.

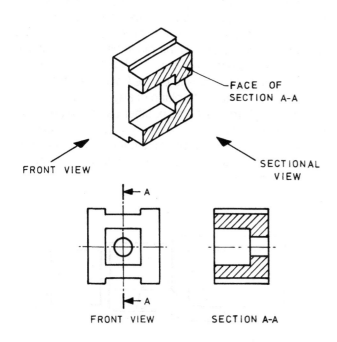

Fig. 1.37 *Representation of a sectional view*

A specific section is identified by letters placed near the arrows, and reference to the sectional view is made by the letters, separated by a short dash, for example section A-A. Where only one cutting plane is used on a drawing, the letters may be omitted.

The chain line may be simplified by omitting the thin part of the line, if clarity is not affected. Arrowheads may also be omitted when indicating symmetrical sectional views or when the sectional view is drawn in the projection indicated on the drawing (see Fig. 1.3).

The identification of a cutting plane may be omitted when it is obvious that the section can only be taken through one location. Figure 1.38 shows a sectional view that is obviously taken on the centre line of the other view.

Sectioning thin areas

Sometimes the section plane passes through very thin areas that cannot be sectioned by normal 45° hatching, for example gaskets, plastic sheet, packing, sheet metal and structural shapes. These areas should be filled in as shown in Figure 1.39(a).

If two or more thin areas are adjacent, a small space should be left between them (Fig. 1.39(b)).

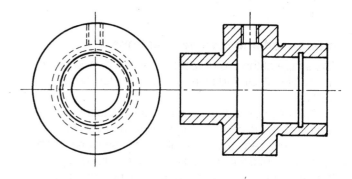

Fig. 1.38 *Sectional view with identification of cutting plane omitted*

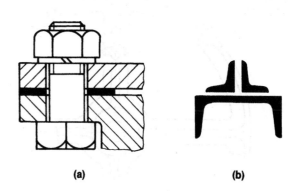

Fig. 1.39 *Sectioning of thin parts*

27

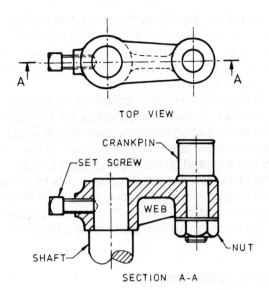

TOP VIEW

CRANKPIN
SET SCREW
WEB
SHAFT
NUT

SECTION A-A

Fig. 1.40 *Exceptions to the general rule of sectioning*

Exceptions to the general rule

As a general rule all material cut by a sectioning plane is cross-hatched in orthogonal views but there are exceptions. When the sectioning plane passes through the centre of webs, shafts, bolts, rivets, keys, pins and similar parts, they are not shown sectioned but in outside view, as in Figure 1.40.

Interposed and revolved sections

The shape of the cross-section of a bar, arm, spoke or rib may be illustrated by a revolved or interposed section.

The interposed section has detail adjacent to it removed and is drawn using a thick line (type A).

The revolved section has the cross-sectional shape revolved in position with adjacent detail drawn against the revolved view. It is drawn using a thin line (type B). Figure 1.41 illustrates these two sections.

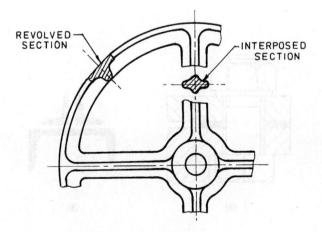

REVOLVED SECTION

INTERPOSED SECTION

Fig. 1.41 *Revolved and interposed sections*

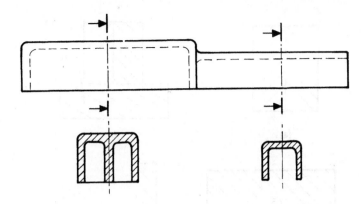

Fig. 1.42 *Removed sections*

Removed sections

These are similar to revolved sections except that the cross-section is removed clear of the main outline for the sake of clarity. The removed section may be located adjacent to the main view (Fig. 1.42) or away from it entirely. In the latter case it must be suitably referenced to the view and section to which it refers. The outline of a removed section is a thick line (type A).

Part or local sections

Part or local sections may be taken at suitable places on a component to show hidden detail. The boundary of the sections is drawn freehand using a type C line, as in Figure 1.43.

Aligned sections

In order to include detail on a sectional view that is not located along one plane, the section plane may be bent to pass through such detail. The sectional view then shows the detail along the line of the bent cutting plane without any indication that the plane has been bent. The principle is illustrated in Figure 1.44(a). Note that when indicating the cutting plane on the front view, heavy lines are used where the plane changes direction.

Figure 1.44(b) illustrates another use of an aligned section, where detail such as holes located on a pitch circle are considered to be rotated into the cutting plane and projected onto the sectional view at their actual distance from the centre line.

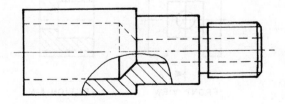

Fig. 1.43 *Part or local sections*

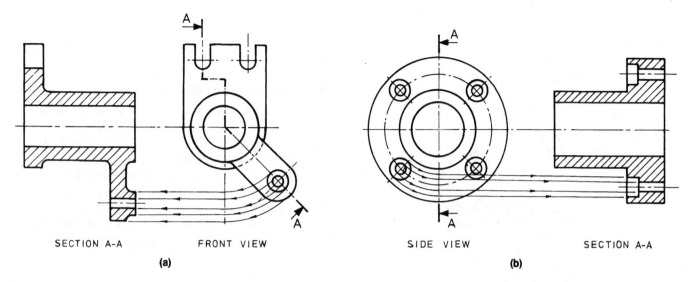

SECTION A-A · FRONT VIEW

(a)

SIDE VIEW · SECTION A-A

(b)

Fig. 1.44 *Aligned sections*

Drawing sectional views

In most cases the normal outside views obtained from orthogonal projection are not sufficient to complete the shape description of an engineering component, both inside and out. Hence other views of a different type must be drawn in conjunction with, or instead of, the normal outside views. These special views are called *sectional views* and the main types used in mechanical drawings are described in this section.

The full sectional view

Figure 1.45 shows an isometric view of a machined block which has been cut through the centre and moved apart. The shape and detail of the counterbored holes are revealed along the face of the cut. This is the purpose of the sectional view—to reveal interior detail. A normal view would be taken from position X.

Figure 1.46 shows the sectional view and a right side view taken from position Y in Figure 1.45. The course of the sectioning plane is indicated by A–A on

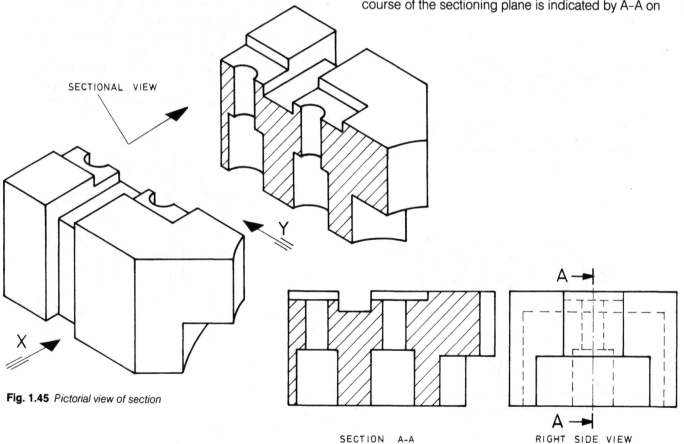

SECTIONAL VIEW

Fig. 1.45 *Pictorial view of section*

SECTION A-A

RIGHT SIDE VIEW

Fig. 1.46 *Orthogonal view of section*

29

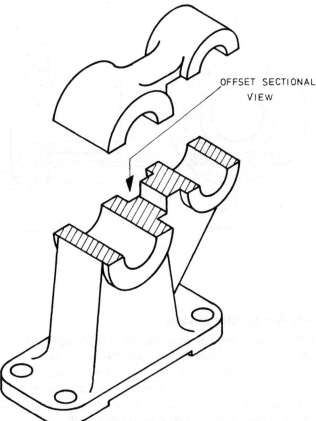

Fig. 1.47 *Pictorial view of offset section*

the side view. The direction of the arrows on the section plane A-A indicates the direction from which the section is viewed.

The offset sectional view

A full sectional view reveals interior detail that lies along one plane only. Sometimes it is desirable to show detail that lies along two or more planes, and this is done by means of an offset sectional view.

Figure 1.47 is an isometric view of a shaft bracket that has been cut by an offset sectioning plane to reveal the detail of the two bosses. The offset sectional view in this case is taken looking down on the bottom piece as shown. Figure 1.48 shows a normal front view and an offset sectional top view of the bracket; the course of the sectional plane is shown by A-A.

Note that there is no line shown on the sectional view where the course of the sectioning plane changes direction.

The half sectional view

This type of view is often used on objects that are symmetrical about a centre line. The cutting plane effectively removes a quarter of the object, as shown in Figure 1.49. The resulting view provides two views in one, as one half shows interior detail and the other half shows external detail. This is illustrated in Figure 1.50.

As with the offset sectional view, the division between the external half and the internal half of the view is not indicated by a full line, but by a centre line. Hidden detail is omitted from the sectioned half of the view, but may be shown on the external half if by so doing the internal shape description is made clearer. This is the case in Figure 1.50, where the hidden detail completes the internal holes revealed in the sectioned half.

SECTION   A-A

A | A

FRONT   VIEW

Fig. 1.48 *Orthogonal view of offset section*

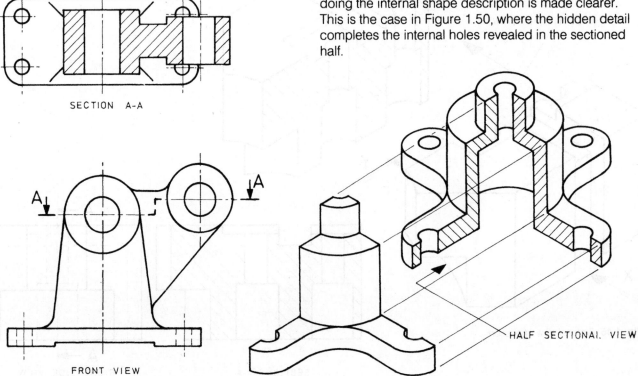

Fig. 1.49 *Pictorial view of half section*

Rules to remember when sectioning

1. A sectional view shows the part of the component in front of the sectioning plane arrows. In third-angle projection the sectional view is placed on the side behind the sectioning viewing plane, and in first-angle projection it is placed on the side in front of the sectional viewing plane.

2. Material that has been cut by the sectioning plane is cross-hatched. Standard exceptions are given on page 28.

3. A sectional view must not have any full lines drawn over cross-hatched areas. A full line represents a corner or edge which cannot exist on a face that has been cut by a plane.

4. As a general rule, dimensions are not inserted in cross-hatched areas, but where it is unavoidable, it may be done as shown in Figure 1.36(c).

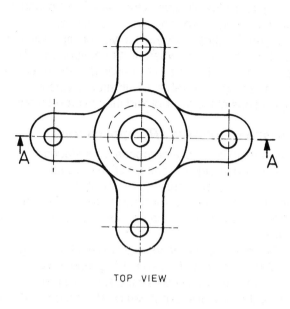

TOP VIEW

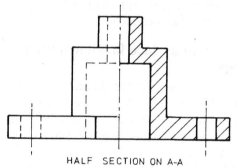

HALF SECTION ON A-A

Fig. 1.50 *Orthogonal view of half section*

Surface texture

Indication on drawings

Symbols indicating the type of surface finish, production methods and/or required roughness of a surface are used on a drawing when this feature is necessary to ensure functionality, and then only on those surfaces that require it. Surface finish specification is not necessary when normal production process finish is satisfactory.

A symbol should be used only once for a given surface, and where possible on a view that shows the size and position of the surface in question.

Surface texture terminology

Figure 1.51 illustrates the standard terminology relating to surface texture. *Surface roughness* (R_a value) is a measure of the arithmetical mean deviation of a short distance of the surface in question.

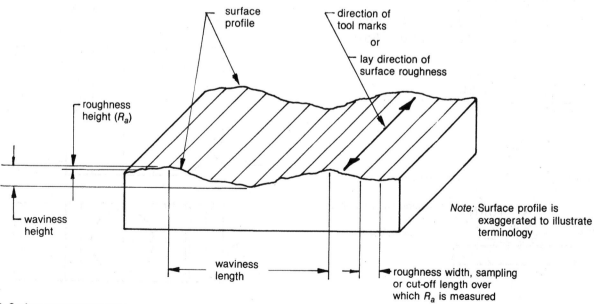

Fig. 1.51 *Surface texture terminology*

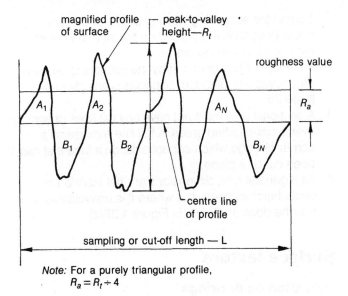

Note: For a purely triangular profile,
$R_a = R_t \div 4$

Fig. 1.52 *Surface roughness profile*

Surface roughness measurement—R_a

The R_a *value* may be defined as the average value of the departure (both above and below) of the surface from the centre line over a selected *sampling* or *cut-off length* (0.08 mm, 0.25 mm, 0.8 mm, 2.5 mm, 8 mm and 25 mm are standard lengths, depending on the production process).

Referring to Figure 1.52, the centre line is positioned so that, over the sampling or cut-off length chosen:

areas above the centre line = areas below the centre line

$$(A_1 + A_2 ... + A_N) = (B_1 + B_2 + ... + B_N)$$

Then to calculate surface roughness:

$$R_a = \frac{(A_1 + A_2 ... + A_N) + (B_1 + B_2 + ... + B_N)}{L}$$

$$= \frac{\text{sum of the areas above and below the centre line}}{\text{sampling or cut-off length}}$$

The preferred or standard values of R_a in micrometres (μm) are as follows:

0.025	0.05	0.1	0.2	0.4	0.8
1.6	3.2	6.3	12.5	25.0	50.0

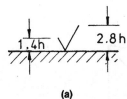

(a)

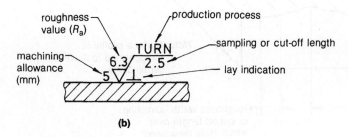

(b)

Fig. 1.53 *Surface roughness symbol*

The standard symbol

Surface finish requirements are indicated on a drawing by a standard symbol, a basic character that may have further information attached to it, depending on the finish requirements of the surface in question.

Figure 1.53(a) illustrates the basic symbol and its size in relation to other drawing characters.

Figure 1.53(b) indicates the type of information that may be required and where it is to be found on the basic symbol. All of this information is seldom required on the one symbol. Table 1.8 illustrates typical applications of the symbol and their interpretations.

Surface roughness (R_a) applications

The ultimate finish of a component's surfaces is determined largely by their function or required appearance characteristics. The ability to produce various finishes is governed by the types of production processes available for component manufacture within a firm.

Designers and drafters must have a knowledge of the above factors before specifying roughness requirements. Table 1.9 lists the standard roughness values, the processes that can produce them, their area of application and some indication of the relative cost associated with their production.

Application of surface finish symbol to drawings

The surface finish symbol should be located so that it can be read from the bottom or right-hand side of the drawing. Symbols should be applied to the edge view of the surface in question, but extension and leader lines may also be used to apply the symbol.

Figure 1.54 illustrates correct methods of applying the symbol.

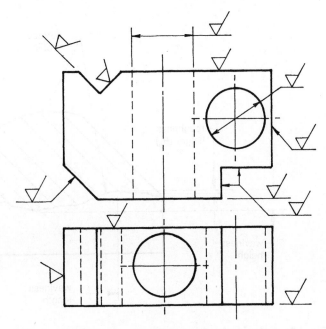

Fig. 1.54 *Application of the surface roughness symbol*

Table 1.8 *Applications of surface finish symbol*

Symbol	Interpretation	Symbol	Interpretation
CONTROLLED SURFACE	the basic symbol—consists of two unequal legs inclined at 60° and resting on the surface to be controlled	0.4/ALL OVER	may be applied in the title block or as a note when a single value applies to all machined surfaces controlled
(machining symbol)	used when machining is necessary to obtain the desired finish	6.3/ ALL OVER EXCEPT WHERE OTHERWISE INDICATED	may be applied in the title block or as a note when a single value applies to the majority of machined surfaces; exceptions should be indicated on the individual surfaces concerned
(symbol with circle)	used when the surface finish is to remain as found from the last process and no material, e.g. a cast or forged part, is to be removed	2 1.6/ TURN	used to specify a turning allowance of 2 mm after which a surface finish of 1.6 μm is required
6.3 3.2/	used to specify maximum and minimum limits of surface roughness obtained by any machining process	N4/	used to specify the roughness value by a standard number equivalent to 0.2 μm or 8 μ ins
25 6.3/ (with circle)	used to specify maximum and minimum limits of surface roughness obtained by any process without machining	3.2/ M	used to specify a roughness value together with a multidirectional lay finish
3.2/ MILL	used to indicate a particular machining process and roughness value	0.8 0.008–4/	used to specify a roughness value together with a waviness height of 0.008 mm and spacing of 4 mm
1.6/2.5	used to indicate a sampling length in millimetres and a machined surface finish	$\sqrt{} = $ 3.2/ MILL	used when a surface texture specification is complicated and is required on a number of surfaces and space on the drawing is limited; the basic symbol is used on the surfaces in question and its meaning is clearly defined in note form on the drawing as shown
CADMIUM PLATE 1.6/ 0.8/	used to indicate roughness before and after surface treatment; note the use of type J line representing the surface after treatment		

Table 1.9 *Standard roughness values*

R_a value (μm)		Process and application
0.025 or 0.025 and 0.05 or 0.05	very fine quality surface finishes, costly to produce	This very smoothly finished surface is produced by fine honing, lapping, buffing or super-finishing machines. It is costly to produce and seldom required. It has a highly polished appearance, depending on the production process, and is normally used on precision instruments such as gauges, laboratory equipment and finely made tools.
0.1 or 0.1		This is similar to the finer grades of finish and has much the same application. Very refined surfaces have this high degree of finish. It is produced by honing, lapping and buffing methods and is costly to produce.
0.2 or 0.2		This fine surface is produced by honing, lapping and buffing methods. This texture could be specified on precision gauge and instrument work, and on high speed shafts and bearings where lubrication is not dependable.
0.4 or 0.4		This fine quality surface can be produced by precision cylindrical grinding, coarse honing, buffing and lapping methods. It is used on high speed shafts, heavily loaded bearings and other applications where smoothness is desirable for the proper functioning of a part.
0.8 or 0.8	medium quality finishes, used where reasonable surfaces are required	This first-class machine finish can be easily produced on cylindrical, surface and centreless grinders but requires great care on lathes and milling machines. It is satisfactory for bearings and shafts carrying light loads and running at medium to slow speeds. It may be used on parts where stress concentration is present. It is the finest finish that it is economical to produce; below this costs rise rapidly.
1.6 or 1.6		This good machine finish can be maintained on production lathes and milling machines using sharp tools, fine feeds and high cutting speeds. It is used when close fits are required but is unsuitable for fast rotating members. It may be used as a bearing surface when motion is slow and loads are light. This surface can be achieved on extrusions, rolled surfaces, die castings and permanent mould castings in controlled production.
3.2 or 3.2		This medium commercial finish is easily produced on lathes, milling machines and shapers. A finish commonly used in general engineering machining operations, it is economical to produce and of reasonable appearance. It is the roughest finish recommended for parts subjected to slow speeds, light loads, vibration and high stress, but it should not be used for fast rotating shafts. This finish may also be found on die castings, extrusions, permanent mould castings and rolled surfaces.
6.3 or 6.3	rough finishes, used where quality surfaces are unimportant	This coarse production finish is obtained by taking coarse feeds on lathes, millers, shapers, boring and drilling machines. It is acceptable when tool marks have no bearing on performance or quality. This texture can also be found on the surfaces of metal moulded castings, forgings, extruded and rolled surfaces, and can be produced by hand filing or disc grinding.
12.5 or 12.5		This surface is produced from heavy cuts and coarse feeds by milling, turning, shaping, boring, disc grinding and snagging. It can also be obtained by sand casting, saw cutting, chipping, rough forging and oxy cutting. This finish is rarely specified and is used only where it is not seen or its appearance is unimportant, e.g. on machinery, jigs and fixtures.
25		This very rough finish is produced by sand casting, torch and saw cutting, chipping and rough forgings. Machining operations are not required as this finish is suitable as found, e.g. on large machinery.

Roughness grade numbers

Where there is a possibility of misinterpretation due to using both metric and imperial units, surface roughness may be indicated by an equivalent surface roughness number, as shown in Table 1.10.

Direction of surface pattern or lay

A production process produces a regular pattern of tool marks on a surface; this feature is called the *lay direction* of the surface.

Table 1.11 illustrates the standard symbols used to represent various lay directions and their interpretations.

Table 1.10 *Roughness grade numbers*

Roughness (R_a)		Roughness number
μm	μ ins	
50.0	2000	N12
25.0	1000	N11
12.5	500	N10
6.3	250	N9
3.2	125	N8
1.6	63	N7
0.8	32	N6
0.4	16	N5
0.2	8	N4

Table 1.11 *Lay symbols*

Lay symbol	Description	Lay symbol	Description
=	direction of tool marks Lay is parallel to the line representing the surface to which the symbol is applied.	C	direction of tool marks Lay is generally circular relative to the centre of the surface to which the symbol is applied.
⊥	direction of tool marks Lay is perpendicular to the line representing the surface to which the symbol is applied.	M	direction of tool marks Lay is multidirectional, but generally has some kind of tool mark pattern.
X	direction of tool marks Lay is slanting in both directions to the line representing the surface to which the symbol is applied.	R	direction of tool marks Lay is approximately radial to the centre of the surface to which the symbol is applied.

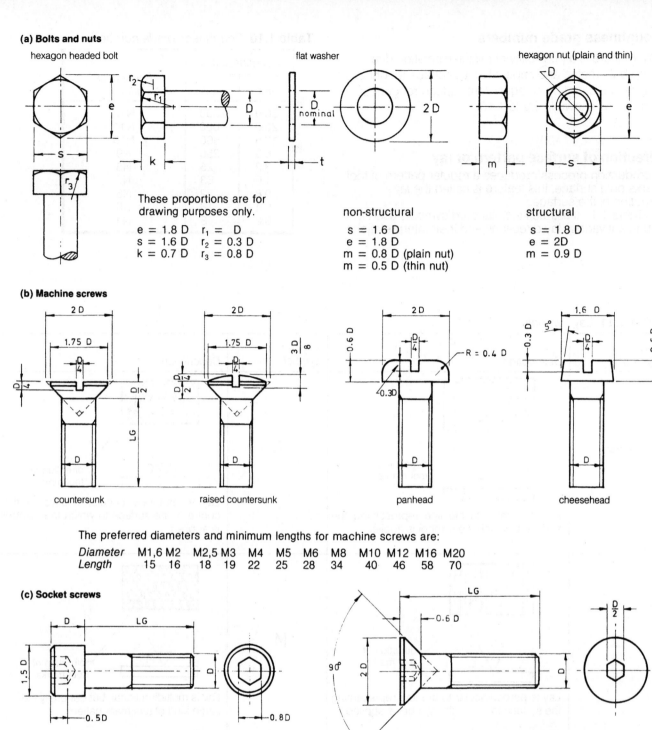

(a) Bolts and nuts

hexagon headed bolt

flat washer

hexagon nut (plain and thin)

These proportions are for
drawing purposes only.

e = 1.8 D   r₁ = D
s = 1.6 D   r₂ = 0.3 D
k = 0.7 D   r₃ = 0.8 D

non-structural

s = 1.6 D
e = 1.8 D
m = 0.8 D (plain nut)
m = 0.5 D (thin nut)

structural

s = 1.8 D
e = 2D
m = 0.9 D

(b) Machine screws

countersunk

raised countersunk

panhead

cheesehead

The preferred diameters and minimum lengths for machine screws are:

Diameter	M1,6	M2	M2,5	M3	M4	M5	M6	M8	M10	M12	M16	M20
Length	15	16	18	19	22	25	28	34	40	46	58	70

(c) Socket screws

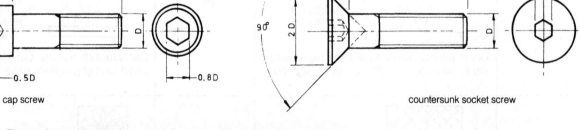

cap screw

countersunk socket screw

The preferred diameters and minimum lengths for socket screws are:

Diameter	M3	M4	M5	M6	M8	M10	M12
Length	8	10	12	16	20	25	30

Fig. 1.55 *Proportions of bolts, nuts and screws*

Fits and tolerances

This section, based on the International Organization for Standardization (ISO) system, introduces the engineering concept of sizing parts before fitting them together to achieve a desirable relative motion between them. Only general applications will be considered.

In manufacture it is impossible to produce components to an exact size, even though they may be classified as identical. Even in the most precise methods of production it would be extremely difficult and costly to reproduce a diameter time after time so that it is always within 0.01 mm of a given basic size. However, industry does demand that parts should be produced between a given maximum and minimum size. The difference between these two sizes is called the *tolerance*, which can be defined as the variation in size that is tolerated. A broad, generous tolerance is cheaper to produce and maintain than a narrow, precise one. Hence one of the golden rules of engineering design is 'always specify as large a tolerance as is possible without sacrificing quality'. There are a number of general definitions and terms used, and these will now be described and illustrated.

Shaft (Fig. 1.56)

A *shaft* is defined as a member that fits into another member. It may be stationary or rotating. The popular concept is a rotating shaft in a bearing. However, when speaking of tolerances, the term shaft can also apply to a member that has to fit into a space between two restrictions, for example a pulley wheel that rotates between two side plates. In determining the clearance fit of the boss between the side plates, the length of the pulley boss is regarded as the shaft.

Hole (Fig. 1.56)

A *hole* is defined as the member that houses or fits the shaft. It may be stationary or rotating, for example a bearing in which a shaft rotates is a hole. However, when speaking of tolerances, the term hole can also apply to the space between two restrictions into which a member has to fit, for example the space between two side plates in which a pulley rotates is regarded as a hole.

Basic size (Fig. 1.56)

This is the size about which the limits of a particular fit are fixed. It is the same for both shaft and hole. It is also called the *nominal size*.

Limits of size (Fig. 1.56)

These are the extremes of size allowed for a dimension. Two limits are possible: one is the *maximum allowable size* and the other is the *minimum allowable size*.

Deviation (Fig. 1.56)

This is the difference between the basic size and the actual size. The extremes of deviations are referred to as the *upper* and *lower deviations*.

Tolerance (Fig. 1.56)

Tolerance is defined as the difference between the maximum and minimum limits of size for a hole or shaft. It is also the difference between the upper and lower deviations.

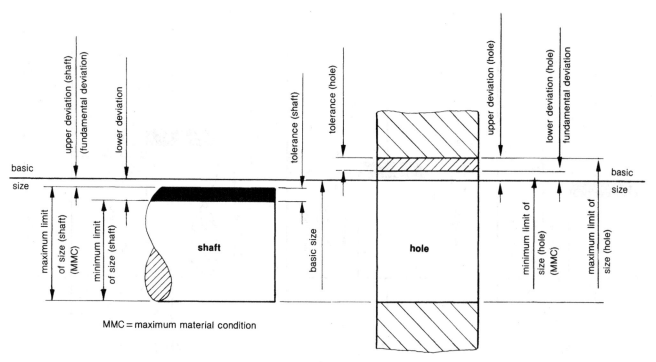

MMC = maximum material condition

Fig. 1.56 *Designation of shaft and hole sizes and limits*

Fit

A *fit* may be defined as the relative motion that can exist between a shaft and a hole (as defined above), resulting from the final sizes achieved in their manufacture. There are three classes of fit in common use: *clearance, transition* and *interference*.

Relative motion between shaft and hole is possible when clearance exists but impossible when interference exists.

Clearance fit (Fig. 1.57(a))

This fit results when the shaft size is always less than the hole size for all possible combinations within their tolerance ranges. Relative motion between shaft and hole is always possible.

The *minimum clearance* occurs at the maximum shaft size and the minimum hole size. The *maximum clearance* occurs at the minimum shaft size and the maximum hole size.

Clearance fits range from coarse or very loose to close precision and locational.

Transition fit (Fig. 1.57(b))

A pure transition fit occurs when the shaft and hole are exactly the same size. This fit is theoretically the boundary between clearance and interference and is practically impossible to achieve, but by selective assembly or careful machining methods, it can be approached within very fine limits.

Practical transition fits result when the tolerances are such that the largest hole is greater than the smallest shaft and the largest shaft is greater than the smallest hole.

Interference fit (Fig. 1.57(c))

This is a fit that always results in the minimum shaft size being larger than the maximum hole size for all possible combinations within their tolerance ranges. Relative motion between the shaft and hole is impossible.

The minimum interference occurs at the minimum shaft size and the maximum hole size.

The maximum interference occurs at the maximum shaft size and minimum hole size.

Allowance (Fig. 1.57)

Allowance is the term given to the minimum clearance (called positive allowance) or maximum interference (called negative allowance) between mating parts. It may also be described as the clearance or interference that gives the tightest possible fit between mating parts.

Grades of tolerance

To give a wide range of control over tolerance, provision has been made in the ISO system for eighteen grades of tolerance, ranging from very fine (the lower numbers) to extremely coarse (the larger numbers). Each grade is approximately 1.6 times as great as the grade below or finer than it. This ratio has been determined after extensive practical investigations and is derived from the relationship $t = kf(d)$, where t is the tolerance and is

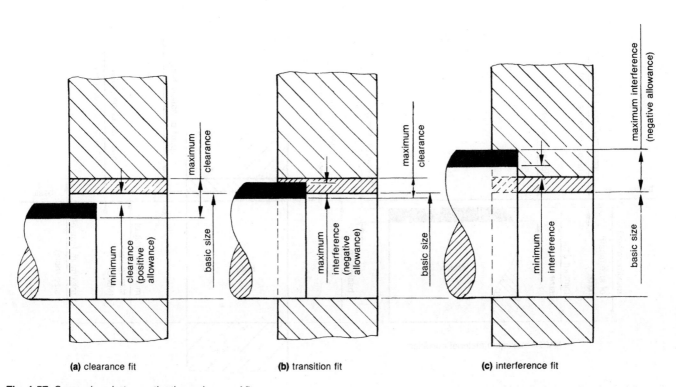

(a) clearance fit   (b) transition fit   (c) interference fit

Fig. 1.57 *Comparison between the three classes of fit*

equal to a function of the diameter multiplied by the constant *k*. Different values of *k* are used to provide a series of tolerance grades for various diameters.

The eighteen grades are designated ITO1, ITO, IT1, IT2, up to IT16. The letters IT (which stand for ISO series of tolerances) are omitted in tables and also when designating fits. The numerical values of these grades of fit for all diameters up to 3150 mm are given in *AS 1654 (Limits and Fits for Engineering)*.

Figure 1.58 illustrates graphically a comparison between some of the grades (IT5 to IT13).

The grade actually represents the size of the tolerance zone and this in turn dictates the degree of accuracy of the machining process required to keep the size within the specified tolerance. Low grades require precision or toolroom machines with highly skilled labour. Coarse grades are much easier to maintain and require cheaper machines and less skilled labour.

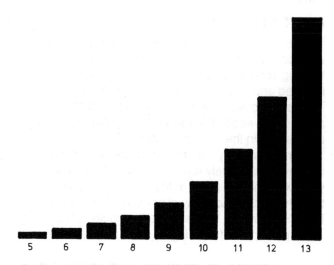

Fig. 1.58 *Comparison of some grades of tolerance. Each grade is 1.6 times as great as the grade lower*

Computer-aided design and drafting/ Computer-aided manufacture

The analytical capabilities of the modern computer are so fast and powerful that the three stages of the manufacturing process—design, drafting and manufacture—can be integrated into the one continuous process, known as CAD/CAM (computer-aided design and drafting/computer-aided manufacture).

Of course, manufacturing forms only a part of the overall manufacturing scene: other functional elements of an enterprise include material purchasing, stores, administration, sales, marketing and so on. All these functions provide information that should be shared for more efficient operation. The modern and most effective way of sharing information is by having it stored in a computer memory (data base) that is readily accessible

by all sections of the organisation. One such system is called a *relational data base management system* (RDBMS). Data base management is truly the key to successful *computer-integrated manufacturing* (CIM), which is a system for co-ordinating the entire spectrum of a company's operations. This text is concerned only with the CAD/CAM process.

Traditional design methods

The manual process of component design entails a laborious succession of steps involving applied mathematical calculations, sketching, prototype drawings, manufacture and trials, modification and sometimes numerous repetitions of the above steps before the final production design is approved; this is followed by detailed production drawings. An engineering designer is not only concerned with the functionality of components; he or she must also be aware of the cost, amount and kind of material used, as well as the type of manufacturing process most suited for production.

The traditional methods of design and drawing are both labour intensive and time consuming—and even then designs that have been approved for production are quite often modified after production. This occurs because the 'trial and error' or 'research and development' phase is not investigated thoroughly or repeated often enough to optimise the design (usually because the huge manual labour input into design calculations and production of drawings is cost prohibitive). In most cases, only two-dimensional drawings are produced, and these do not provide the 'finished look' perspective that a three-dimensional drawing can provide. If a component is used in stress or load applications, a designer must undertake many hours or even days of complex mathematical calculations to determine critical dimensions and tolerances. As already stated, the designer must also be aware of the economic need to use as little material as possible and at the same time be sure that the design is strong enough to withstand the loads imposed upon it. These two requirements are in opposition, and so the 'perfect design' must lie on a very narrow path between the two.

How can a modern-day designer best design a component while keeping the cost down so that pricing of the component is keen in a competitive market?

Principles of CAD/CAM

The introduction of the computer to facilitate mathematical calculations has revolutionised the design process. The extension of the computer's capabilities into the world of graphics has allowed the engineer to optimise design. A continuous process of trial and rejection is used until the 'perfect design' is found.

The system comprises two components—*hardware* and *software*. Hardware refers to the equipment used, such as the computer, graphics screen, keyboard, printer, plotter and so on. Software refers to the computer programs that allow the user to automatically carry out the calculations and drawing operations designated within the program. Printers and plotters produce the typed output or engineering drawings at the conclusion of the design process.

Computer programs tailormade to carry out certain calculations or complete drawing operations on graphics screens are very costly. However, the hardware system is useless without them, and one of the main criteria to be assessed when purchasing a CAD system is the quantity and quality of the software packages that will run on the system. In fact, this is often the major factor when analysing the various systems on offer.

CAD/CAM hardware

The main item of hardware in any CAD/CAM system is the computer, and over the past three decades the most desirable configurations have changed considerably. During the 1960s the large mainframe computer was used, in the 1970s mini-computers served multiple graphics screens and the 1980s saw the emergence of the personal computer (PC) and stand-alone workstations. The modern trend is to provide every design engineer with his or her own workstation. A typical workstation set-up is shown in Figure 1.59—a computer with up to 4 megabytes of main memory, a display monitor and two methods of inputting commands (a keyboard for printed commands; a mouse or puck moved on a digitising plate to control a screen cursor, a graphics tablet or a digitiser).

Moving the mouse over the plate automatically moves the cursor over a menu displayed on the side of the screen, and by depressing a button on the mouse, the designer can execute the relevant function on the screen. In this way, given the appropriate software, two-dimensional and three-dimensional drawings can be prepared line by line on the screen, and when this is accomplished a hard copy of the drawing may be printed out on a printer or plotter.

Other methods of controlling the screen cursor include a light pen, joystick and rollers. Some systems have the menu displayed on a digitising tablet instead of on the screen. A digitising tablet is a special plate in which each co-ordinate corresponds to a screen co-ordinate, and returning the puck to a co-ordinate always brings the cursor to the same point on the screen. Some digitisers (graphics tablets) incorporate the menu for selection of drawing functions. Drafting competence with all computer graphics systems is directly related to the time spent at the terminal and will certainly differ from person to person.

CAD/CAM software

The range of CAD/CAM software available to users is enormous and increases dramatically year by year. Existing software packages are continually being updated and improved, then released as new versions. The continuous revision and improvement of software packages is absolutely essential if they are to remain competitive in this dynamic market.

It may be safely said that computer-aided drafting will never completely replace manual drafting. The latter will always have a place in the design office, even though such a place may eventually be very small. Drafters, however, will still have to acquire a knowledge of the principles of drawing and attain the ability to read

Fig. 1.59 *Typical CAD/CAM work station*   *SUN MICROSYSTEMS AUSTRALIA PTY LTD*

and analyse a drawing. These skills are necessary to provide the correct input commands into the computer graphics system and to be able to interpret and analyse the output.

In simple terms, a specific software package can be described as a large number of standard commands stored within the computer's memory. Knowing how to access each of these commands by the name provided, the operator can execute each command visually on a screen monitor. Thus, command by command, a two-dimensional or three-dimensional drawing may be built up in much the same way as a drafter manually constructs a drawing in the conventional manner.

Packages that provide two-dimensional and three-dimensional drawings are easily mastered by a drafter, and many educational institutions teach the use of these packages to apprentices, technicians, technical officers and professional engineers as part of their undergraduate course requirements.

Just as a drafter has to acquire a knowledge of drawing principles to be able to master two-dimensional and three-dimensional computer drawing, so too does the designer have to attain a knowledge of drawing principles as well as the principles of design of engineering elements before he or she can make use of the many software packages available to facilitate the design process.

The CAD process

Consider the computer design process for a critical component, one that has to be designed to withstand stresses or loads during its working life. The process consists of a number of quite specific steps from the initial concept to the manufacturing stage, as outlined in Figure 1.60. Simple components that do not require mathematical modelling or stress analysing need only steps 1, 8 and 9 to complete their premanufacturing phase.

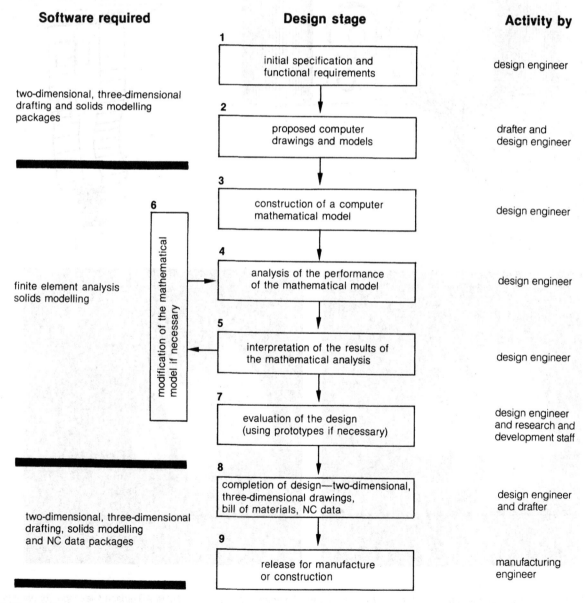

Software required	Design stage	Activity by
	1 initial specification and functional requirements	design engineer
two-dimensional, three-dimensional drafting and solids modelling packages	**2** proposed computer drawings and models	drafter and design engineer
	3 construction of a computer mathematical model	design engineer
finite element analysis solids modelling	**6** modification of the mathematical model if necessary — **4** analysis of the performance of the mathematical model	design engineer
	5 interpretation of the results of the mathematical analysis	design engineer
	7 evaluation of the design (using prototypes if necessary)	design engineer and research and development staff
two-dimensional, three-dimensional drafting, solids modelling and NC data packages	**8** completion of design—two-dimensional, three-dimensional drawings, bill of materials, NC data	design engineer and drafter
	9 release for manufacture or construction	manufacturing engineer

Fig. 1.60 *Steps in the computer design process*

The interaction of the computer and its associated software packages with the design process greatly assists the designer and the drafter throughout all the design stages. Initially the component may be drawn line by line in two and/or three dimensions using the computer's graphics capability, or it may be modelled using one of three systems:

1. *Wireframe modelling* constructs a model as a combination of frames added together to give the three-dimensional shape (Fig. 1.61). This method shows no mass representation, hence it is not suitable for sectioning and mass/volume calculations.

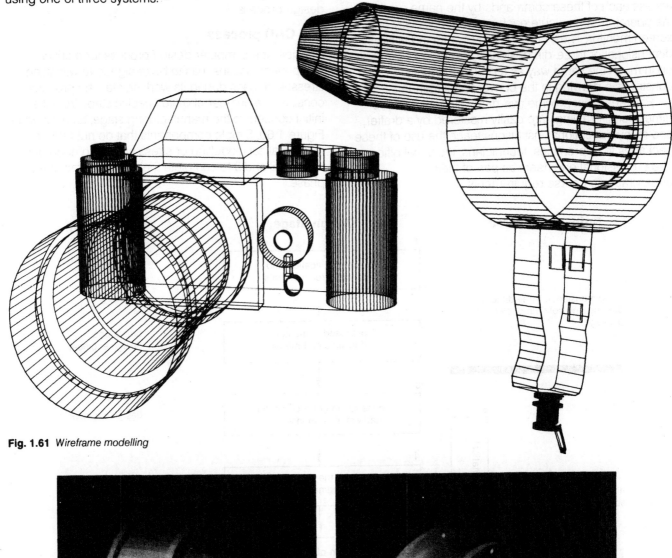

Fig. 1.61 *Wireframe modelling*

Fig. 1.62 *Surface modelling of complex shapes (a) solid shaded model of detector component of STARLAB (space telescope) (b) photograph of detector component   HAWKER DE HAVILLAND LIMITED (ENGINEERING DIVISION)*

2. *Surface modelling* also creates the three-dimensional shape of an object but has no mass representation. However, its prime use is to represent complex surfaces, such as those found on vehicle body panels, aircraft bodies and ship hulls (Fig. 1.62). The surface is divided into patches and is controlled mathematically so that the shape or contour of the surface is free of ripples. The strongest feature of this system is that it will allow the interactive modification of a surface after an initial design. Also, local areas may be modified and blended into the overall surface without altering the surface outside the local area concerned.

3. *Solids modelling*, as its name implies, includes the mass of the object as well as its surface shape and hence is suitable for mass and volume calculations. The model is created by one of three techniques.

First, it may be developed in various ways by the combination of two or more primitive solids, which are included as a standard part of the software package (Fig. 1.63(a) and (b)).

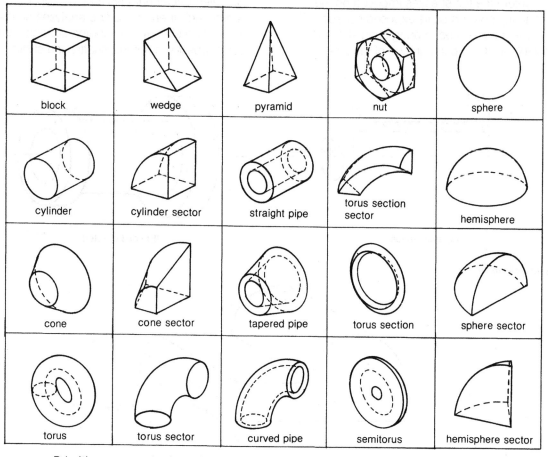

Primitive geometrical solids used in Pafec's Boxer solids modelling software

(a)

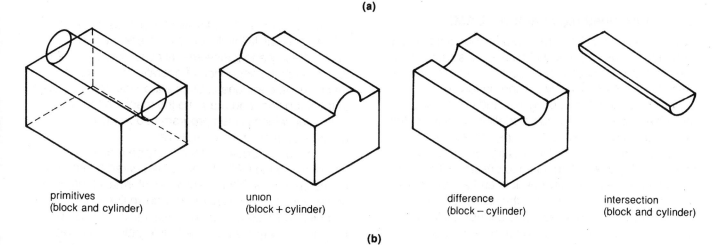

primitives
(block and cylinder)

union
(block + cylinder)

difference
(block − cylinder)

intersection
(block and cylinder)

(b)

Fig. 1.63 *Solids modelling*

Second, spinning a given two-dimensional contour (Fig. 1.64(a)) about a specified axis will create a solid model (Fig. 1.64(b)).

Third, a two-dimensional shape can be drawn and then 'dragged' normal to the plane of the shape to produce an extruded solid (Fig. 1.65).

These three techniques, used singly or in combination, enable any solid shape to be modelled in its exact geometrical configuration. This allows the designer to apply *finite element analysis*, an exact mathematical modelling technique that allows the designer to determine the stresses imposed on the component at any position on its exposed surfaces. The design can then be modified to bring such stresses within allowable limits. The integration of the three

systems—wireframe, surface and solids modelling—within the one software package is highly desirable as each has its own particular value to the designer.

The process of remodelling, analysing and modifying can be carried out over and over again at great speed and at a fraction of the cost of the conventional methods described earlier. Thus, the 'ideal' model is designed for strength, appearance and economy of construction.

The above process is represented in Figure 1.60 by steps 3, 4 and 5 and back through loop 6 if necessary. Finite element techniques may be applied to static, dynamic and thermal loading analyses on many engineering elements, including springs, masses, beams, trusses, columns, shafts, and shells.

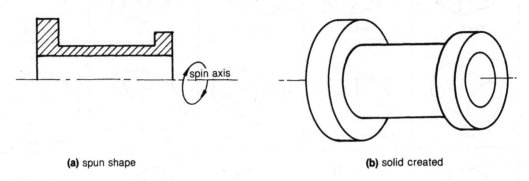

(a) spun shape

(b) solid created

Fig. 1.64 *Spun solid model*

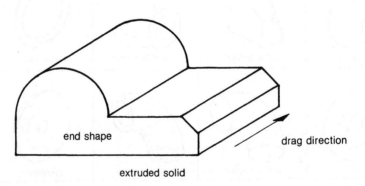

end shape

drag direction

extruded solid

Fig. 1.65 *Extruded solid model*

Computer-aided manufacture (CAM)

Once a design has been released for manufacture, the computer modelling does not cease; the simulation of the production process will enable the optimisation of operations prior to actually making prototypes, or purchasing expensive tooling, jigs and fixtures. It is possible to carry out a solids modelling exercise of the machine tool table, fixtures, clamps, tool pieces and workpiece in position and to actually *move* the tool through its operation sequence to ensure that such movements are physically possible and are those that will produce the component in the shortest time.

Having verified tool paths and set-ups, instructions are sent to the machine tool to produce the component in one of two ways. First, by using what is known as a numerical control (NC) software package, a tape can be

produced from the CAD terminal and its associated printer. This will faithfully record in a special coded format the *x*, *y* and *z* movements necessary to move the tool through the sequence of operations required to machine the component. Such a tape can then be taken to the shop floor and read into the machine tool's computer, which is then programmed ready to commence machining operations.

Second, the above process may be modified in cases where the CAD station is hard-wired to machines on the shop floor. Then machining information created at the workstation is sent directly down line to the machine tool's computer, thus eliminating the use of tape. In this way it is possible to integrate the operations of a whole factory of machining processes from a central computer.

Geometrical constructions, charts and diagrams 2

Drafting is a skill not easily obtained and a high degree of efficiency can only be achieved after a drafter has become proficient in the use of the instruments of the art. A sound knowledge of basic geometrical constructions is needed to complete drawings in a skilful manner.

This section introduces these instruments with a series of exercises and provides a large number of basic geometrical constructions for reference. It includes graded practical exercises to test instrument skill and geometrical construction knowledge.

Exercises using drawing instruments

The following exercises are designed to give some degree of efficiency in the use of drawing instruments. It is suggested that these exercises be used in the first practical sessions or as a first assignment.

Students should attempt to reproduce the following drawings, maintaining the neatness and clarity shown in the examples.

2.1

Practise drawing types of line A to H from pages 5–7. Full lines are outlines, type A. Dash lines are hidden detail, type B.

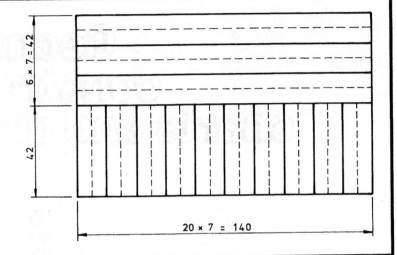

2.2

Mark off the 140 mm and 84 mm sides into 28 mm segments with dividers to ensure perfect squares.

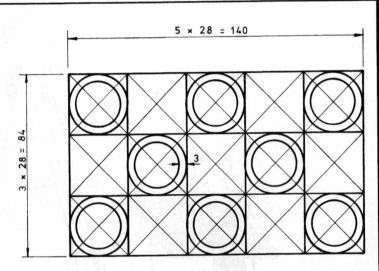

2.3

Commence at centre A
with 6 mm radius,
work out to B,
then in to C.

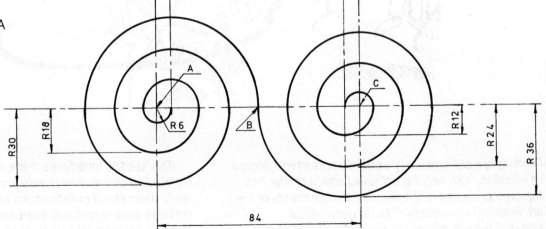

2.4

Locate centres of 70 mm diameter circles on the corners of a 70 mm square, then draw circles lightly. Construct triangles using the 30°-60° set square.

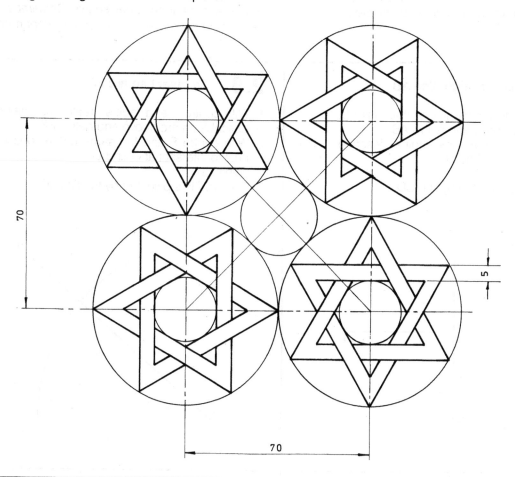

2.5

Divide the horizontal diameter into twelve 7 mm segments and use these points as centres for the semicircles.

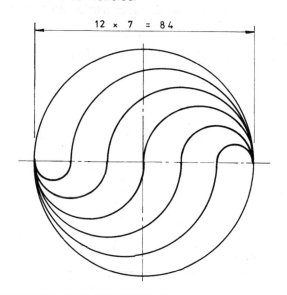

12 × 7 = 84

2.6

Divide the circle into twenty-four sectors using only the 45°and 30°-60°set squares, singly and in combination.

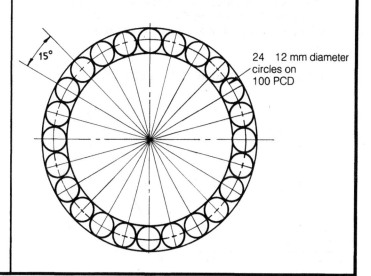

15°

24   12 mm diameter circles on 100 PCD

Exercises in geometrical constructions

In technical drawing, it is often necessary to make geometrical constructions in order to complete an outline.

The following basic constructions are given for reference. Practise each example until you become familiar with the techniques shown. *Remember*, neatness and accuracy are extremely important.

2.7 To bisect a straight line AB

1. With centre A and radius greater than half AB, describe an arc above and below AB.
2. With centre B and the same radius, describe arcs to intersect at C and D.
3. Join C and D.
4. AB is bisected at point O by line CD.

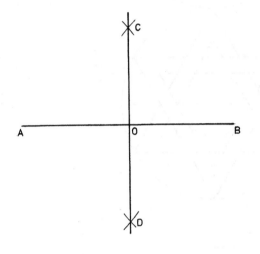

2.8 To bisect an arc EF

1. With centre E and radius greater than half EF, describe an arc above and below EF.
2. With centre F and the same radius, describe arcs to intersect at G and O.
3. Join OG.
4. The arc EF is bisected by the line OG.

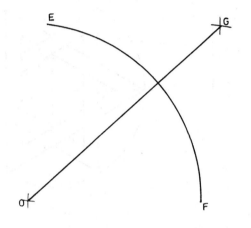

2.9 To construct an angle equal to a given angle

1. Let ABC be the given angle.
2. Draw a line EF.
3. With B as centre and any radius, describe an arc GH.
4. With E as centre and the same radius, describe an arc to intersect EF at K.
5. With centre K and radius GH, describe an arc at J.
6. Join EJ.
7. The angle DEF is equal to the angle ABC.

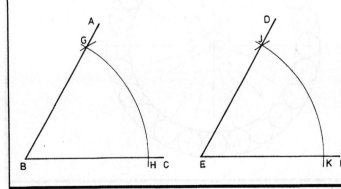

2.10 To bisect a given angle

1. Let ABC be the given angle.
2. With centre B and any radius describe an arc ED.
3. With centres E and D draw similar arcs to intersect at F.
4. Join BF.
5. The required bisector of the angle ABC is BF.

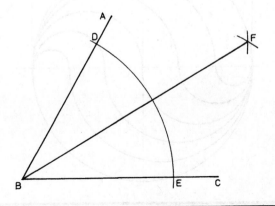

2.11 To trisect a right angle

1. Let ABC be the given angle.
2. With centre B and any radius, describe an arc DE.
3. With D and E as centres and the same radius, mark off G and F respectively on this arc.
4. Join FB and GB.
5. Angle ABC is trisected by FB and GB.

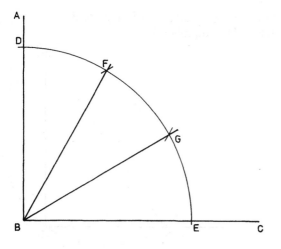

2.12 To construct an equilateral triangle on a given side

1. Let AB be the given side.
2. With centre A and radius AB, describe an arc.
3. With centre B and the same radius, describe an arc to intersect the first at point C.
4. Join AC and BC.
5. ABC is the required equilateral triangle.

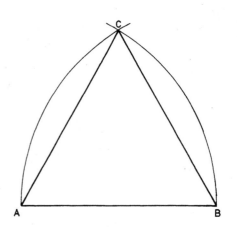

2.13 To construct an isosceles triangle, given the base and altitude

1. Let AB be the base of the triangle and a the altitude.
2. Bisect AB by the perpendicular line EF at point C.
3. From point C, mark off the altitude CD equal to a.
4. Join AD and BD.
5. ABD is the required isosceles triangle.

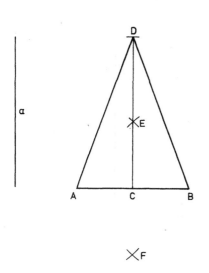

2.14 To construct a triangle, given the three sides

1. Let a, b, and c be the given sides.
2. Draw AB equal to the side a.
3. With centre A and radius equal to side c, describe an arc.
4. With centre B and radius equal to side b, describe an arc to intersect the first at point C.
5. Join AC and BC.
6. ABC is the required triangle.

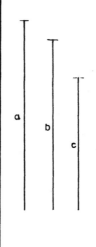
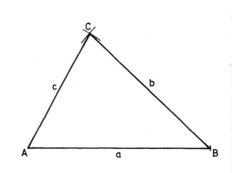

2.15 To construct an isosceles triangle, given the base and the angle at the vertex

1. Let AB be the given base and DEF the given angle.
2. Bisect AB at H and produce the bisector.
3. At A, construct angle PAQ equal to angle DEF.
4. Bisect angle QAB.
5. Produce the bisector to meet the perpendicular from H at point C.
6. Join BC.
7. ABC is the required triangle.

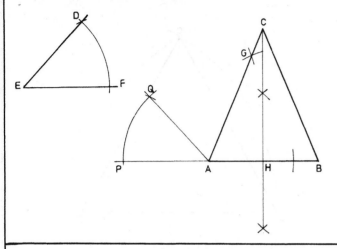

2.16 To draw a perpendicular to one end of a given line

1. Let AB be the given line.
2. Mark point O any convenient distance above AB.
3. With O as centre and radius OB, describe an arc cutting AB at point C and passing through B.
4. Join CO and produce to intersect the arc at point D.
5. Join DB.
6. DB is perpendicular to AB at point B.
   Note: ABD equals 90°.

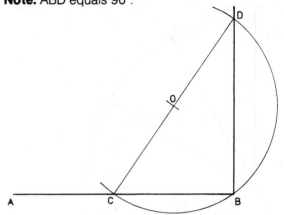

2.17 To draw a perpendicular to a line from a point outside the line

1. Let AB be the given line and P the given point.
2. With P as centre and any radius, describe an arc to intersect AB at C and D.
3. With C as centre and any radius, describe an arc below AB.
4. With D as centre and the same radius, describe an arc to intersect the first at point Q.
5. Join QP.
6. PE is perpendicular to the line AB.
   Note: AEP and PEB are both equal to 90°.

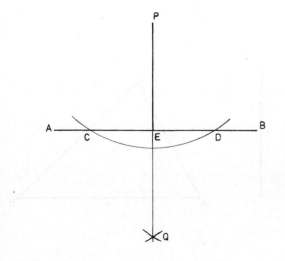

2.18 To draw a perpendicular from a point in a line

1. Let AB be the given line and P a point in the line.
2. With P as centre and any radius, describe an arc to intersect AB at points C and D.
3. With C and D as centres and any radius, describe arcs to intersect at point E.
4. Join EP.
5. EP is the required perpendicular.
   Note: APE and EPB are both equal to 90°.

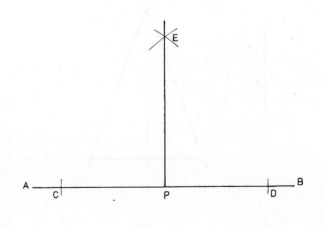

2.19 To divide the circumference of a circle into twelve parts

1. Let O be the centre of the given circle.
2. Draw diameter AOB perpendicular to diameter COD (see Exercise 2.18).
3. With centres A, C, B and D and radius AO, describe arcs to intersect the circumference of the circle.
4. The circle is divided into twelve equal parts by points A, J, E, C, G, K, B, M, H, D, F and L.

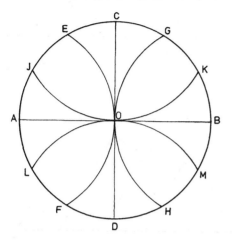

2.20 To construct a square on a given side

1. Let AB be the given side.
2. At points A and B construct perpendiculars equal in length to AB (see Exercise 2.18).
3. Join CD.
4. ABDC is the required square.

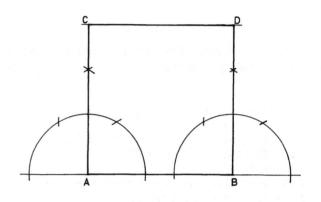

2.21 To construct a square, given the diagonal

1. Let AB be the given diagonal.
2. Bisect the diagonal AB at point O (see Exercise 2.7).
3. With centre O and radius OA, describe arcs to intersect the bisector at points C and D.
4. Join AC, CB, BD and DA.
5. ACBD is the required square.

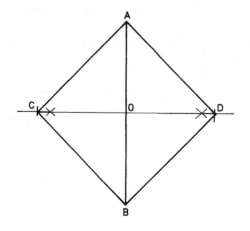

2.22 To construct a rectangle, given the length of its adjacent sides

1. Let a and b be the adjacent sides.
2. Draw AB equal to the length of side a.
3. At B erect BD perpendicular to AB and equal to side b (see Exercise 2.16).
4. With centre A and radius BD, describe an arc.
5. With centre D and radius AB, describe an arc to intersect at point C.
6. Join AC and CD.
7. ABDC is the required rectangle.

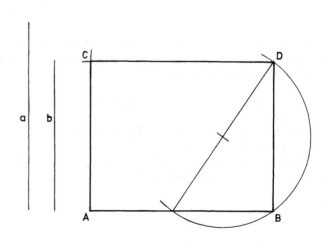

2.23 To construct a rhombus, given the side and a base angle

1. Let BC be the given side and MON the given angle.
2. At B construct the angle ABC equal to the given angle MON (see Exercise 2.9).
3. Set off BA equal to BC.
4. With A and C as centres and radius BC, describe arcs to intersect at point D.
5. Join AD and CD.
6. ABCD is the required rhombus.

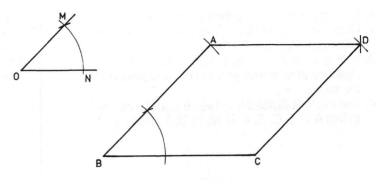

2.24 To divide a line into any number of equal parts, e.g. six equal parts

1. Let AB be the given line.
2. Draw AC at an angle of approximately 30° to AB.
3. With any convenient radius, set off from A six equal divisions along the line AC.
4. Join 6 to B.
5. With the aid of two set squares draw lines parallel to 6B from the other points.
6. The parallel lines cut AB into six equal parts.

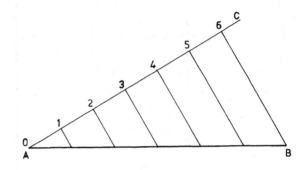

2.25 To draw a line parallel to a given line through a given point

1. Let AB be the given line and P the given point.
2. From P and with a radius less than AP, describe an arc to cut AB at D.
3. With centre D and radius DP, describe an arc to cut AB at E.
4. With centre D and radius EP mark off DF.
5. Join FP to give the required line.

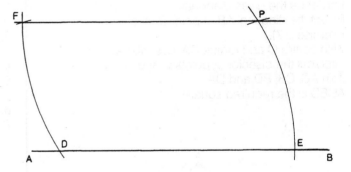

2.26 To draw a line parallel to a given line at a given distance from it

1. Let AB be the given line and c the given distance.
2. From two points well apart on AB draw two arcs of radius equal to c.
3. Draw a line tangential to these two arcs.
4. The line CD is parallel to AB.

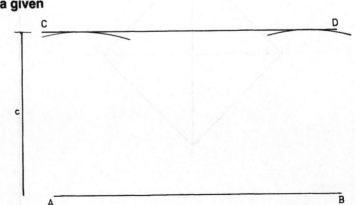

2.27 To divide a line into any number of equal parts, e.g. five equal parts

1. Let AB be the given line.
2. From A draw AC at any convenient angle to AB.
3. From B, draw BD parallel to AC.
4. With any convenient radius, set off from A five equal divisions along AC.
5. With the same radius, set off from B five equal divisions along BD.
6. Join 0 and 5, 1 and 4, and so on.
7. The parallel lines cut AB into five equal parts.

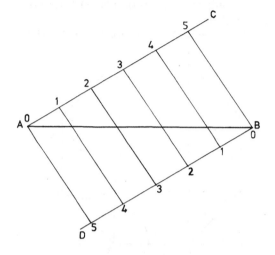

2.28 To construct a regular hexagon on a given side

1. Let AB be the given side.
2. With centre A and radius AB, describe an arc.
3. With centre B and the same radius, describe an arc to cut the first at O.
4. With O as centre and OA as radius, describe a circle.
5. From A or B and using radius OA, step off arcs around the circle at FEDC.
6. Join all points to complete the hexagon.

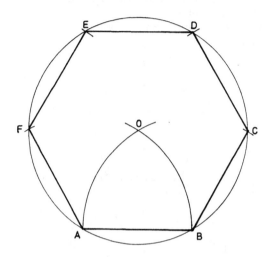

2.29 To construct a regular pentagon on a given side

1. Let AB be the given side.
2. Bisect AB at point C (see Exercise 2.7).
3. Erect a perpendicular at B (see Exercise 2.18). Make BD equal to AB.
4. With C as centre and CD as radius, describe an arc to cut AB produced at point E.
5. From A and B, with radius AE, draw arcs to intersect at F.
6. With A, B and F as centres and radius AB, draw arcs to intersect at G and H.
7. Join FG, GA, FH and HB to complete the pentagon.

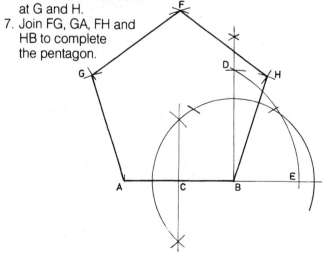

2.30 To inscribe a regular pentagon in a given circle

1. Let the given circle have centre O.
2. Draw the horizontal diameter AB and the vertical diameter CD to intersect at O.
3. Bisect OB at E (see Exercise 2.7).
4. With E as centre and radius EC, describe an arc to intersect AB at F.
5. With C as centre and radius CF, describe an arc to intersect the given circle at G and K.
6. Join CG. This is one side of the pentagon.
7. With G and K as centres and CG as radius, set off points H and J.
8. Join GH, HJ, JK and KC to complete the pentagon.

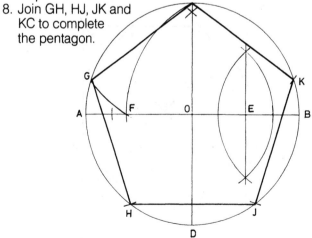

2.31 To construct a regular pentagon on a given side

1. Let AB be the given side.
2. With centres A and B and radius AB, draw complete circles.
3. With C, the point of intersection, as centre and the same radius, draw an arc to cut both circles at D and E.
4. Join C to F and produce.
5. From D and E draw lines through F to obtain G and H.
6. With centres G and H and radius AB, describe arcs to intersect at K.
7. Join AG, GK, KH and HB to complete the pentagon.

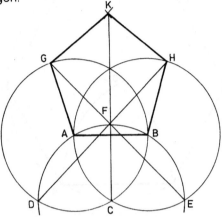

2.32 To construct a regular octagon on a given side

1. Let AB be the given side.
2. At A and B erect perpendiculars (see Exercise 2.18).
3. At A and B construct angles of 45° (see Exercise 2.10).
4. Mark off AE and BF equal to AB.
5. Join EF. Erect perpendiculars from E and F (see Exercise 2.16).
6. Mark off EG and FH equal to AB.
7. With G and H as centres and AB as radius, describe arcs to meet perpendiculars from A and B at C and D.
8. Join GC, CD and DH to complete the octagon.

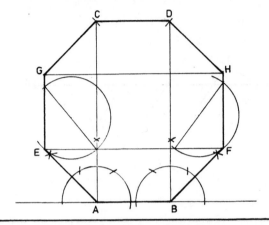

2.33 To inscribe a regular octagon in a given square

1. Let ABCD be the given square.
2. Draw the diagonals AC and BD to intersect at O.
3. With centres A, B, C and D, and radius equal to AO, draw arcs to intersect the sides of the square.
4. Letter all points of intersection.
5. Join FG, HJ, KL and ME to complete the octagon.

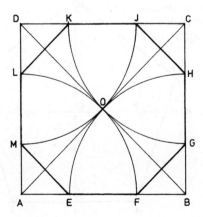

2.34 To inscribe a regular heptagon in a circle

1. Draw the horizontal diameter AB. From A draw AP at any convenient angle to AB.
2. From A step off seven equal divisions along AP and join 7 to B.
3. Draw a line parallel to 7B from point 2 to cut AB at C.
4. With A and B as centres and AB as radius, describe arcs to intersect at O.
5. Join OC and produce to intersect the circumference of the circle at D.
6. Join AD. This is one side of the heptagon.
7. Step off AD around the circumference of the circle. Join DE, EF, FG, GH, HJ and JA.

2.35 To construct a regular polygon on a given side, e.g. a heptagon

1. Let AB be the given side.
2. Bisect AB at point C (see Exercise 2.7).
3. With C as centre and CA as radius, describe an arc to intersect the bisector at 4.
4. With A as centre and AB as radius, describe an arc to intersect the bisector at 6.
5. Bisect the distance 6–4 at 5.
6. Add distance 4–5 to 6 to give 7.
7. With centre 7 and radius 7A, draw a circle.
8. Step AB around seven times to give a heptagon.

Note: to construct an octagon, add distance 4–5 to 7 giving 8. With centre 8 and radius 8A, draw a circle. Step AB around eight times to give the octagon.

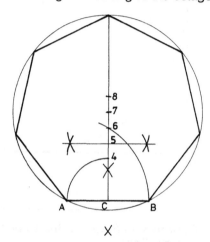

2.36 To construct a circle to pass through three given points, A, B and C, not in the same straight line

1. Join AB and BC.
2. Bisect AB and BC (see Exercise 2.7).
3. The bisectors intersect at point O, the centre of the required circle.
4. With centre O and radius OA (or OB or OC), describe a circle to pass through points A, B and C.

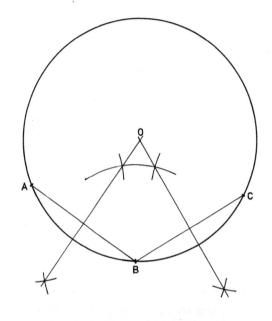

2.37 To construct a circle of given radius to pass through two given points

1. Join points A and B.
2. With centres A and B and radius r, describe arcs to intersect at O.
3. Then O is the centre of the required circle.
4. With O as centre and r as radius, describe a circle to pass through A and B.

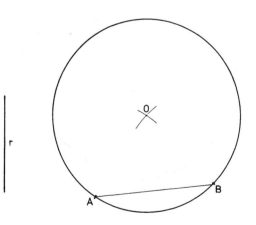

2.38 To construct a tangent to a circle from a point on its circumference

1. Let O be the centre of the circle and P the given point.
2. Join OP.
3. Erect a perpendicular APB to OP (see Exercise 2.16).
4. APB is the required tangent.

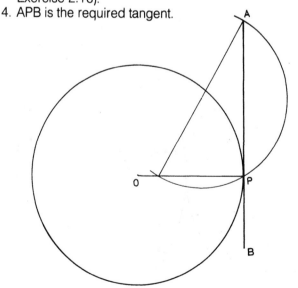

2.39 To construct a tangent to a circle from a point outside the circle

1. Let O be the centre of the circle and P the given point.
2. Join OP.
3. Bisect OP at point X (see Exercise 2.7).
4. With X as centre and XP as radius, describe a semicircle to intersect the circle of T.
5. Join PT. PT is the required tangent.

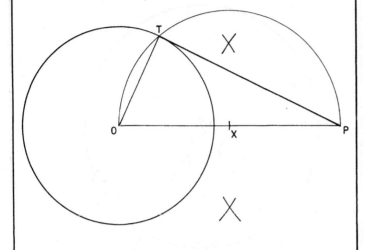

2.40 To draw an arc tangential to two straight lines at right angles

1. Let AB and BC be the two straight lines at right angles and c the radius of the arc.
2. From B mark off BE and BF equal to the radius c of the given arc.
3. With E and F as centres and radius c, describe arcs to intersect at O.
4. From O with radius c, draw an arc which will be tangential to AB and BC.

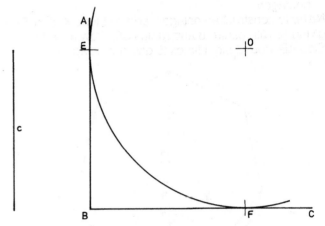

2.41 To draw an arc tangential to a given arc and centred on the opposite side to the point of tangency

1. Let A be the centre of the given arc.
2. Let a be the radius of the given arc and b the other radius.
3. From A and radius a + b, describe an arc.
4. With centre B anywhere on this arc and radius b, describe the required arc tangential to the given arc.

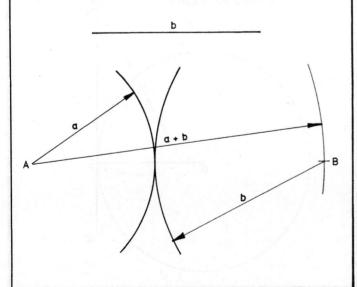

2.42 To draw an arc tangential to two straight lines

1. Let AB and BC be the given lines and c a given distance.
2. Draw two lines parallel to the given lines AB and BC at the given distance c from them.
3. Let the two lines intersect at point D.
4. With D as centre and radius c describe an arc which will be tangential to AB and BC.

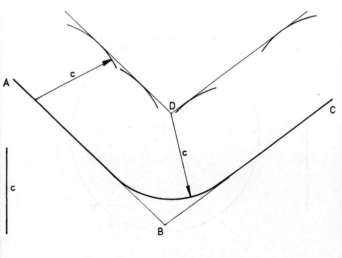

2.43 To draw an arc tangential to two arcs— externally

1. Let a, b and c be the radii.
2. With centres A and B, draw two arcs of given radii a and b.
3. With A as centre and a + c as radius, describe an arc.
4. With B as centre and b + c as radius, describe an arc to intersect at C.
5. With C as centre and radius c, describe an arc which will be tangential to the other two.

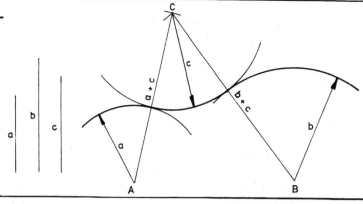

2.44 To draw an arc tangential to two arcs— internally

1. With centres A and B, draw two arcs of given radii a and b.
2. With A as centre and c−a as radius, describe an arc.
3. With B as centre and c−b as radius, describe an arc to intersect the first at C.
4. With C as centre and radius c, describe an arc which will be tangential to the other two.

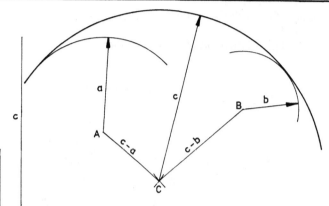

2.45 To draw an arc tangential to a line and another arc

1. Let BC be the given line and a and b given radii.
2. With centre A and radius a, describe an arc.
3. Draw a line parallel to BC and distance b from it.
4. With A as centre and radius a + b, describe an arc to intersect the parallel line at D.
5. With D as centre and b as radius, describe an arc which will be tangential to the given line BC and the arc with radius a.

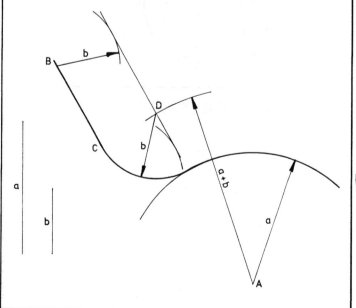

2.46 To draw an arc tangential to two arcs and enclosing one of them

1. Let A and B be centres of two arcs of radius a and b respectively and let c be the radius of the required arc.
2. With A and B as centres, describe arcs a + c and c − b respectively to intersect at C.
3. With centre C and radius c describe the required arc.
4. Join AC to intersect the curve at E and produce CB to intersect the curve at F. Then E and F are the points of tangency of the three arcs.

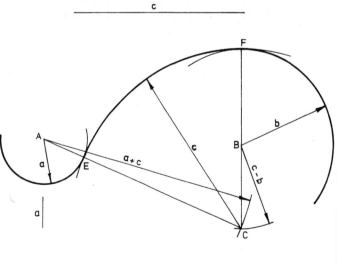

2.47 To draw a direct common tangent to two given circles

1. Let A and B be the centres of two circles of radii r and R respectively.
2. Join A and B and bisect AB at X (see Exercise 2.7).
3. With X as centre and AX as radius, draw a semicircle.
4. With centre B and radius R – r, describe a circle to intersect the semicircle at C.
5. Join BC and produce to intersect the larger of the given circles at D.
6. Join CA. With CA as radius and D as centre mark point E.
7. Join ED to give the required tangent.

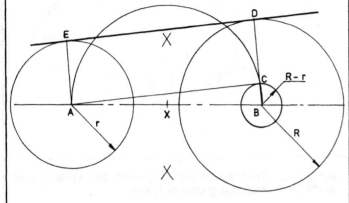

2.48 To draw a transverse common tangent to two given circles

1. Let A and B be the centres of two circles of radii R and r respectively.
2. Join A and B and bisect AB at X (see Exercise 2.7).
3. With X as centre and AX as radius, draw a semicircle.
4. With centre B and radius R + r, describe a circle to intersect the semicircle at C.
5. Join BC to intersect the circle centre B at point D.
6. Join CA. With CA as radius and D as centre mark point E.
7. Join ED to give the required tangent.

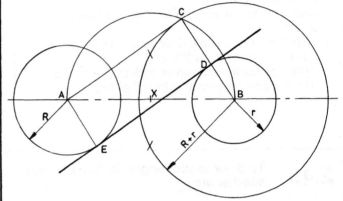

2.49 To obtain the length of the circumference of a circle

1. Let O be the centre of the given circle and AOC its diameter.
2. Draw a horizontal line tangential to point B (see Exercise 2.38).
3. Draw lines from points A and C to intersect the tangent at 60° and let the points of intersection be E and F. The line EF equals the semicircumference of the circle.
4. Mark length EF twice along a line to produce GH. GH equals the circumference of the given circle.

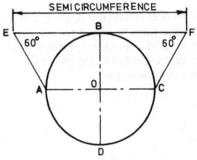

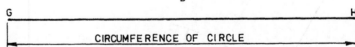

2.50　To draw an involute to a given circle

1. Draw a straight line AB tangential to the base circle and equal to its circumference (see Exercise 2.38).
2. Divide AB into twelve equal parts (see Exercise 2.27) and divide the base circle into twelve equal parts. This means that each part of the line AB is equal in length to each part of the circle.
3. From the points of division on the base circle, draw tangents using the 60°-30° set square.
4. Commencing at the next tangent to A, transfer length A1 to this tangent.
5. Transfer A2 to the next tangent, then A3 and so on until AB is reached.
6. Draw a smooth curve through the points A, 1, 2, ... 11, B, and this curve is the involute of the base circle.

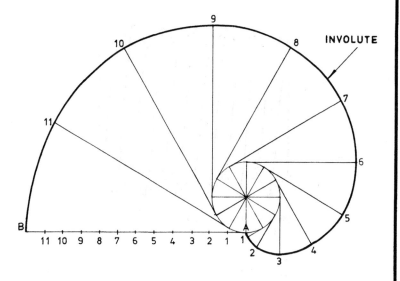

The involute curve

An *involute* can best be described as the path traced out by the end of a piece of string as it is unwound from the circumference of a circle called the base circle. One of the most useful applications of the involute curve in engineering is on the profile of gear teeth. Figure 2.1 illustrates an involute gear tooth in which that part of the tooth between the top and the base circle is of involute form.

The sides of the tooth are generated by two separate involutes from a common base circle and are spaced so that the tooth thickness at the pitch circle is a known value depending on the circumference of the gear and the number of teeth.

An involute may be generated from a straight line, polygon, circle or indeed any closed figure. Figure 2.2 illustrates involutes formed from various shapes.

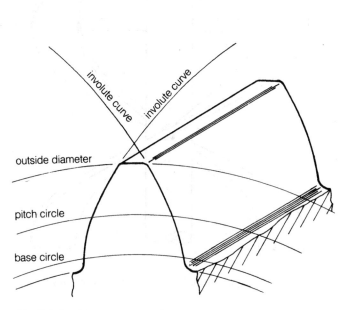

Fig. 2.1 *Involute spur gear tooth*

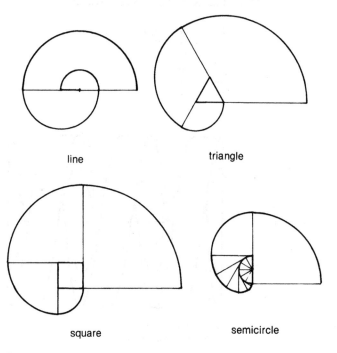

Fig. 2.2 *Involutes formed from various shapes*

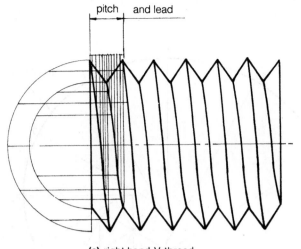

(a) right-hand V thread
(single start)

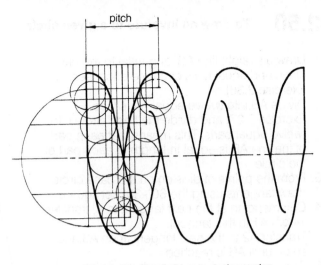

(d) right-hand round compression spring

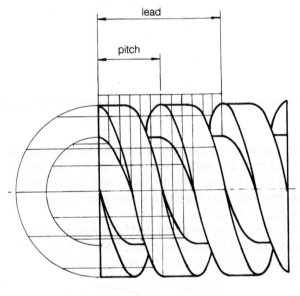

(b) right-hand square thread
(double start)

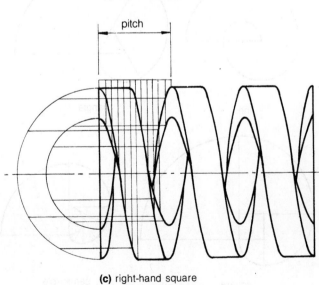

(c) right-hand square
compression spring

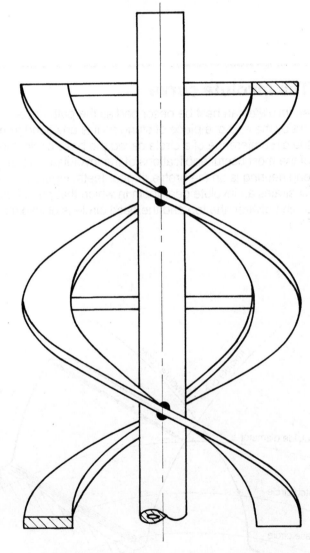

(e) left-hand double-flight ribbon-type screw conveyor

Fig. 2.3 *Practical applications of the helix*

The cylindrical helix

A *helix* is the path traced out by a point as it moves along and around the surface of a cylinder with uniform angular velocity and, for each circumference traversed, moves a constant length (called the *lead*) in a direction parallel to the axis.

The helix angle can be found by constructing a right-angled triangle, the base of which is the circumference of the cylinder and the vertical height of which is equal to the lead, as shown in Figure 2.4.

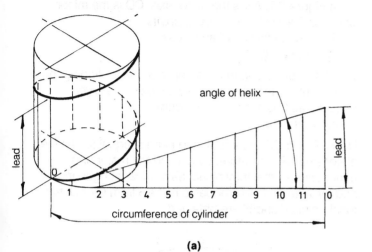

(a)

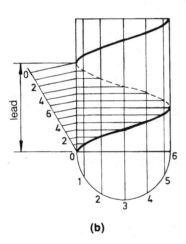

(b)

Fig. 2.4 *Right-hand helix*

The helix finds many applications in industry: screw threads, springs and conveyors are typical examples of its use. Drawings of these items are illustrated in Figure 2.3

Conic sections

When a cone is intersected by a plane, one of four well-known geometrical curves is obtained, depending on the angle of intersection. Figure 2.5 shows the side view of a cone and the curves that are relevant to a given plane of intersection.

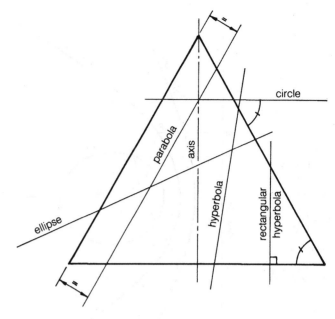

Fig. 2.5 *Various conic sections*

1. When the intersecting plane is perpendicular to the axis, the section outline is a *circle*.
2. When the intersecting plane makes a greater angle with the axis than does the sloping surface, the section outline is an *ellipse*.
3. When the intersecting plane makes the same angle with the axis as does the sloping surface, the section outline is a *parabola*.
4. When the intersecting plane makes a lesser angle with the axis than does the sloping surface, the section outline is a *hyperbola*.

The true shape of these four sections can be found by projecting an auxiliary view from the edge view of the sectioning plane (see Exercise 2.51).

The ellipse, parabola and hyperbola may also be constructed by considering the geometrical definition governing them: *An ellipse, parabola or hyperbola is the locus of a point which moves so that its distance from a fixed point (called the* focus*) and its perpendicular distance from a fixed straight line (called the* directrix*) bear a constant ratio to each other (called the* eccentricity*).*

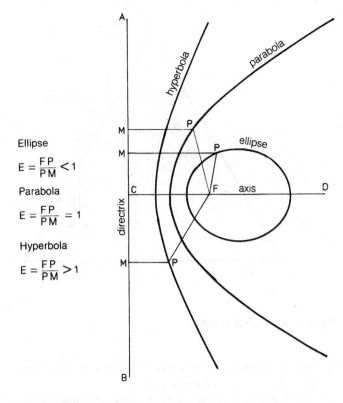

Ellipse

$$E = \frac{FP}{PM} < 1$$

Parabola

$$E = \frac{FP}{PM} = 1$$

Hyperbola

$$E = \frac{FP}{PM} > 1$$

Fig. 2.6 *Construction of conic sections by geometric ratios*

Figure 2.6 illustrates the three curves constructed in this manner. AB and CD are two lines at right angles and are referred to as the directrix and axis respectively. A point F, which can be positioned anywhere on the axis, is called the focus. The shape of the three curves is governed by a ratio called the eccentricity (E), which is the ratio:

$$E = \frac{\text{distance from a point (P) on the curve to focus (F)}}{\text{perpendicular distance from point (P) on the curve to the directrix}}$$

Hence, if the distance of the focal point F from the directrix and the eccentricity of the curve are known, the curve may be constructed quite easily by plotting a series of points that conform to the eccentricity ratio. For the three conics the ratios are as follows:

ellipse—any ratio less than unity, for example $\frac{1}{2}$
parabola—a ratio equal to unity
hyperbola—any ratio greater than unity, for
example $\frac{2}{1}$

The ellipse

An ellipse is a closed symmetrical curve with a changing diameter that varies between a maximum and minimum length. These two lengths are known as the *major axis* and *minor axis* respectively. The lengths of the axes may vary greatly, and it is upon their relative sizes that the shape of the ellipse depends.

An ellipse may be defined geometrically as the curve traced out by a point (P) which moves so that the sum of its distances from two fixed points (F and F′) is constant and equal in length to the major axis.

In Figure 2.7, AB is the major axis, CD is the minor axis, and F and F′ are the *focal points*.

From the definition of an ellipse,

FP + PF′ = AB

The definition also leads to a construction for finding the focal points F and F′, when only the axes are given, because as C is a point on the curve,

CF + CF′ = AB

Now CF and CF′ are equal, and each is equal to half the major axis AB. Therefore, by placing the major and minor axes so that they bisect each other at right angles and by taking a radius equal to half AB from C or D, the focal points F and F′ are obtained.

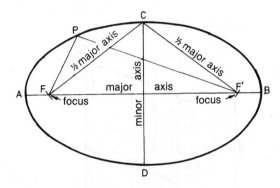

Fig. 2.7 *Definition of the ellipse geometry*

Exercises on conic sections and ellipses

The following constructions are given for reference. You should practise them and become familiar with the method and techniques.

2.51   To draw the conic sections by auxiliary projection

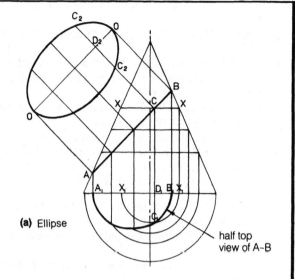

(a) Ellipse

The first curve, the *circle*, is drawn by determining the diameter of the cone at that section through which the intersecting plane passes.

The *ellipse* is drawn as shown in diagram (a) by projecting an auxiliary view perpendicular to the line of intersection AB. This gives the true shape of the section, which is elliptical.

1. Draw the side view of the cone and a half top view (semicircle) attached to the base.
2. Let AB be the edge view of the cutting plane.
3. To draw the half top view of this cutting plane, project A and B to intersect the base of the cone at A_1 and B_1, these two points being the extreme points on the half top view of the cutting plane A–B.
4. Draw any horizontal section X–X to intersect AB at C.
5. Draw the half top view of X–X, that is, the semicircle X_1–X_1.
6. Project C on to X_1–X_1 to intersect at C_1, which is one point on the half top view of A–B. Let this projector intersect the base at D_1.
7. Other points on this view are similarly found by taking other horizontal sections through A–B. Two are shown on the diagram.
8. Draw a smooth curve $A_1 C_1 B_1$ passing through the points determined to give the half top view of A–B.
9. Project OO parallel to AB. OO is the long axis of the ellipse.
10. Project C at right angles to AB to intersect OO at D_2 and mark off $D_2 C_2$ on either side of OO equal to $D_1 C_1$; that is, $C_2 C_2$ is the true width of the section A–B at point C.
11. The points C_2 are on the required ellipse. The four other points are similarly found to give eight points all told. More should be determined if the ellipse is larger.
12. A freehand curve is drawn through the points to give the required ellipse.

The *parabola*, diagram (b), and the *hyperbola*, diagram (c), are drawn in a similar manner except that they have a flat base because the intersecting plane A–B cuts through the base of the cone.

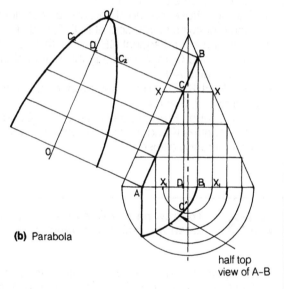

(b) Parabola

half top view of A-B

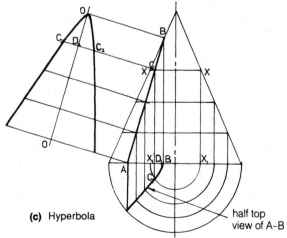

(c) Hyperbola

half top view of A-B

2.52 To construct an ellipse (approximate methods)

The four-centre method is used when given the major axis AB and the minor axis CD.

1. Draw the major and minor axes to intersect at O. Join AC.
2. With centre O and radius OC, describe an arc to intersect AO at E.
3. With centre C and radius AE, describe an arc to intersect AC at F.
4. Draw the perpendicular bisector of AF to intersect AO at G and CD (produced) at H. G and H are the centres of two arcs for forming half of the ellipse.
5. Make OJ equal OH and OK equal OG to give two more centres for the other half of the ellipse.
6. Join JG, HK and JK, and produce all three.
7. With centres H, J, G and K, and radii HC, JD, GA and KB respectively, describe arcs to form the ellipse. The tangent points of the four arcs are at points 1, 2, 3 and 4.

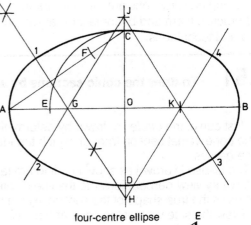

four-centre ellipse

The isometric ellipse is used when given the centre O and the radius of the circle AO required for isometric representation.

1. Draw the isometric axes (any two lines 60° apart) to intersect at O.
2. With centre O and radius AO, describe a circle to intersect the isometric axes at A, D, B and C.
3. Through these four points draw lines parallel to the isometric axes to intersect at E, F, G and H.
4. Draw the long diagonal GE to pass through O.
5. From E and G, mark off along this diagonal distances EJ and GK equal to AO.
6. Join FK, HK, FJ and HJ, and produce these to intersect EFGH at points 1, 2, 3 and 4.
7. With centres H and F, describe arcs 1–2 and 3–4 respectively.
8. With centres K and J, describe arcs 2–3 and 1–4 to complete the ellipse.

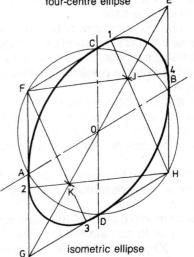

isometric ellipse

2.53 To draw an ellipse by concentric circles method, given the major and minor axes

1. Draw two concentric circles having diameters AB and CD, respectively equal to the major and minor axes of the ellipse.
2. Divide the circles into twelve equal parts with either the compass or a 30°–60° set square (see Exercise 2.19).
3. Draw perpendicular lines from points of division on the outside circle to intersect horizontal lines drawn from corresponding points of division on the inside circle.
4. The intersection of these lines together with points A, B, C and D gives twelve points on the ellipse.
5. Draw a freehand curve through these points to give the required ellipse.

This method of constructing an ellipse is probably the easiest to remember and the one generally recommended when the lengths of the axes are given.

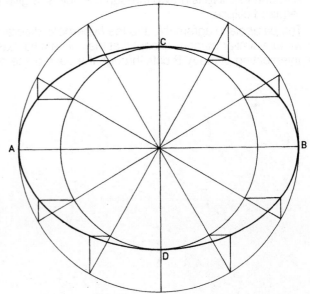

2.54 To draw an ellipse by the intersecting arcs method, given the major and minor axes

1. Draw the major and minor axes, AB and CD, to bisect each other at right angles.
2. With centre C and radius equal to half AB, describe arcs to cut AB at F and F′.
3. Select points 1, 2, 3, 4 and 5 between F and the centre of the ellipse. Place point 1 fairly close to F.
4. With radius of length A–1 and centre F and F′ describe four arcs, one in each quadrant of the ellipse.
5. With radius of length B–1 and centres F and F′, describe four more arcs to give points 1′ in each quadrant.
6. With radius of length A–2 and centres F and F′ describe four arcs, one in each quadrant of the ellipse.
7. With radius of length B–2 and centres F and F′ describe four more arcs to give points 2′ in each quadrant.
8. Repeat this procedure for points 3′, 4′ and 5′.
9. Draw a freehand curve through the points to give the required ellipse.

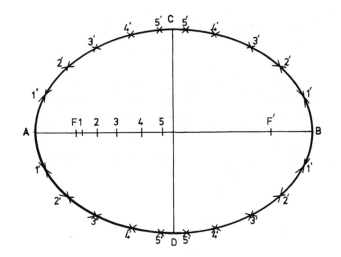

2.55 To draw an ellipse by the intersecting lines method

1. Draw the major and minor axes, AB and CD, to bisect each other at right angles at O.
2. Draw a rectangle EFGH of length AB and width CD (see Exercise 2.22).
3. Divide AO and AE into four equal parts (see Exercise 2.24).
4. Join C to the points of division on AE.
5. Join D to the point of division on OA and produce these lines to meet C1, C2 and C3 to give three points on the ellipse.
6. Using horizontal and vertical ordinates from these points, obtain three points in each of the other three quadrants.
7. Draw a freehand curve through these points to give the required ellipse.

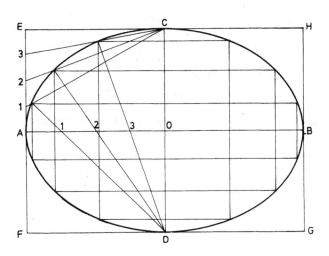

2.56 To draw an ellipse with the aid of a trammel, given the major and minor axes

1. Make a paper trammel from a piece of paper, as shown.
2. Draw the major and minor axes, AB and CD, to bisect each other at right angles.
3. Using the trammel as shown and taking care to keep F on the minor axis and E on the major axis, plot a series of points, using D as the marker. The trammel is rotated through 360° to give a full range of points on the curve.
4. Draw a freehand curve through the points to give the required ellipse.

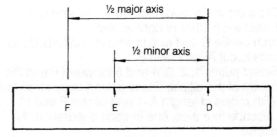

paper trammel

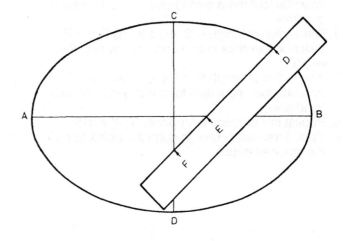

2.57 To draw the tangents to an ellipse from a point P outside the circumference of the ellipse

1. Let ADBC be the ellipse with focal points F and F′ and let P be the point outside the ellipse.
2. With centre P and radius PF′ describe an arc.
3. With centre F and radius AB describe another arc to cut the first arc at M and N.
4. Join FM and FN to cut the ellipse at T and T′ respectively.
5. Join PT and PT′ to complete the required tangents.

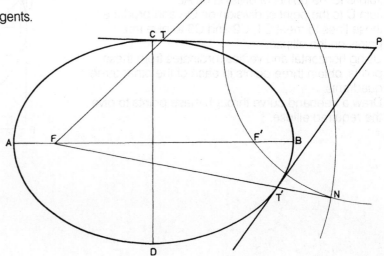

2.58 To draw a tangent and a normal to an ellipse from a point P on the circumference of the ellipse

1. Let ADBC be the ellipse with focal points F and F' and let P be the point on the ellipse.
2. Join FP and F'P and produce them to H and G.
3. Bisect angle FPG (see Exercise 2.10).
4. The bisector TP is the required tangent to the ellipse at point P.
5. Bisect angle GPH.
6. The bisector MN is the required normal to the ellipse at point P.

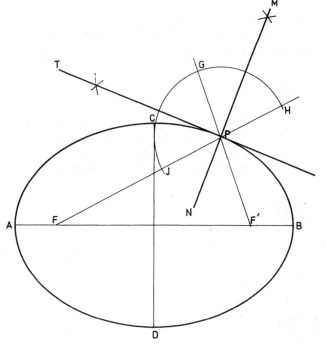

Exercises on plane geometry

The following exercises have been devised to test knowledge of plane geometry.

2.59

By bisection, divide a line 75 mm long into four equal parts.

2.60

Divide an arc into four equal parts.

2.61

Given a line AB 50 mm long, at point B erect a perpendicular making BC equal to AB. Divide the angle ABC into three parts.

2.62

Let AB, BC and AC measure 75 mm, 48 mm and 65 mm, respectively. Using these measurements, construct a scalene triangle.

2.63

Let the line RS, 65 mm long, represent the diagonal of a square. Construct a square RTSV on this diagonal.

2.64

The diagonal of a rectangle measures 100 mm. Construct a rectangle ACBD given that one side is 90 mm.

2.65

Construct a rhombus with sides of 50 mm and two internal angles of 60°.

2.66

Construct a triangle ABC, with AB 86 mm, BC 70 mm and the angle ABC 60°. From point B on the triangle construct a perpendicular to the side AC.

2.67

Construct an isosceles triangle given that the base AB is 45 mm and the base angles are 75°.

2.68

Draw two circles with diameters 30 mm and 88 mm, respectively, with their centres 100 mm apart. Draw direct common tangents to them.

2.69

Draw a line AC 106 mm long. Divide the line into two parts AB and BC, such that AB:BC is 2:3. At C erect a perpendicular CD 56 mm long and join AD and BD. Describe a circle to pass through points A, B and D.

2.70

Draw a line AB 100 mm long. At A construct an angle BAC of 75°. At B construct an angle ABC of 60°. On AB as diameter, draw a circle. Construct tangents to this circle from point C.

2.71

Construct a quadrilateral ABCD which has the following dimensions:

$$
\begin{aligned}
AB &= 62 \text{ mm} \\
\text{angle } BAD &= 52\frac{1}{2}° \\
CD &= 53 \text{ mm} \\
AD &= 70 \text{ mm} \\
BC &= 53 \text{ mm}
\end{aligned}
$$

2.72

From point P in a line AB, construct a line PR, 30 mm long such that angle RPB is 60°. Construct a circle tangent to AB at P and passing through R. At R construct a tangent to this circle. Construct a second circle tangent to AB and to the first circle.

2.73

The foci of an ellipse are 88 mm apart and its minor axis is 62 mm long. Construct the ellipse using the intersecting arcs method.

2.74

On a diagonal of 120 mm, construct a rectangle with one side 56 mm long. Using the intersecting lines method, construct an ellipse within the rectangle.

Scales

A scale indicates the ratio between the actual length of lines in a drawing and the true measurements of the object drawn. The scales recommended in the metric system are:

1. full size   1:1
2. enlargement   2:1, 5:1, 10:1, 20:1
3. reduction   1:2, 1:5, 1:10, 1:20

The scale should always be stated and inserted in the title block or bottom right-hand corner of the drawing.

Plain scales

These may be either reduction or enlargement scales. They generally represent either two separate units, or one unit and parts of that unit. When the larger unit is easily divided into the parts or second units, a plain scale may be constructed as shown in Exercises 2.59–2.61.

Enlargement scales

These scales are used to determine lengths that bear a constant ratio to the true size length, although they are always *larger* than the true size length, for example a 2:1 scale would show lengths twice their true size.

Drawings made using this scale would be twice the true size. Similarly a 5:1 scale drawing would be five times the true size drawing.

Reduction scales

These scales are used to determine lengths that bear a constant ratio to the true length, although they are always *smaller* than the true size, for example a 1:2 scale would show lengths half their true size. Drawings made using lengths taken from this scale would be half the true size. Similarly a 1:5 scale drawing would be one-fifth of the true size drawing.

Diagonal scales

Sometimes it is desirable to divide a length into a large number of small parts that cannot successfully be plotted or marked accurately. To overcome this difficulty a diagonal scale is used, the principle of which is shown in Figure 2.8.

Suppose a length has to be divided into, say, forty parts. The diagonal scale to achieve this, Figure 2.8(a) is drawn as follows. Divide the length into five equal parts (choose a number that will divide wholly into 40, for example 40/5 = 8). Eight horizontal lines are now drawn, equally spaced about 2 or 3 mm above the length line and numbered as shown. Diagonals are now drawn across the five divisions from the bottom right-hand corner to the top left-hand corner. For each horizontal line crossed by a diagonal from bottom to top, the diagonal moves 1/8 of the width of the rectangle to the left, that is, 1/8 of 1/5 = 1/40 of the length. Therefore, the total distance one diagonal moves to the left is 8/40 of the length. A distance that represents, say, 26/40 of the length is found by counting three of the diagonals (24/40) and going up the fourth diagonal to the second horizontal line. The distance required is between the two marks shown on the scale.

Figure 2.8(b) shows an alternative diagonal scale which also divides the length into forty parts. However, it uses four diagonals and ten horizontal lines. The distance representing 26/40 is the same as in Figure 2.8(a).

Note: The number of parts required must always be equal to the number of diagonals multiplied by the number of horizontal lines, that is, 4 × 10 or 5 × 8.

Figure 2.9 demonstrates the use of a diagonal scale. It shows a reduction scale of 1:20, representing metres and centimetres to a maximum of 3 m. Distances of 1.37 m and 2.82 m have been marked off.

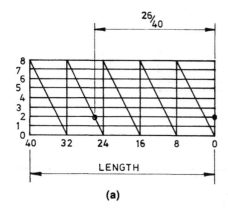

(a)

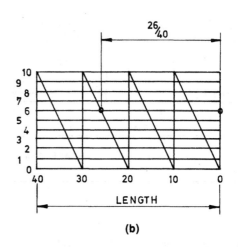

(b)

Fig. 2.8 *Principle of diagonal scale*

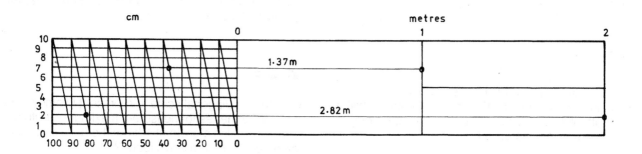

SCALE 1:20

Fig. 2.9 *Use of the diagonal scale to measure distances of 1.37 m and 2.82 m*

Exercises on scale construction

Exercises 2.75–2.77 should be followed so that students become familiar with scale problems.

Exercises 2.78–2.82 can then be undertaken to test the knowledge gained.

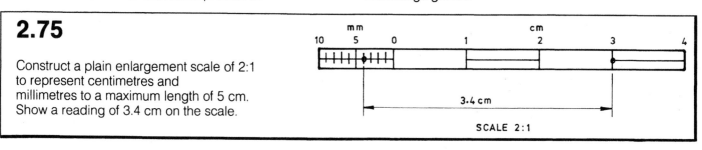

2.75

Construct a plain enlargement scale of 2:1 to represent centimetres and millimetres to a maximum length of 5 cm. Show a reading of 3.4 cm on the scale.

3.4 cm

SCALE 2:1

2.76

Construct a plain reduction scale of 1:5 to represent centimetres up to a length of 60 cm. Show a reading of 46 cm on the scale.

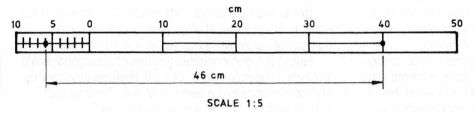

2.77

If a length of 50 mm represents 2 m, construct a plain reduction scale to show metres and tenths of a metre up to 5 m. Show a reading of 2.7 m on your scale.

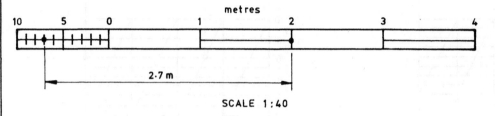

2.78

Construct a plain scale of reduction 1:10 to measure 140 cm in units of 10 cm and 1 cm. Show a length of 97 cm on your scale.

2.79

Construct a plain enlargement scale of 2:1 to measure 4 cm in units of cm and mm. Construct a pentagon with a scale length of side of 1.3 cm. Measure the scale length of its diagonal to the nearest mm.

2.80

Given that a measurement of 100 mm represents 5 m, construct a plain scale to show metres and tenths of a metre up to 7 m.

2.81

Construct a diagonal scale of 1:10 to measure metres and centimetres to a length of 2 m. Using true lengths of 13 cm, 9 cm and 8 cm, construct a triangle. Determine the area of the triangle to scale.

2.82

Construct a diagonal scale to represent kilometres and units of 20 metres to measure a maximum length of 5 kilometres. Use a scale of 1 km = 4 cm. To what scale length does 93 mm correspond?

Construction of geometric shapes and templates

The construction of geometrical shapes and templates demands a high degree of accuracy from the drafter. It is also important that a procedure be followed.

Figure 2.10 has been constructed using the following stages:

1. Accurately position all centre lines (Fig. 2.10(a)).

2. Circles and arcs that, when positioned, require no further construction can be drawn in heavy detail. Additional arcs or circular constructions are left in light detail (Fig. 2.10(b)).
3. The centre points for tangential arcs can be found (Fig. 2.10(c)).
4. Draw the tangential arcs and lines to complete the shape of the required template (Fig. 2.10(d)) and dimension (Fig. 2.10(e)).

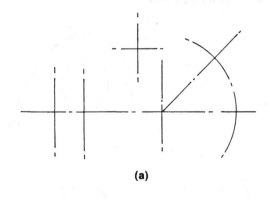

(a)

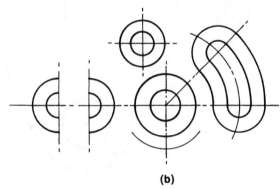

(b)

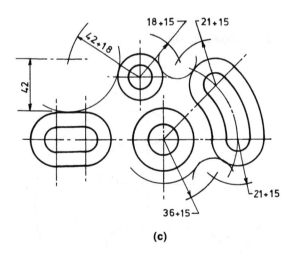

(c)

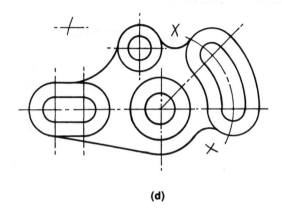

(d)

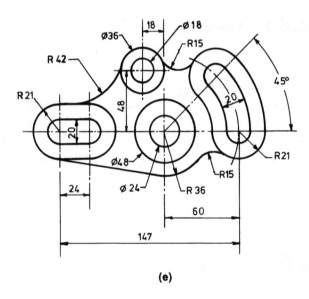

(e)

Fig. 2.10 *Construction of geometric shapes and templates*

Exercises on geometric shapes and templates

The following exercises should be constructed with the aid of compasses and set squares. Refer to the geometrical constructions described earlier in this chapter to ensure the use of correct and accurate methods.

All construction lines to locate radii centres should be shown. Indicate all points of tangency with a neat cross. Do not dimension the shapes. A uniform thickness and darkness of outline is required throughout.

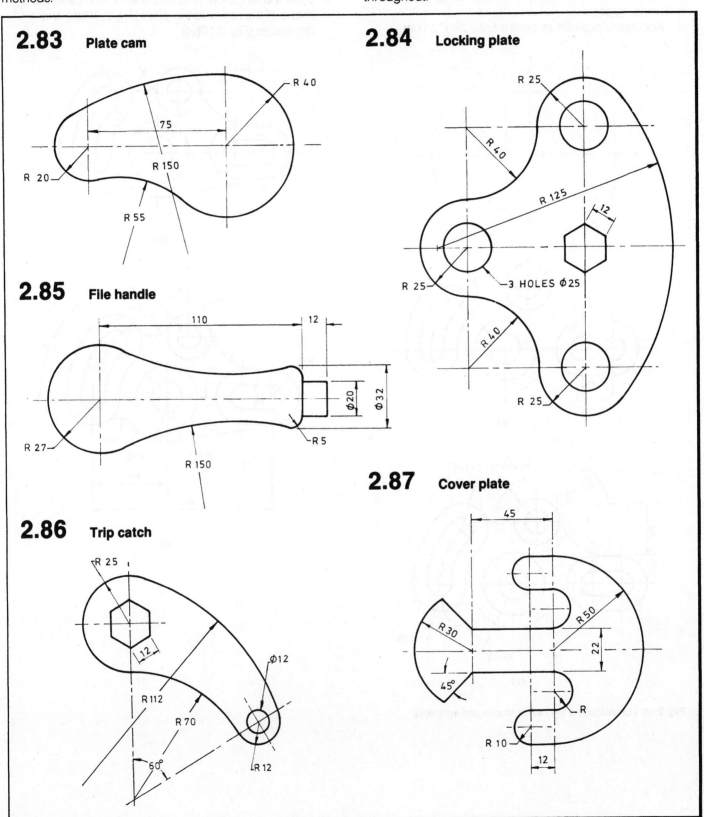

2.83 Plate cam

2.84 Locking plate

2.85 File handle

2.86 Trip catch

2.87 Cover plate

2.88 Trip lever

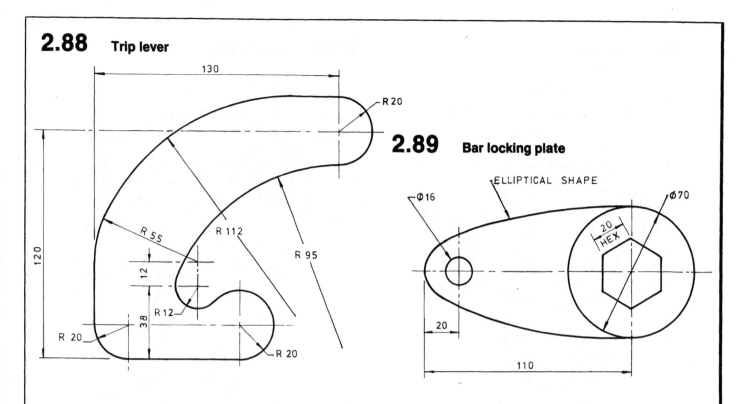

2.89 Bar locking plate

2.90 C wrench

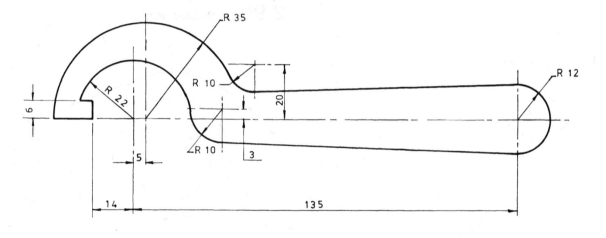

2.91 Spanner

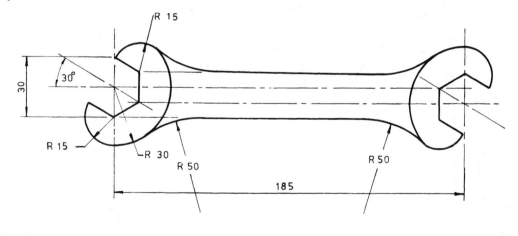

2.92 Cam plate

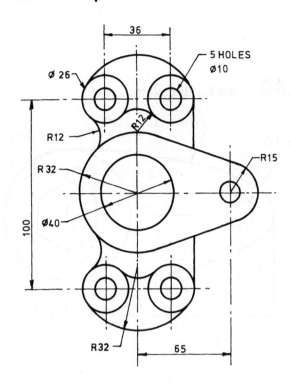

2.93 Lever bracket

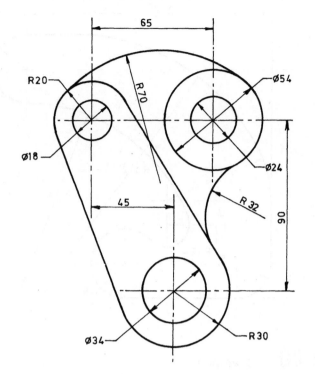

2.94 Radius link

2.95 Rocker arm

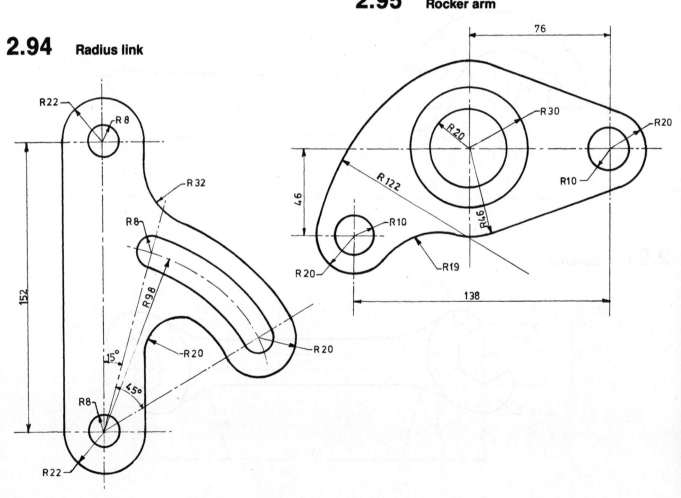

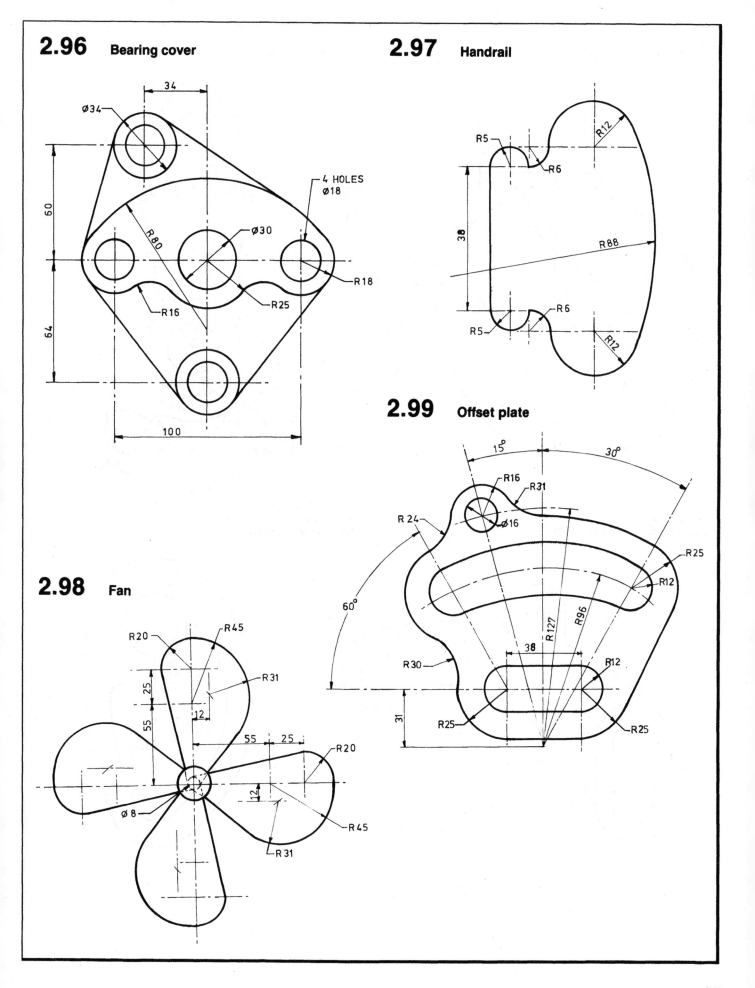

2.96 Bearing cover

2.97 Handrail

2.99 Offset plate

2.98 Fan

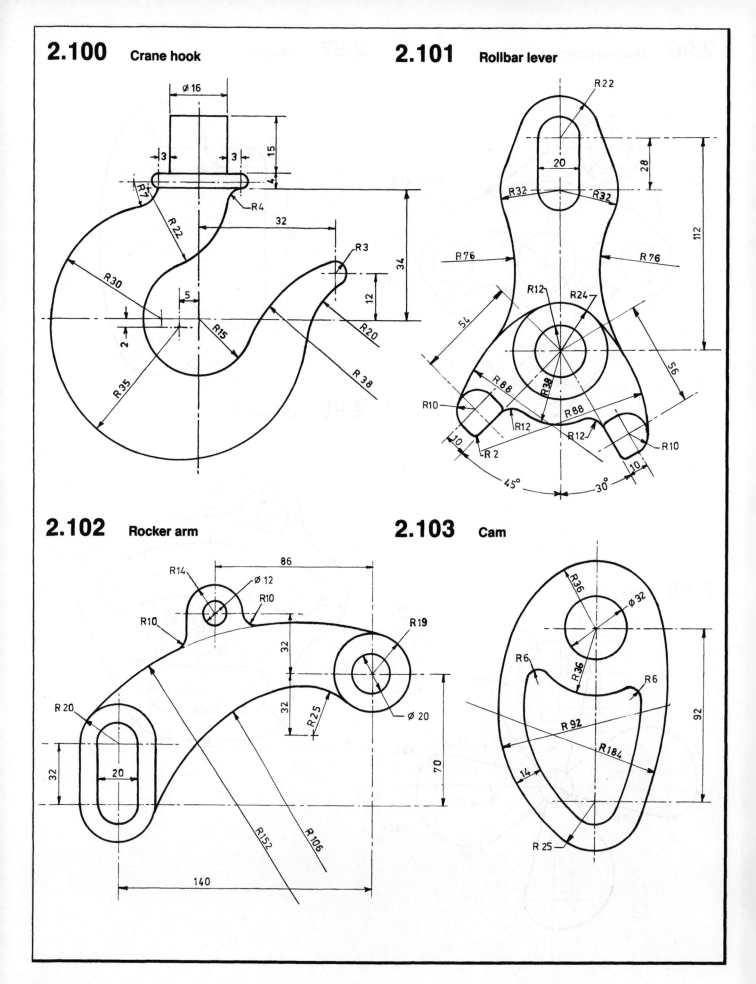

2.100 Crane hook

2.101 Rollbar lever

2.102 Rocker arm

2.103 Cam

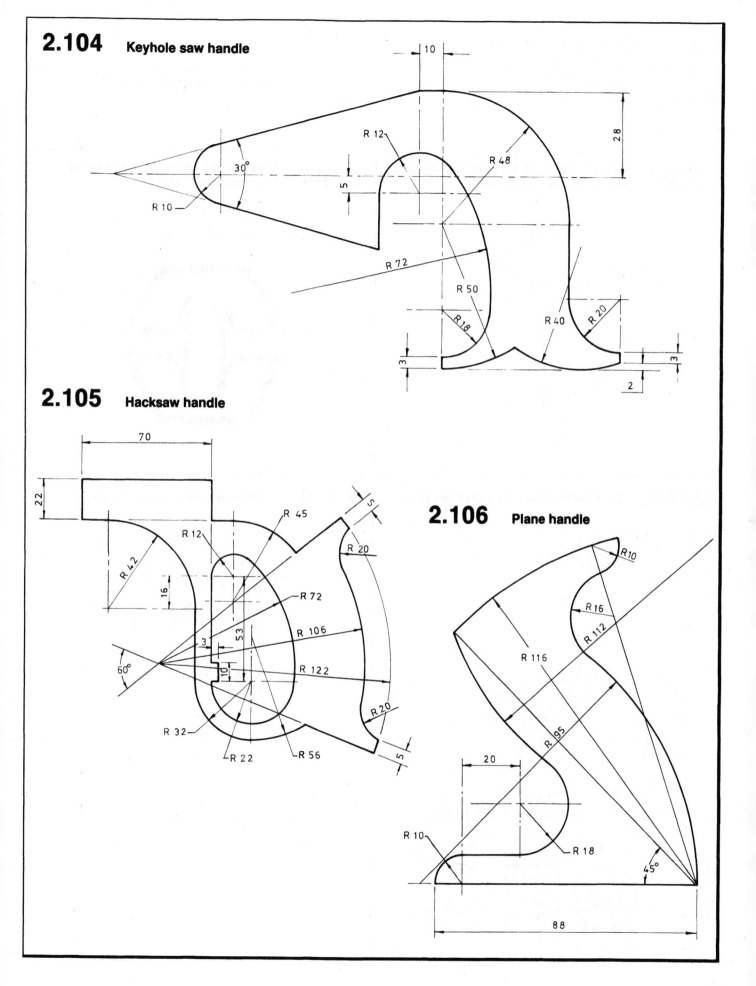

2.104 Keyhole saw handle

2.105 Hacksaw handle

2.106 Plane handle

Trade symbols, logograms and pictographs

Trade symbols, logograms and pictographs are often constructed using geometric shapes and templates, and the following examples will provide an interesting addition to this section.

Students should be encouraged to select other examples. Colouring-in could also be a part of the project.

2.107 John P. Young & Associates Pty Ltd Management Consultants

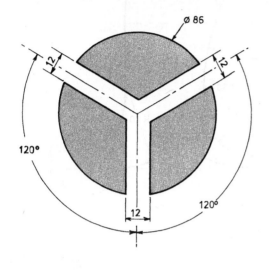

2.108 Northern Territory, Department of Transport and Works

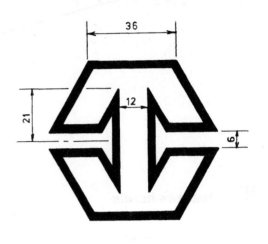

2.109 Queensland Electricity Commission

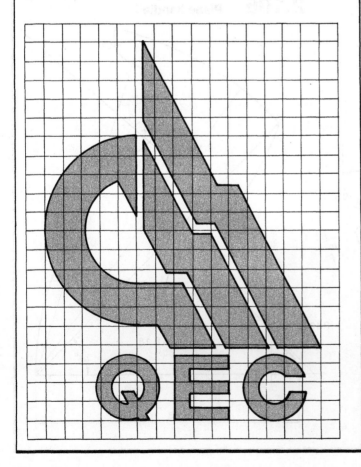

2.110 Legal Aid Commission, Victoria

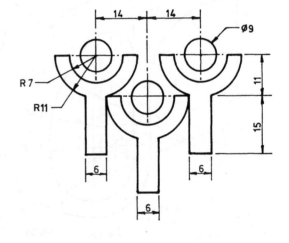

2.111 Worksafe Australia (National Occupational Health and Safety Commission)

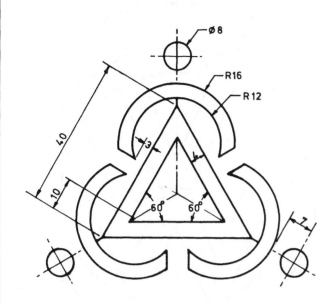

2.112 Alcoa of Australia Limited

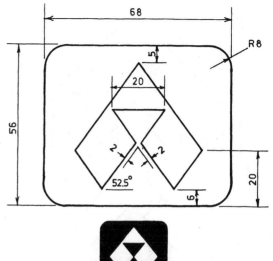

ALCOA
AUSTRALIA

2.113 Chandler and MacLeod Consultants Pty Ltd

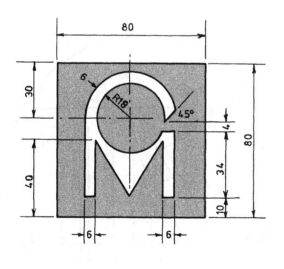

2.114 Reark Research Pty Ltd

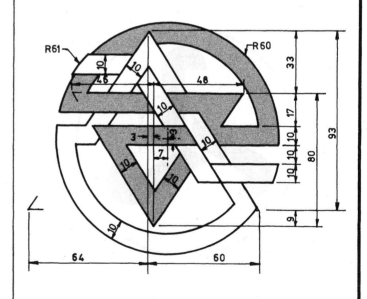

2.115 World Travel Headquarters Pty Ltd

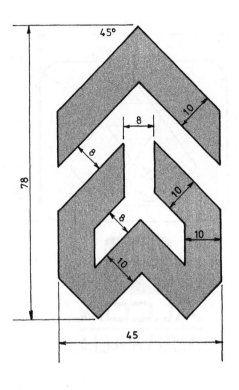

2.116 Volkswagen Australia Pty Limited

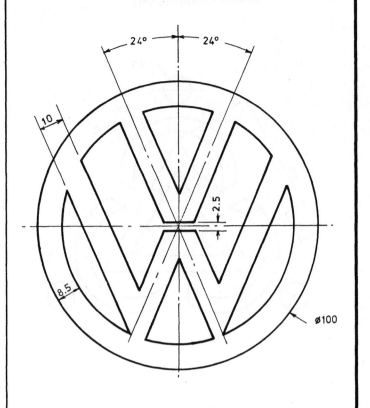

2.117 Ausonics Pty Ltd

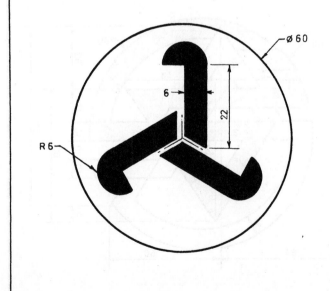

2.118 Tait Electronics (Aust.) Pty Ltd, Brisbane

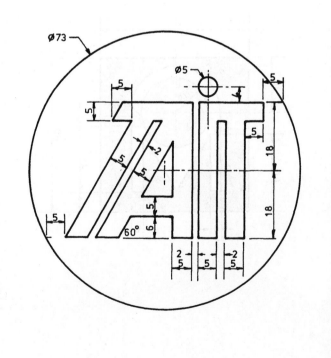

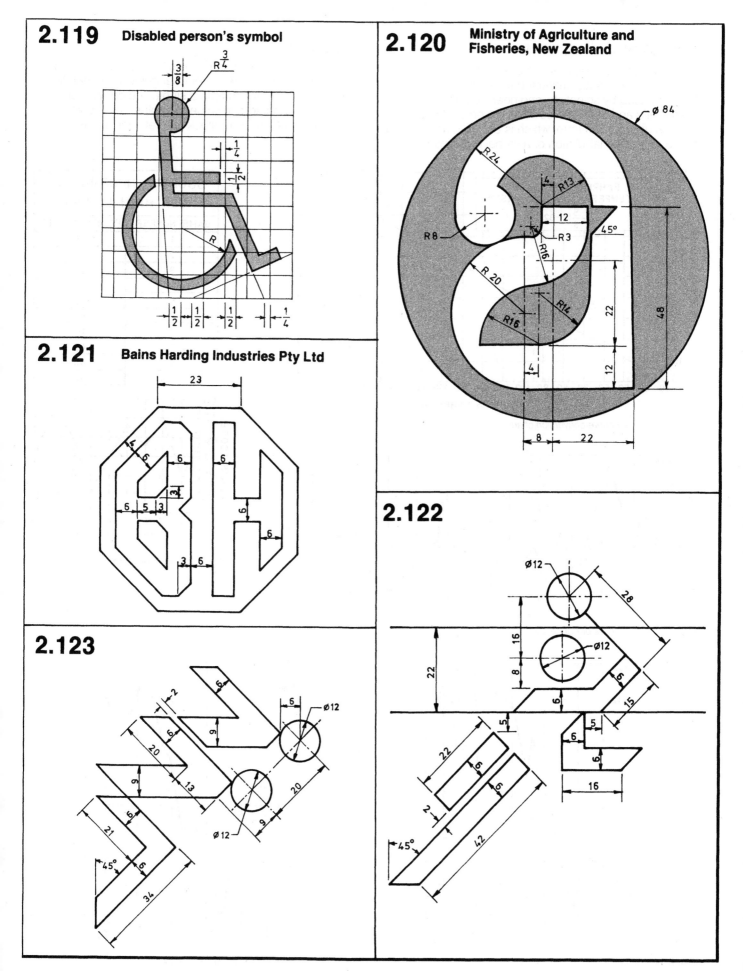

2.119 Disabled person's symbol

2.120 Ministry of Agriculture and Fisheries, New Zealand

2.121 Bains Harding Industries Pty Ltd

2.122

2.123

Charts and diagrams

Graphic charts and diagrams are used to record and study research results presented in numerical form. They are used by scientists, engineers, economists, statisticians and other professionals to present data in a pictorial or graphical form, which is easier to understand than a numerical tabulation or a verbal description.

Line charts

Line charts, as shown in Figures 2.11–2.15, are most often used to show trends or changes over a specific period of time, for example a single line conversion chart for changing one value to another, or showing surging metal prices or stock market fluctuations.

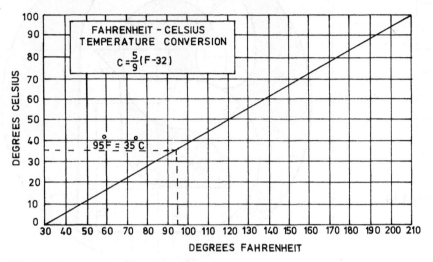

Fig. 2.11 *A temperature conversion chart*

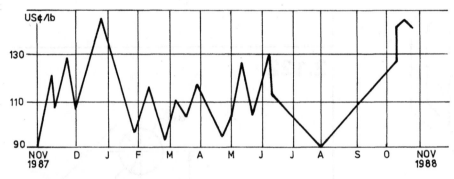

Fig. 2.12 *Graph of copper price surges*

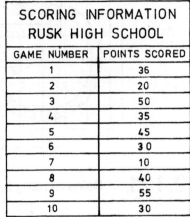

GAME NUMBER	POINTS SCORED
1	36
2	20
3	50
4	35
5	45
6	30
7	10
8	40
9	55
10	30

SCORING INFORMATION RUSK HIGH SCHOOL

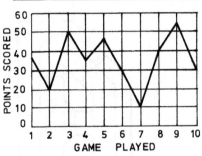

Fig. 2.13 *Graph of variation in scoring rates over time*

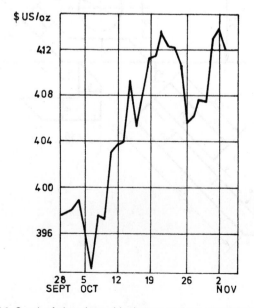

Fig. 2.14 *Graph of changing gold prices*

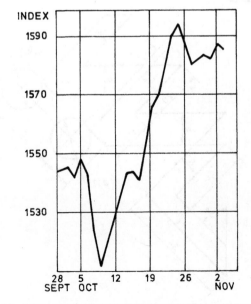

Fig. 2.15 *Graph of the all-ordinaries index*

Bar graphs

Bar charts or column charts, as shown in Figures 2.16-2.19, are probably the most familiar and most easily read and understood graphic charts. They provide a means of comparing similar items and may be presented vertically or horizontally. Vertical bar charts may be drawn so as to show an oblique view to increase the visual impact of the chart by the addition of depth.

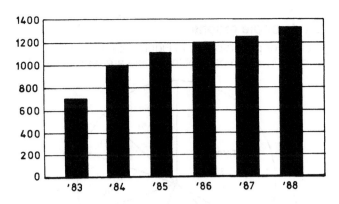

Fig. 2.16 *Bar chart showing school enrolments, 1983-88*

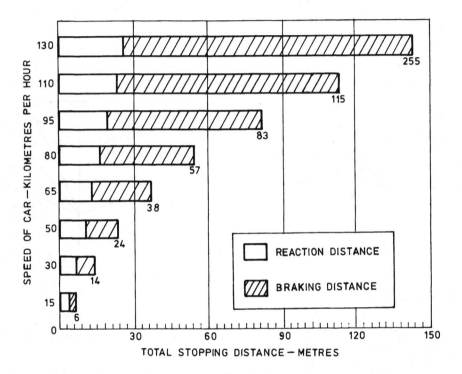

Fig. 2.17 *Horizontal bar chart of stopping distances by speed of car*

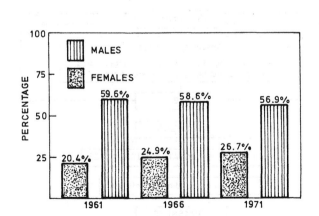

Fig. 2.18 *Bar chart showing porportion of population in the workforce, 1961-71*

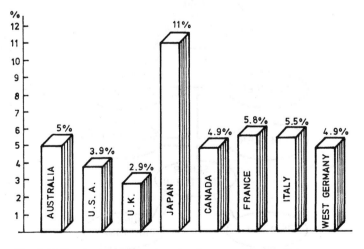

Fig. 2.19 *Bar chart in oblique view showing economic growth by percentage of GDP average annual rate for ten years (1960-61 to 1970-71)*

Circle or pie graphs

A circle or pie graph, as in Figures 2.20–2.23, is a form of chart in which parts of the whole are represented by sectors totalling 360°. This is equivalent to 100%, and, therefore, 3.6° equals 1%, 90° equals 25% and so on. Pie charts may also be drawn as a pictorial view to increase the visual impact.

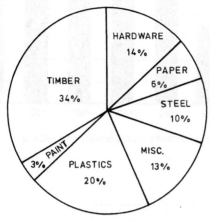

Fig. 2.20 *Pie chart of manual arts budget estimate*

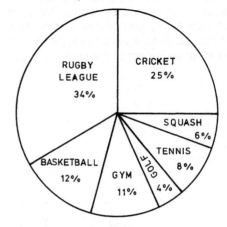

Fig. 2.21 *Pie chart of sport allocation*

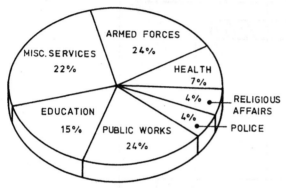

Fig. 2.22 *Pie chart in isometric view showing estimated government expenditure*

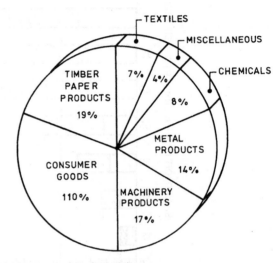

Fig. 2.23 *Pie chart in oblique view showing employment distribution*

Organisation and flow charts

Organisation charts, Figures 2.24–2.25, illustrate an established procedure for processing work through various stages of design, development or manufacture. Flow charts, Figures 2.26 and 2.27, show the path of a series of operations through which material or a product goes during manufacture.

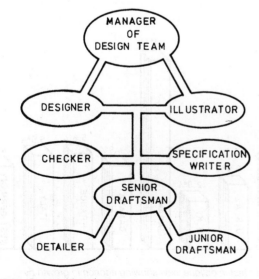

Fig. 2.24 *Organisation chart*

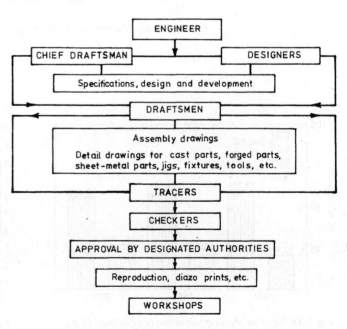

Fig. 2.25 *Organisation chart*

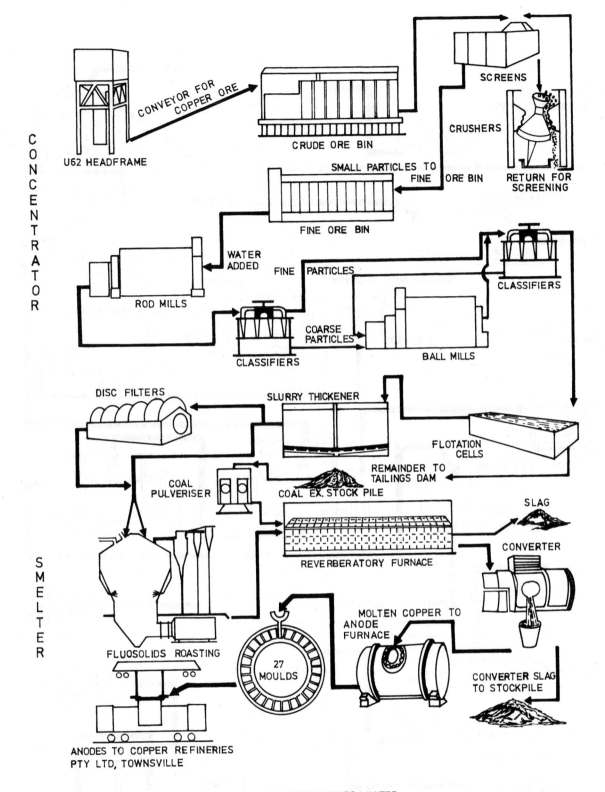

CONCENTRATOR

U62 HEADFRAME

CONVEYOR FOR COPPER ORE

CRUDE ORE BIN

SCREENS

CRUSHERS

RETURN FOR SCREENING

SMALL PARTICLES TO FINE ORE BIN

FINE ORE BIN

WATER ADDED

ROD MILLS

FINE PARTICLES

CLASSIFIERS

CLASSIFIERS

COARSE PARTICLES

BALL MILLS

DISC FILTERS

SLURRY THICKENER

FLOTATION CELLS

REMAINDER TO TAILINGS DAM

COAL PULVERISER

COAL EX. STOCK PILE

SLAG

CONVERTER

SMELTER

FLUOSOLIDS ROASTING

REVERBERATORY FURNACE

27 MOULDS

MOLTEN COPPER TO ANODE FURNACE

CONVERTER SLAG TO STOCKPILE

ANODES TO COPPER REFINERIES PTY LTD, TOWNSVILLE

Fig. 2.26 *Flow chart showing stages in copper production*   *MOUNT ISA MINES LIMITED*

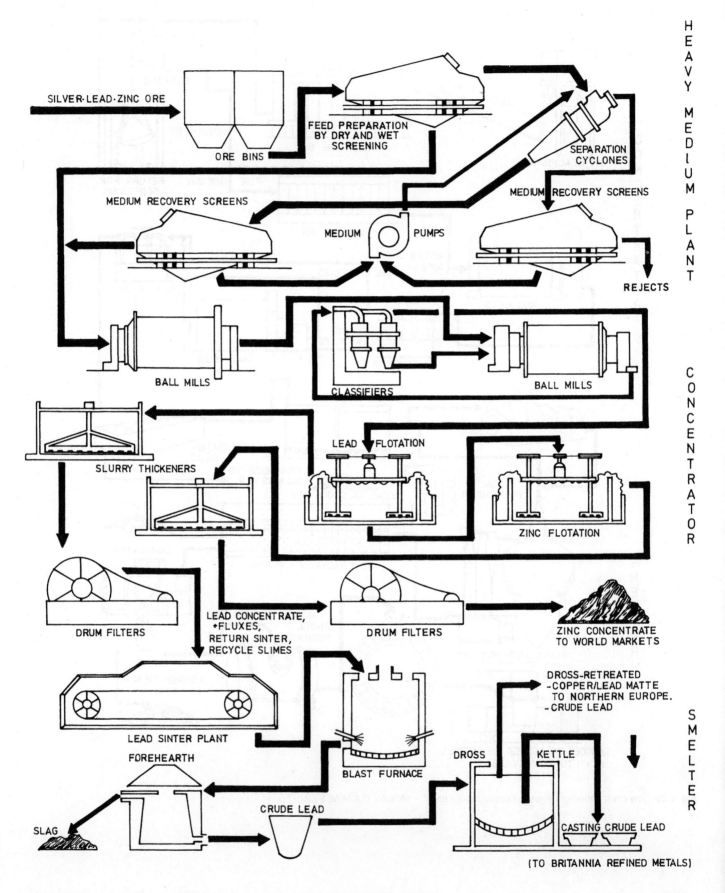

SILVER·LEAD·ZINC ORE

ORE BINS

FEED PREPARATION BY DRY AND WET SCREENING

SEPARATION CYCLONES

MEDIUM RECOVERY SCREENS

MEDIUM RECOVERY SCREENS

MEDIUM PUMPS

REJECTS

BALL MILLS

CLASSIFIERS

BALL MILLS

SLURRY THICKENERS

LEAD FLOTATION

ZINC FLOTATION

DRUM FILTERS

LEAD CONCENTRATE, +FLUXES, RETURN SINTER, RECYCLE SLIMES

DRUM FILTERS

ZINC CONCENTRATE TO WORLD MARKETS

DROSS-RETREATED
-COPPER/LEAD MATTE TO NORTHERN EUROPE.
-CRUDE LEAD

LEAD SINTER PLANT

FOREHEARTH

BLAST FURNACE

DROSS

KETTLE

SLAG

CRUDE LEAD

CASTING CRUDE LEAD

(TO BRITANNIA REFINED METALS)

Fig. 2.27 *Flow chart showing stages in lead production*   *MOUNT ISA MINES LIMITED*

Descriptive geometry

Descriptive geometry is a branch of drawing that involves the study of the spatial relationships of points, lines, plane figures and solids. The understanding of such relationships is the theoretical basis for all architectural and engineering drawing, making it possible to represent three-dimensional objects as two-dimensional drawings.

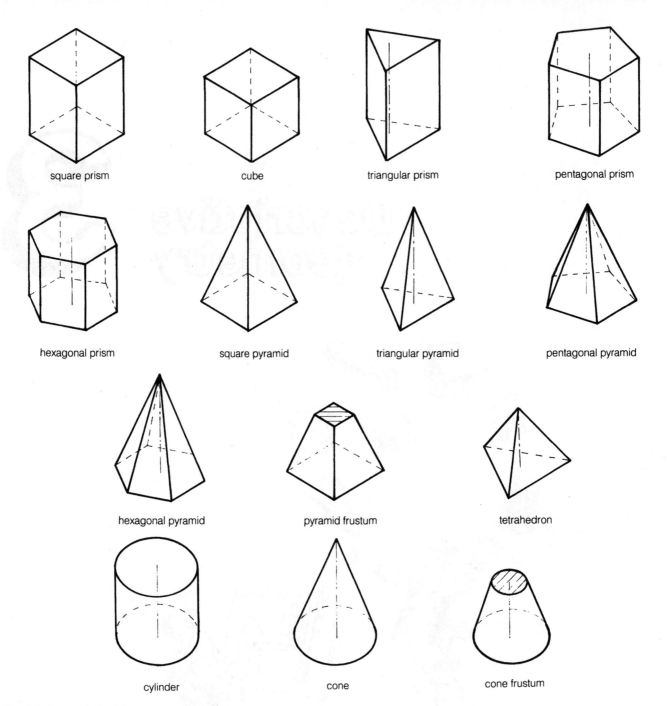

Fig. 3.1 *Geometrical solids*

square prism

cube

triangular prism

pentagonal prism

hexagonal prism

square pyramid

triangular pyramid

pentagonal pyramid

hexagonal pyramid

pyramid frustum

tetrahedron

cylinder

cone

cone frustum

Projection

Solids are represented in solid geometry by views known as the top view and the front view. The top view is drawn on the horizontal plane and the front view on the vertical plane.

Figure 3.2 shows a rectangular-based prism suspended below the horizontal plane and behind the vertical plane. By projecting vertically from the top four corners of the prism, we obtain the rectangular shape of the top of the prism, marked ea, fb, gc and hd on the horizontal plane (HP).

By projecting forward from the front four corners of the prism, we obtain the rectangular shape of the front of the prism, marked f'e', b'a', c'd' and g'h' on the vertical plane (VP). By joining the two planes and opening them to represent a sheet of drawing paper, we can see the top view and front view of the prism (Fig. 3.3). The front view is drawn directly below and in line with the top view.

The light lines joining these two views are called projectors. We say that one view is projected from the other onto its relevant plane of projection. Each part of the solid will retain its shape and size no matter how far from each plane it is positioned.

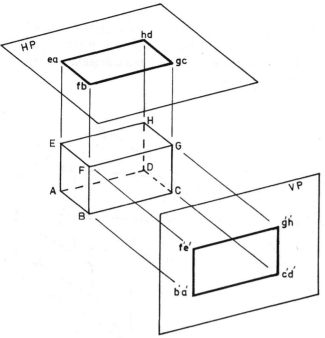

Fig. 3.2 *Projection of rectangular-based prism*

Fig. 3.3 *Projections of rectangular-based prism on a sheet of drawing paper*

Projections of a line

The drawings in Figure 3.4 illustrate the projection of a line onto the planes of projection.

Figure 3.4(a) shows a pictorial and a plane view of a line AB parallel to the HP and VP. Figure 3.4(b) shows a pictorial and a plane view of a line AB, parallel to the VP but inclined at an angle to the HP. Figure 3.4(c) shows a pictorial and a plane view of a line AB, parallel to the HP but inclined at an angle to the VP.

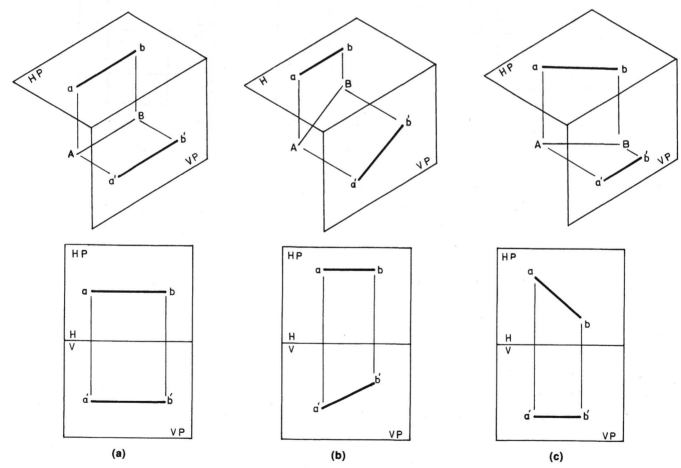

(a) (b) (c)

Fig. 3.4 *Projections of a line*

Projections of a surface

The drawings in Figure 3.5 illustrate the projection of a surface onto the planes of projection.

Figure 3.5(a) shows the pictorial and a plane view of a surface ABCD parallel to the HP. Figure 3.5(b) shows the pictorial and a plane view of the same surface ABCD at an angle to the HP. Figure 3.5(c) shows the pictorial and a plane view of the same surface ABCD at an angle to the VP.

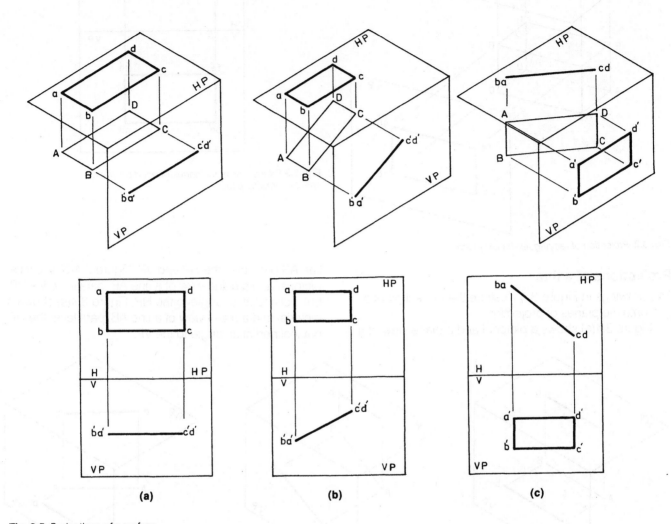

Fig. 3.5 *Projections of a surface*

Exercises on solid geometry

In the following exercises the solution is provided for the first part and you should study the method used and then attempt to solve the rest of the exercises.

3.1(a)

Draw the top, front and right side view of a rectangular-based prism, base 50 mm × 34 mm, height 60 mm. A rectangular end touches the HP and a 50 mm × 60 mm face is parallel to and 20 mm behind the VP.

Solution
1. Draw the HV line.
2. To obtain the top view, draw a rectangle fb, gc, hd and ea with the line fb–gc parallel to and 20 mm from the HV line.
3. To obtain the front view, project all points from the top view below the HV line.
4. From the HV line mark the height of the prism, 60 mm.
5. Letter the points b'a', c'd', g'h' and f'e' to complete the front view.
6. To obtain the right side view, draw a 45° line from g'h'. Project horizontal lines from the top view to intersect the 45° line.
7. Draw vertical lines from these points to intersect horizontal lines from the front view.
8. Letter all points d"a", c"b", g"f" and h"e" to complete the right side view.

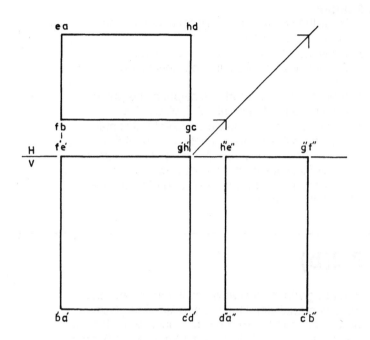

3.1(b)

You are given the front view of a right square prism, side of square 40 mm, axis 55 mm. The 55 mm edges are inclined at 60° to the HP.
  Draw the following views in third-angle projection:
1. the given front view
2. a top view
3. a left side view

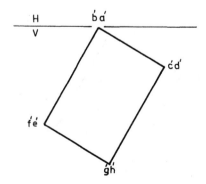

3.1(c)

You are given the top view of a right square prism, side of square 40 mm, axis 60 mm. The 60 mm edges are inclined at 45° to the VP.
  Draw the following views in third-angle projection:
1. the given top view
2. a front view
3. a right side view

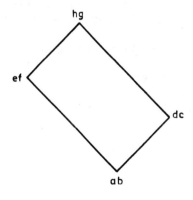

3.2(a)

An equilateral triangular-based prism, side of base 45 mm, axis 60 mm, has a triangular end touching the HP and a rectangular surface parallel to and 20 mm behind the VP. Draw the top, front and left side views of the prism.

Solution

1. Draw the top view of the prism and letter the points ad, be and cf.
2. Draw the HV line 20 mm below ad–be.
3. Project all points from the top view below the HV line.
4. From the HV line mark the height of the prism, 60 mm. Letter the points d', f', e', b', c', a'.
5. To obtain the left side view, project horizontal lines from the top view to intersect the 45° line.
6. Draw vertical lines from these points to intersect horizontal lines from the front view.
7. Letter all points f", d"e", a"b", and c" to complete the left side view.

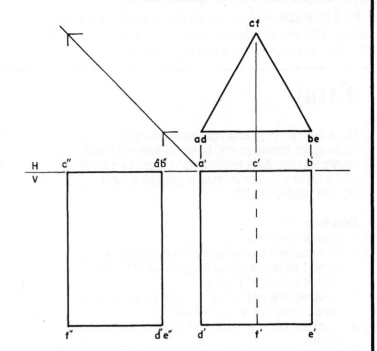

3.2(b)

You are given the top view and the front view of a truncated equilateral triangular prism. The prism is now inclined to the HP until the truncated surface is parallel to the HP. Side of end 40 mm, axis 60 mm.
   Draw the following views in third-angle projection:
1. the given top and front views
2. the top and front views of the inclined prism

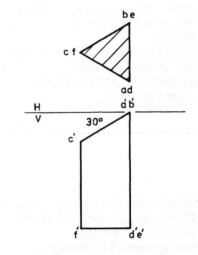

3.2(c)

You are given the front view of an equilateral triangular prism, side of end 36 mm, axis 45 mm. The longer edges of the prism are inclined at 60° to the HP.
   Draw the following views in third-angle projection:
1. the given front view
2. the top view
3. a left side view

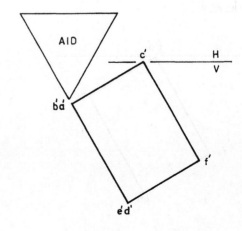

3.3(a)

Draw the top and front view of a right square pyramid, side of base 50 mm, axis 50 mm, when it is positioned 20 mm behind the VP with its apex touching the HP. Two edges of the base are parallel to the VP.

Solution
1. Draw the top view of the pyramid abcd and draw the diagonals to intersect at o.
2. Draw the HV line 20 mm below bc.
3. Project all points from the top view below the HV line.
4. Set off 50 mm, the height of the pyramid. Join o′ to b′a′ and o′ to c′d′ to complete the elevation.

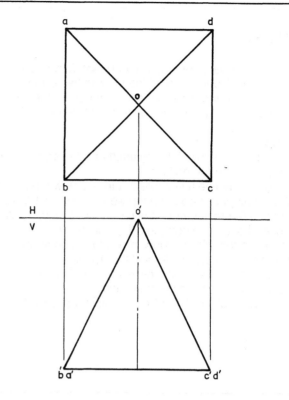

3.3(b)

You are given the front view of a right square pyramid, side of base 60 mm, axis 60 mm, with its axis inclined at 45° to the HP.

Draw the following views in third-angle projection:
1. the given front view
2. the top view
3. a left side view

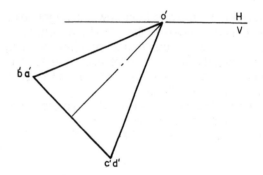

3.3(c)

You are given the elevation of a truncated right square pyramid, side of base 60 mm, axis 70 mm. The pyramid has been truncated by an inclined plane at the mid-point of the axis.

Draw the following views in third-angle projection:
1. the given front view
2. a top view

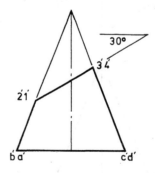

3.4(a)

A right hexagonal pyramid, side of base 35 mm, axis 70 mm, is positioned behind the VP with two edges of the base parallel to the VP and the apex touching the HP. Draw the top, front and left side view of the pyramid.

Solution
1. Draw the top view of the pyramid abcdef, and draw the diagonals to intersect at o.
2. Draw the HV line a suitable distance below the line bc.
3. Project all points from the top view below the HV line.
4. Set off 70 mm, the height of the pyramid. Join a', b'f', c'e' and d' to o' to complete the front view.
5. To obtain the left side view, project horizontal lines from the top view to intersect the 45° line.
6. Draw vertical lines from these points to intersect horizontal lines from the front view.
7. Join f"e", a"d", b"c", to o" to complete the left side view.

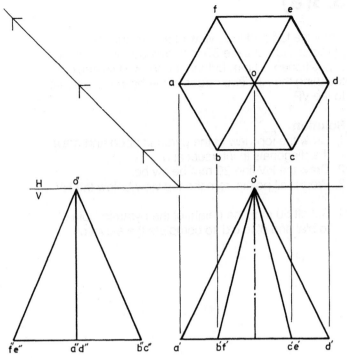

3.4(b)

You are given the top and front views of a right hexagonal pyramid positioned behind the VP with two edges of the base at right angles to the VP and the apex touching the HP. The pyramid is tilted to the left about the edge b'a' until the base makes an angle of 45° to the HP. Side of base is 30 mm, axis 55 mm.

Draw the following views in third-angle projection:
1. the given top and front views
2. the top and front views of the pyramid in the tilted position

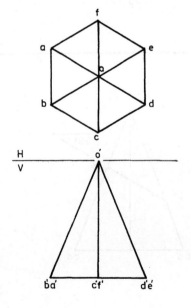

3.4(c)

You are given the top and front views of a truncated right hexagonal pyramid positioned behind the VP with two edges of the base parallel to the VP. The pyramid has been truncated at the mid-point of the axis. The truncated pyramid is tilted to the right about point d' until the base makes an angle of 60° to the HP. Side of base is 35 mm, axis 70 mm.

Draw the following views in third-angle projection:
1. the given top and front views
2. the top and front views of the frustrum in the tilted position

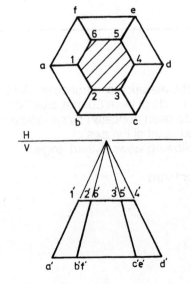

3.5(a)

Draw the top and front views of a right cylinder, diameter of end 56 mm, axis 70 mm, when it is positioned behind the VP with the axis parallel to the VP. An end of the cylinder touches the HP.

Solution
1. Draw the top view of the cylinder. Position both centre lines and letter the points ad, o and bc.
2. Position the HV line a suitable distance below the top view.
3. Project points ad and bc below the HV line and mark out 70 mm, the length of the axis.
4. Join points a', b', c', d' to complete the front view of the cylinder.

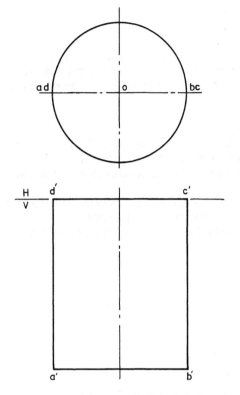

3.5(b)

You are given the front view of a right cylinder with its axis inclined at 45° to the HP and parallel to the VP. Diameter of base is 50 mm, axis 75 mm.

Draw the following views in third-angle projection:
1. the given front view
2. the top view

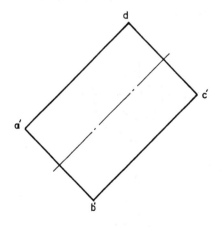

3.5(c)

A right cylinder, diameter of end 50 mm, axis 60 mm, is positioned behind the VP with the ends parallel to the VP and the axis parallel to the HP. The cylinder is pivoted about its curved surface until the end ab makes an angle of 30° with the VP.

Draw the following views in third-angle projection:
1. the given top and front views
2. the top and front views of the cylinder in the new position

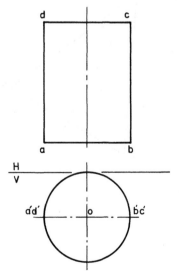

3.6(a)

Draw the top and front view of a right cone, diameter of base 60 mm, axis 65 mm when it is positioned behind the VP with its axis at right angles to the HP and parallel to the VP.

Solution

1. Draw the top view of the cone. Position both centre lines and letter the points a, o and b.
2. Position the HV line a suitable distance below the top view.
3. Project points a, o and b below the HV line and mark off 65 mm, the height of the cone.
4. Join points a' and b' to o' to complete the front view of the cone.

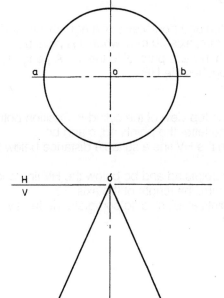

3.6(b)

You are given the front view of a right cone, diameter of base 54 mm, axis 60 mm. The axis of the cone makes an angle 60° to the HP.

Draw the following views in third-angle projection:
1. the given front view
2. a top view

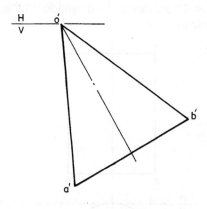

3.6(c)

You are given the top and front views of a truncated right cone, diameter of base 60 mm, axis 60 mm. The cone has been truncated at the mid-point of its axis. The truncated cone is tilted to the left about point a' until the base makes an angle of 60° to the HP.

Draw the following views in third-angle projection:
1. the given top and front views
2. the top and front views of the truncated cone in the tilted position

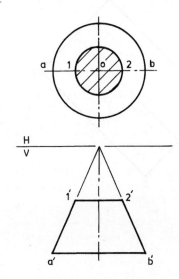

Auxiliary views and planes

An auxiliary view is a projection onto an auxiliary plane inclined to one of the principal planes of projection (i.e. HP or VP). The view is at right angles to the inclined surface.

Figure 3.6(a) is a pictorial view of a point P behind the horizontal, vertical and auxiliary planes. The projections of this point are shown pictorially: p, the top view, p', the front view and p″, the auxiliary view.

Figure 3.6(b) shows the three planes opened out with the auxiliary plane attached to the vertical plane. The projections of the point P are also shown.

Figure 3.6(c) shows the projections of the point P with the projection lines at right angles to their respective reference lines. The distances marked X are equal.

Figure 3.7(a) shows a point P behind the horizontal, vertical and auxiliary planes. The projections of this point are shown pictorially: p, the top view, p', the front view and p″, the auxiliary view.

Figure 3.7(b) shows the planes opened out with the auxiliary plane attached to the horizontal plane. The projections of the point P are also shown.

Figure 3.7(c) shows the orthogonal projection of the point P with the projection lines at right angles to their respective reference lines. The distances marked X are equal.

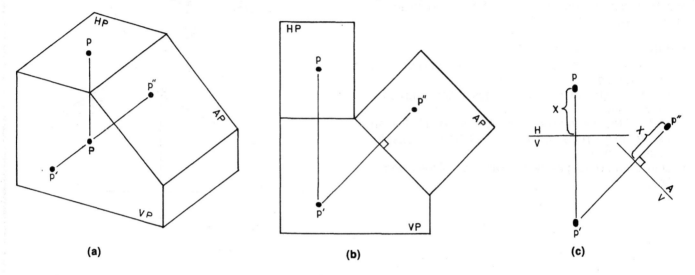

Fig. 3.6 *Projections of a point P to the horizontal, vertical and auxiliary inclined planes*

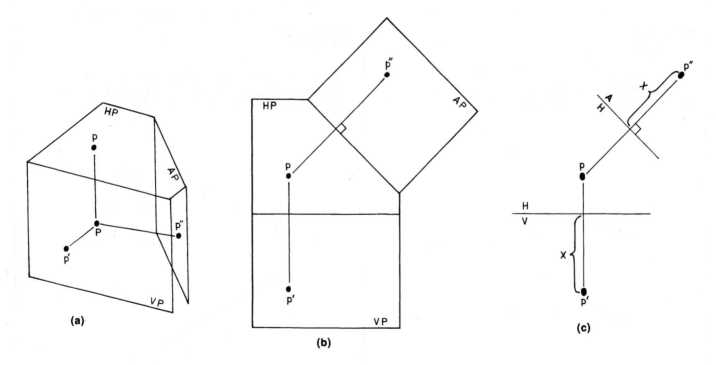

Fig. 3.7 *Projections of a point P to the horizontal, vertical and auxiliary vertical planes*

Exercises on auxiliary views

Up to this point, the exercises have been solved using three planes of projection: horizontal, front vertical and end vertical. These regular planes can offer solutions to numerous problems, but it is necessary to introduce auxiliary planes of projection to describe solutions for inclined or slanted surfaces.

These exercises with inclined surfaces will be drawn on auxiliary planes that are parallel to the inclined surface to enable the true shape of the surface to be shown. There are also several exercises that include the projection of simple solids onto an auxiliary plane. These exercises have been included to test knowledge and the principles of projection. Solutions are given for some to demonstrate the method.

3.7(a)

You are given the top view ab and the front view a'b' of a line. Project an auxiliary front view of the line onto an auxiliary vertical plane at 30°.

Solution
1. At a suitable distance from the top view, draw a line at 45° to represent the auxiliary plane.
2. From a and b, draw projection lines at right angles to the H'V' line.
3. Mark off the distances of a' and b' from the HV line and transfer to the projectors above the H'V' line. Letter the points a" and b".
4. Join a" and b" to give the auxiliary view of the given line.

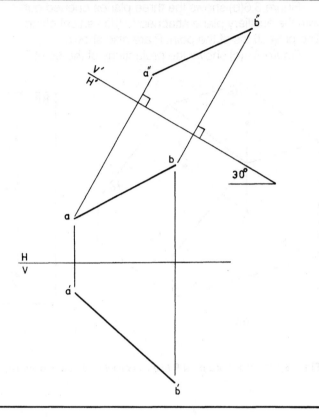

3.7(b)

You are given the top view ab and the front view a'b' of a line. Project an auxiliary top view of the line onto an auxiliary horizontal plane at 45°.

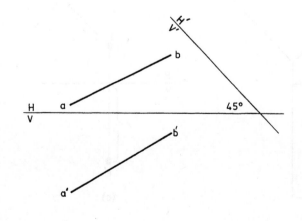

3.7(c)

You are given the top view ab and the front view a'b' of a line. Project an auxiliary top view of the line onto an auxiliary horizontal plane at 60°.

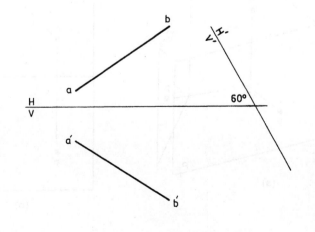

3.8(a)

You are given the top view abcd and the front view a′d′–b′c′ of a rectangle. Project an auxiliary top view to represent the true shape of the rectangle on an auxiliary horizontal plane at 30°.

Solution

1. At a suitable distance from the front view, draw a line at 30° to represent the auxiliary plane.
2. From a′d′ and b′c′, draw projection lines at right angles to the H′V′ line.
3. Mark off the distances of a and b from the HV line and transfer to the projectors above the H′V′ line. Letter the points a″ and b″.
4. Mark off the distances of d and c from the HV line and transfer to the projectors above the H′V′ line. Letter the points d″ and c″.
5. Join a″, d″, c″ and b″ to obtain the true shape of the rectangle.

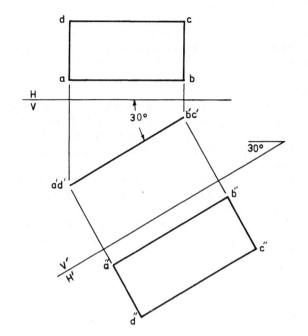

3.8(b)

You are given the top view ab and the front view a′b′ of a line. Find the true length of the line by projecting an auxiliary view.

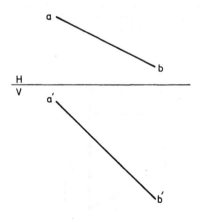

3.8(c)

The line ba–cd represents the top view of a square with side 50 mm.
1. Draw the front view of the square.
2. Project the true shape of the square onto an auxiliary vertical plane at 45°.
3. Project an auxiliary top view of the square onto an auxiliary horizontal plane at 60°.

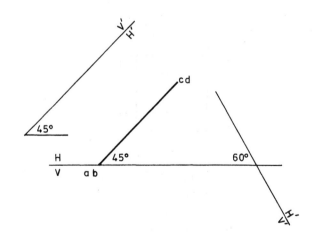

3.9(a)

You are given the top view and the front view of a machine wedge. Draw the auxiliary view of the wedge so that the true shape of the oblique surface is obtained.

Solution
1. At a suitable distance from the top view, draw a line parallel to the oblique surface in the front view. This line will represent the auxiliary plane.
2. Project all points from the front view to the auxiliary plane.
3. Step off the distance from the top view to the HV line.
4. Step off the width of the top view and transfer to the auxiliary view.
5. Join all points to complete the auxiliary view. The true shape of the oblique surface is shown shaded.

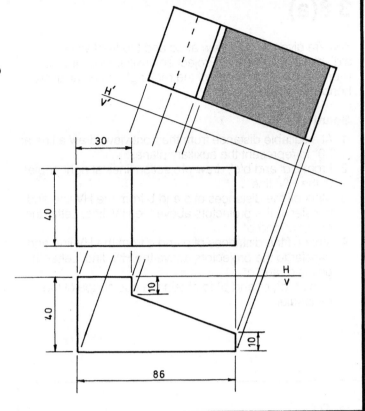

3.9(b)

You are given the top and front views of a shaped block.
1. Draw the given views.
2. Project the true shape of the sloping surface or project a complete auxiliary view of the block on an auxiliary plane parallel to the sloping surface.

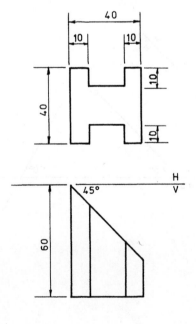

3.9(c)

You are given the top and front views of a machined block.
1. Draw the given views.
2. Project the true shape of the sloping surface or project a complete auxiliary view of the block on an auxiliary plane parallel to the sloping surface.

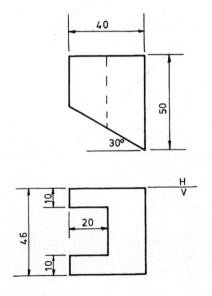

3.10(a)

You are given the top and front views of a right square prism. Project an auxiliary front view of the prism onto an auxiliary vertical plane at 60°.

Solution

1. At a suitable distance from the top view, draw a line at 60° to represent the auxiliary plane.
2. From the top view, draw projectors from all points at right angles to the H'V' line.
3. Mark off the height of the prism along the projectors.
4. Join all points and letter them as shown.

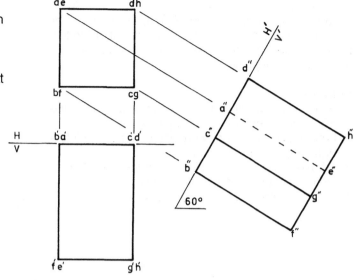

3.10(b)

You are given the front view of a right square prism, side of end 45 mm, axis 60 mm, with its shorter edges at 30° to the HP.
1. Draw the given view and project the top view.
2. Project an auxiliary front view of the prism onto an auxiliary vertical plane at 45°.

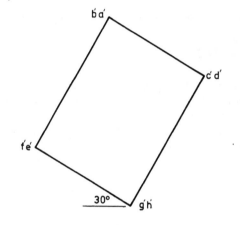

3.10(c)

You are given the top view of a right square prism, side 45 mm, axis 60 mm, with its longer edges inclined to the VP at 30°.
1. Draw the given view and project a front view.
2. Project an auxiliary top view of the prism onto an auxiliary horizontal plane at 60°.

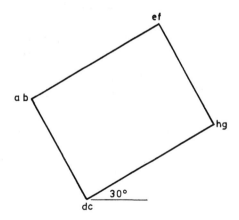

3.13(a)

You are given the top view and front view of a cylinder, diameter of end 48 mm, axis 70 mm. The cylinder has been truncated at an angle of 45°. Draw a partial auxiliary view of the cylinder showing the true shape of the truncated surface.

Solution

1. Draw a line parallel to the oblique surface of the cylinder.
2. From points 0, 1, 2, 3, 4, 5 and 6 in the top view, project lines at right angles to the H'V' line.

3. Set off on these projectors the vertical distances obtained from the front view.
4. Join all points with a smooth curve to complete the true shape.

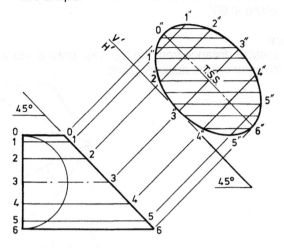

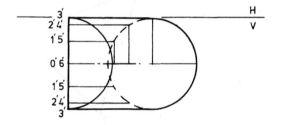

3.13(b)

You are given the top and front views of a cylinder truncated at an angle of 45°. The diameter of the end is 36 mm and the length of bc is 56 mm.
1. Draw the given views.
2. Project the true shape of the sloping surface or project a complete auxiliary view of the truncated cylinder on an auxiliary plane parallel to the sloping surface.

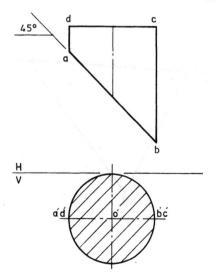

3.13(c)

You are given the front and auxiliary views of a right cylinder, diameter of end 48 mm, axis 60 mm.
1. Draw the given views.
2. Project a top view of the cylinder.

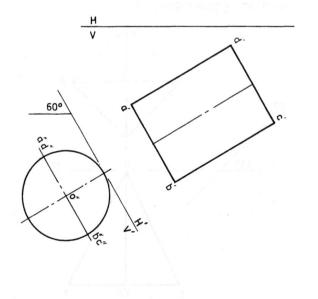

3.14

You are given the top and front views of a shaped block.
1. Draw the given views.
2. Project the true shape of the sloping surface or project a complete auxiliary view of the block on an auxiliary plane parallel to the sloping surface.

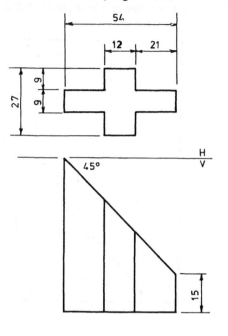

3.15

You are given the front view and the incomplete view of a truncated hexagonal pyramid.
1. Draw the given front view.
2. Project and complete the top view.
3. Draw an auxiliary view of the pyramid on an auxiliary plane parallel to the truncated surface.

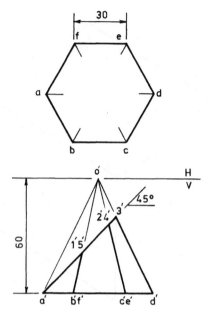

3.16

You are given the top and front views of an anchor block.
1. Draw the given views.
2. Project the true shape of the sloping surface or project a complete auxiliary view of the block on an auxiliary plane parallel to the sloping surface.

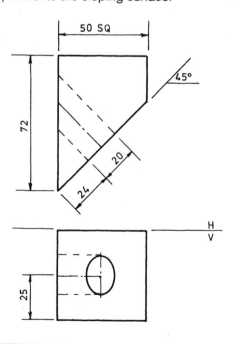

3.17

You are given the top and front views of a casting.
1. Draw the given views.
2. Project the true shape of the sloping surface or project a complete auxiliary view of the casting on an auxiliary plane parallel to the sloping surface.

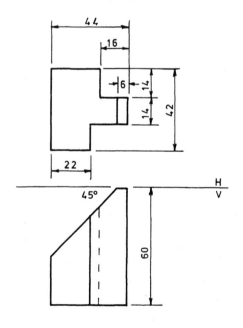

The inclined plane

Figure 3.8 shows the pictorial view of an inclined plane (IP) rising to the right and making an angle of 45° to the HP. The intersections of the plane with the VP and HP are shown as VT, the vertical trace, and HT, the horizontal trace. T on the HV line is the intersection of the traces. Figure 3.9 shows the orthogonal projection of the traces. The characteristics of the inclined plane are:

1. The HT is perpendicular to the HV line.
2. The true angle between the HT and VT is a right angle.
3. The angle the VT makes with the HV line is the angle of inclination of the inclined plane (IP) to the HP.
4. The VT is the front view of the inclined plane.

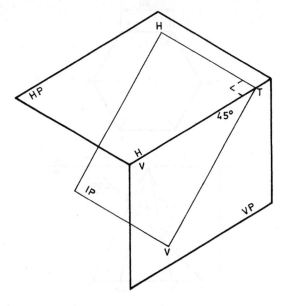

Fig. 3.8 *Pictorial view of an inclined plane*

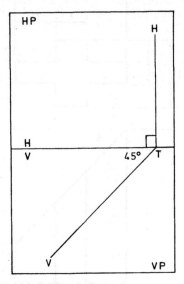

Fig. 3.9 *Orthogonal projection of the traces*

Exercises on the inclined plane

The following exercises include demonstrations of the method to be followed and problems for the student to solve.

3.18(a)

A cube, side 20 mm, is resting on an inclined plane rising to the right at an angle of 30° to the HP. The cube is positioned so that a face is parallel to and 10 mm behind the VP and another face is parallel to and 40 mm from the HT. Draw the front and top views of the cube.

Solution
1. Draw the VT and HT at 30° and 90° respectively to the HV line.
2. Draw the front view of the cube with the corner f'g' 40 mm from point T along the vertical trace. Letter the remaining points.
3. Project all points vertically from the front view.
4. Mark off 10 mm from the HV line to obtain point f, and a further 20 mm to obtain point g.
5. Join all points and letter them as shown.

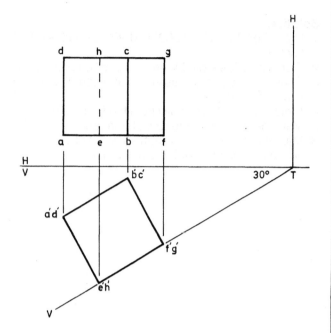

3.18(b)

A right square pyramid, side of base 25 mm, axis 25 mm, rests on an inclined plane rising to the left at 45° to the HP. A side of base is parallel to and 24 mm from the HT. The apex of the pyramid is 30 mm behind the VP. Draw the front and top views of the pyramid.

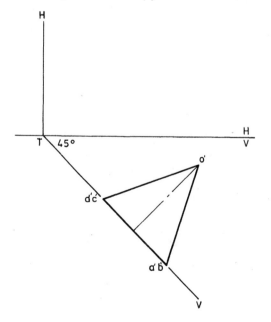

3.18(c)

A right hexagonal prism, side of end 12 mm, axis 30 mm, rests on a hexagonal end on an inclined plane rising to the right at 60° to the HP. A corner of this end is 20 mm from the HT. The axis of the prism is 25 mm behind the VP. Draw the front and top views of the prism.

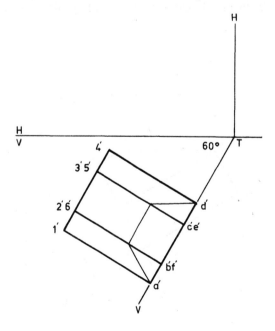

107

3.19(a)

A right pentagonal pyramid, side of base 15 mm, axis 25 mm, rests on its base on an inclined plane rising to the right at 45° to the HP. The axis of the pyramid is 34 mm from the HT and 20 mm behind the VP. Draw the front and top views of the pyramid.

Solution

1. Draw the VT and HT at 60° and 90° respectively to the HV line.
2. From point T, set off 34 mm, the position of the axis.
3. Construct a pentagon as an aid and complete the front view. Letter all points.
4. Project all points vertically from the front view.
5. Mark the position of the axis 20 mm above the HV line.
6. Mark the positions of the remaining points.
7. Join all points and letter them as shown.

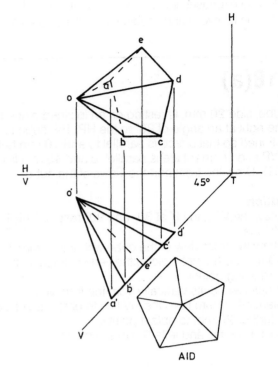

3.19(b)

A right hexagonal prism, side of end 15 mm, axis 30 mm, rests on a hexagonal end on an inclined plane rising to the right at 60° to the HP. An edge of this end is 20 mm from the HT. The axis of the prism is 30 mm behind the VT. The prism has been truncated by a cutting plane parallel to the HP. Draw the front and top views of the prism.

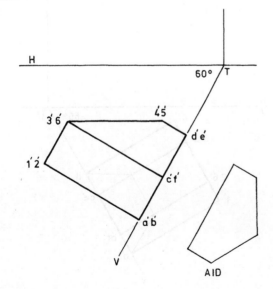

3.19(c)

A right cone, diameter of base 30 mm, axis 35 mm, rests on its circular base on an inclined plane rising to the left at 45° to the HP. The axis of the cone is 40 mm from the HT and 40 mm behind the VP. Draw the front and top views of the cone.

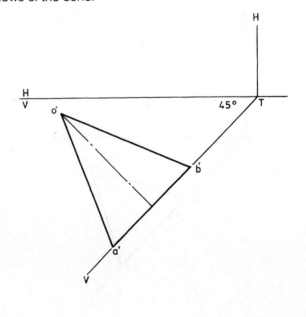

Orthogonal projection

First and third angle

Engineers and drafters throughout the world use the orthogonal system of projection for illustrating the shape and dimensions of many types of features. It is a multiview system in which the principal views are 90° apart in the horizontal and vertical planes, giving a total of six possible views: front, back, top, bottom and both sides.

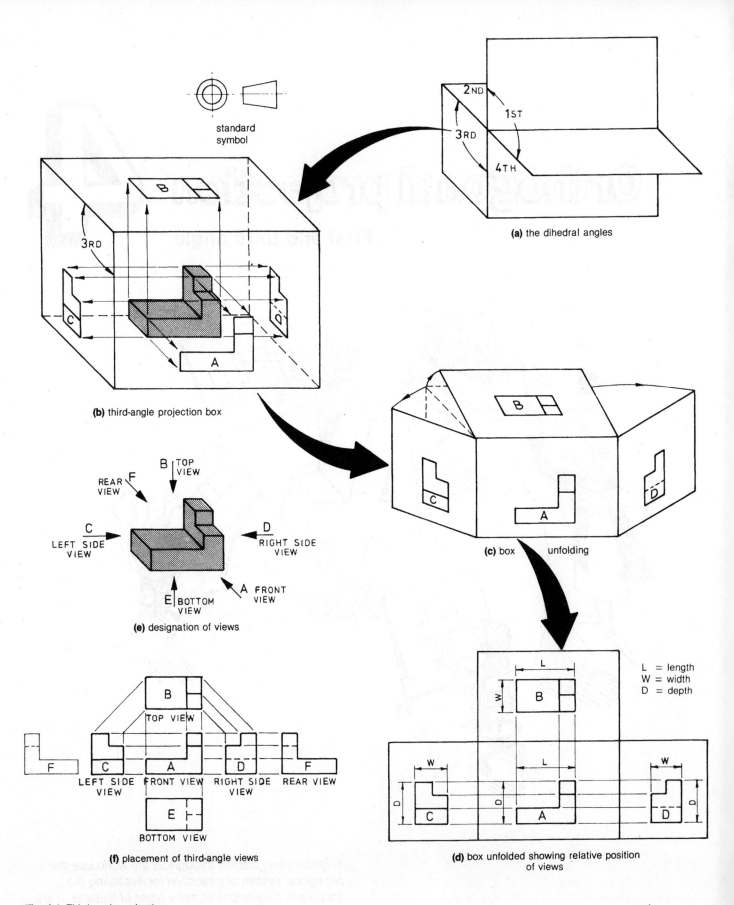

standard symbol

(a) the dihedral angles

(b) third-angle projection box

(c) box unfolding

(e) designation of views

REAR VIEW F

B TOP VIEW

C LEFT SIDE VIEW

D RIGHT SIDE VIEW

E BOTTOM VIEW

A FRONT VIEW

(d) box unfolded showing relative position of views

L = length
W = width
D = depth

(f) placement of third-angle views

B
TOP VIEW

F

C LEFT SIDE VIEW

A FRONT VIEW

D RIGHT SIDE VIEW

F REAR VIEW

E BOTTOM VIEW

Fig. 4.1 *Third-angle projection*

Orthogonal projection

Engineering drawings are normally intended to indicate the shape and size of an object. However, all objects have three dimensions, namely length, breadth and depth, and the problem of representing these on a drawing as well as conveying an impression of shape to the reader is overcome by the use of a technique called *orthogonal projection*.

Orthogonal projection is a method of viewing an object so that a number of plane views may be obtained, each of which includes two of the object's three dimensions of length, breadth and depth.

When a horizontal and vertical plane intersect at right angles, four right angles (known as *dihedral angles*) are formed and are numbered conventionally as shown in Figure 4.1(a). Orthogonal projections are commonly based on the first and third angles and so are known as *first-angle projection* and *third-angle projection*, respectively. In the interests of standardisation, Standards Australia has recommended that the third-angle method of projection be used.

Third-angle projection

The third dihedral angle forms the basis of a six-sided transparent box (Fig. 4.1(b)), in which the object is imagined to be placed so that two of its three principal dimensions (length, breadth and depth) are contained in each of six possible views reflected onto the sides of the box. (Only four views are shown in Fig. 4.1(b); the views from underneath and from the rear are omitted for clarity.) So that the views reflected onto the sides of the box may be represented on one plane, the box is unfolded, as shown in Figure 4.1(c) and (d). Unfolding the box in this way positions the views in a unique manner with respect to each other. Such relative positioning applies universally to all views obtained by this method.

It is essential that drawings made by the third-angle method be identified, preferably by the use of the standards symbol illustrated in Figure 4.1 or by the words 'third-angle projection' printed in a conspicuous place on the drawing, usually in the title block.

Designation of third-angle views

Figure 4.1(e) illustrates the six possible viewing positions of third-angle projection, and the preferred designation of each view. However, when the method of projection is indicated by the standard symbol, the principal views shown in Figure 4.1(f) require no further identification.

Figure 4.1(f) shows the relative placement of the six designated views, which include the rear view F and the bottom view E not illustrated on the previous figures. The rear view F may be positioned as shown or on the left of the left side view C, as indicated by the light outline.

Number of views

Although six possible views may be drawn, all six are very rarely required. The number used should be just sufficient to indicate the shape of the object and to enable a clear definition of the size of all features. For most drawings, three views are adequate. However, the front view is always provided, and whatever number and combination is decided on, they should all be adjacent views. Examples of three-view, two-view and one-view drawings are shown in Figure 4.2(a), (b) and (c), respectively. In Figure 4.2(c) one view only is required because the diameter symbol defines the shape at right angles to the axis.

Other views, such as section, auxiliary, partial and revolved views, may be used in conjunction with the six principal views to more satisfactorily describe an object. They are illustrated in Chapter 1, pages 27 and 28.

Projection of orthogonal views

Because orthogonal views bear a standard relationship to each other according to the unfolding of the projection box, details such as edges, surfaces and holes located on one view may be transferred to other views by projection methods. Projecting horizontally between the front, rear and side views with the aid of a T square enables height measurements to be transferred quickly and accurately from one view to another. The front view is normally drawn first, and from it detail may be projected horizontally to the side and rear views or vertically to the top and bottom views, and vice versa.

Figure 4.3 illustrates the principle for third-angle projection, showing how detail may be projected between the two side, front and top views.

There are three methods of projecting between the top and side views: Figure 4.4(a) uses a 45° set square, Figure 4.4(b) uses compasses and Figure 4.4(c) combines horizontal and vertical projection lines from a 45° line. In Figure 4.4(a), (b) and (c), the distances between views is the same; however, the distance may be varied by moving the projection quadrant to the side, as in Figure 4.4(d). The top view may be moved further from the front view without altering the side view in a similar manner. The ability to vary the distances between views at will is necessary for proper layout of the views on the drawing sheet.

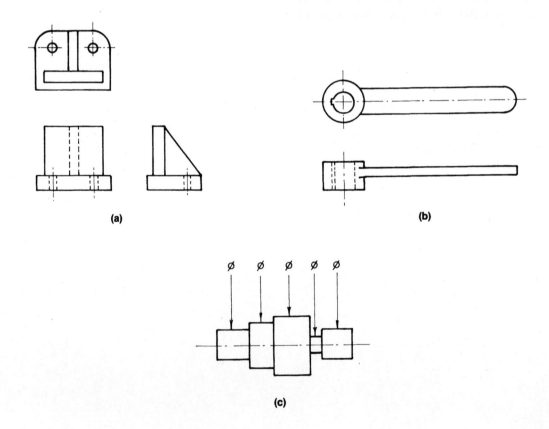

(a)

(b)

(c)

Fig. 4.2 *Choosing the number of views*

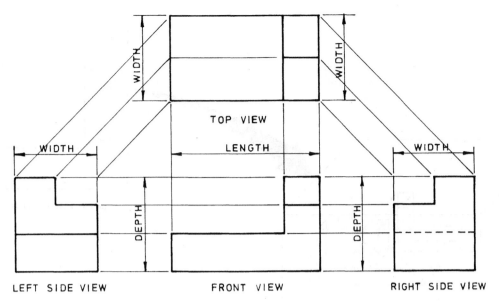

Fig. 4.3 *Relationship of orthogonal views*

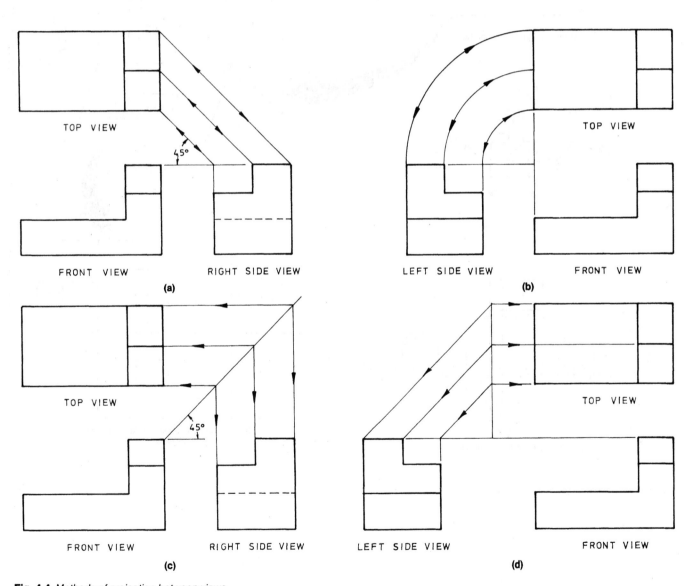

TOP VIEW

45°

FRONT VIEW   RIGHT SIDE VIEW

(a)

TOP VIEW

LEFT SIDE VIEW   FRONT VIEW

(b)

TOP VIEW

45°

FRONT VIEW   RIGHT SIDE VIEW

(c)

TOP VIEW

LEFT SIDE VIEW   FRONT VIEW

(d)

Fig. 4.4 *Methods of projection between views*

113

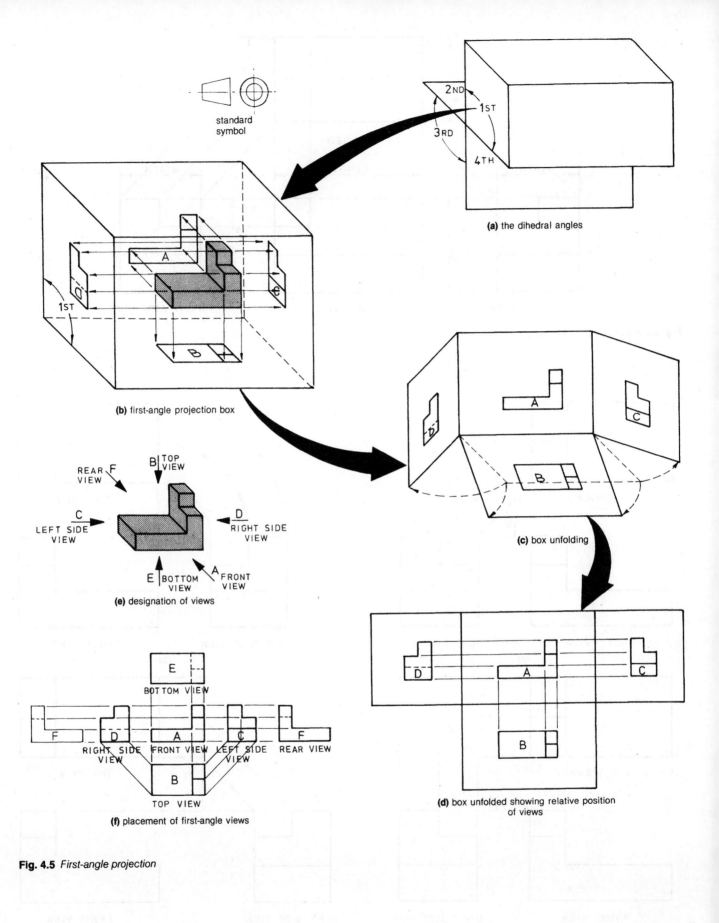

standard
symbol

(a) the dihedral angles

(b) first-angle projection box

(c) box unfolding

(d) box unfolded showing relative position of views

REAR
VIEW F

B TOP
VIEW

LEFT SIDE
VIEW C

D
RIGHT SIDE
VIEW

E BOTTOM
VIEW

A FRONT
VIEW

(e) designation of views

E
BOTTOM VIEW

F | D | A | C | F
RIGHT SIDE VIEW | FRONT VIEW | LEFT SIDE VIEW | REAR VIEW

B
TOP VIEW

(f) placement of first-angle views

Fig. 4.5 *First-angle projection*

114

First-angle projection

The second method of projecting plane views, known as first-angle projection, is illustrated in Figure 4.5. In the interest of standardisation, Standards Australia has recommended that this method *not* be used and that third-angle projection be the preferred method. However, first-angle protection is still used by many firms, and it is essential for the student of engineering drawing to understand the principles of both methods.

Relationship between first-angle and third-angle views

As illustrated in Figure 4.5(e), the designation of views in first-angle projection is identical to that in third-angle projection (Fig. 4.1(e)). However, a comparison between the two methods of unfolding the dihedral box will show that the relative positions of the views are different. The difference may be stated simply as follows: *A view in third-angle projection is placed so that it represents the side of the object nearest to it on the adjacent view (Fig. 4.1(f)). A view in first-angle projection is placed so that it represents the side of the object furthest from it on the adjacent view (Fig. 4.5(f)).*

In all other respects the rules of projection for the two methods are identical.

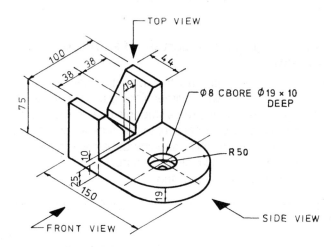

Fig. 4.6 *Component to be drawn*

Production of a mechanical drawing

After deciding on a selection of views, the production of a mechanical drawing can be divided into five stages, as follows:

1. drawing of borderline and location of views on the drawing sheet
2. light construction of views
3. lining in of views
4. dimensioning and insertion of subtitles and notes
5. drawing of title block, parts list and revisions table

Drawing of borderline and location of views

Consider the component shown by the isometric view in Figure 4.6, the orthogonal projection of which is to be drawn on an A2 size sheet (594 mm × 420 mm). The views to be drawn are indicated by the arrows.

It will be observed that the overall length of the front view is 150 mm and its height is 75 mm. The top view is 150 mm long and its width is 100 mm, and the side view is 100 mm wide and 75 mm high.

When the number and designation of views have been decided, their correct layout within the available working space is necessary to give the drawing an overall balanced and pleasing appearance. The available working space is that portion of the drawing sheet remaining after allowances have been made for the insertion of such items as the title block, parts list and revisions table. An indication of the dimensions of available working space on various types of drawing is shown in Figure 4.7.

Assuming that a title block 35 mm high is to be provided in the bottom right-hand corner of the drawing frame, the available working space is equal to 566 × (400 − 35) = 566 × 365. (Dimensions of drawing frames and border widths for various sheet sizes are given in Table 1.4.)

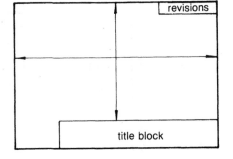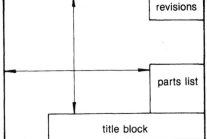

Fig. 4.7 *Dimensions of working space*

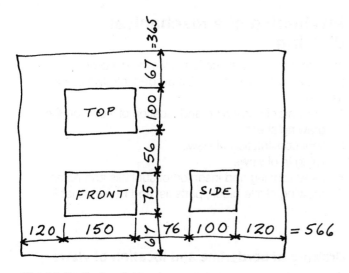

Fig. 4.8 *Positioning of views*

On a piece of rough working paper, sketch a rectangle representing the working space (Fig. 4.8). Place the views in their relative positions on the sheet and draw two lines, one horizontal and one vertical, along which dimensions are placed. Mark in the dimensions of the three views. Decide on distances to be allowed between views, and mark these as well. If possible, at least 40 mm should be allowed between views to make room for the insertion of dimensions. Now mark on the sketch the equal distances remaining between the views and the borderline. Finally, add up the dimensions along the two dimension lines and make sure they total the dimensions of the working space, for example 566 × 365 mm.

When making the rough sketch for the location of views, always see that the rectangles representing the views are in their correct relative positions, otherwise the views will be wrongly placed on the drawing sheet.

Light construction of views

Figure 4.9 illustrates this stage. Draw a light horizontal line 67 mm up from the title block. This line will pass through the bottom of the front and side view. Lightly draw in the front view, taking measurements from the rough sketch (Fig. 4.8). Hidden detail lines are drawn finally at this stage, as lining them in later is difficult. Next, lightly construct either the top or side view (say the top). Wherever possible, use vertical projection lines from the front view for the location of detail on the top view. Finally, using horizontal projection lines from the front view and 45° projection lines from the top view, lightly construct the side view. Note that arcs and circles should be lined in at this stage.

It is important that construction lines are ruled lightly, as they may have to be erased or altered as the drawing progresses.

Lining in of views

Lining in should be done systematically for the three views. Commencing with horizontal lines and at the top of the top view, line in progressively working down the page with the T square. Starting at the left-hand side and working across the page, line in all vertical lines using a T square and set square combined. A chisel point pencil should be used for lining in to give consistent line thickness. Construction and projection lines may be left on the drawing, provided they are very light. If any construction lines need to be erased, this should be done before lining in commences. The views, when lined in, are shown in Figure 4.10.

Dimensioning and insertion of subtitles and notes

At this stage it is necessary to explain the main principles involved when dimensioning a mechanical drawing. There are two rules students must remember:

1. Each dimension necessary to describe a component should be given, and it should not be necessary to deduce an essential dimension from other dimensions on the drawing.

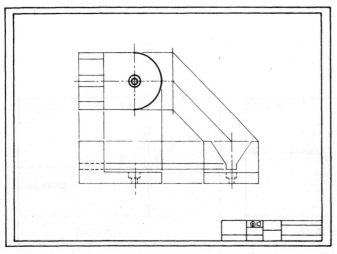

Fig. 4.9 *Construction of views*

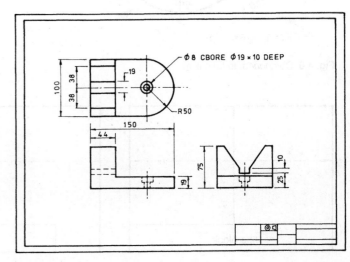

Fig. 4.10 *Completed views*

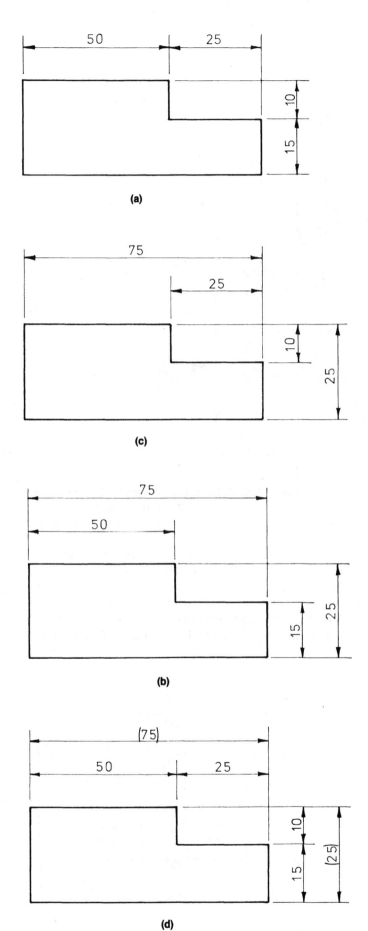

Fig. 4.11 *Principles of linear dimensioning*

2. There should be no more dimensions than are necessary to define a component. That is, it should not be possible to deduce a dimension from other dimensions if that dimension has already been given directly.

Three correct methods of dimensioning lengths that conform to the above rules are given in Figures 4.11(a), (b) and (c). The following points should be noted:

1. Where two dimensions together give the length of an object, as in Figure 4.11(a), the overall dimension is omitted.
2. When an overall length is shown, as in Figures 4.11(b) and (c), a non-functional intermediate dimension is omitted.

Sometimes it is useful to show an overall dimension, even though all intermediate dimensions are supplied. The overall dimension is shown as an auxiliary dimension, and this is indicated by placing it in brackets, as shown in Figure 4.11(d). An auxiliary dimension is in no way binding as far as machining operations are concerned.

Dimensions that govern the working of a component are essential and are called *functional dimensions*. All other dimensions are called *non-functional dimensions*.

The above rules show that a drafter must fully understand the working of a component to be able to indicate functional dimensions; by correct dimensioning, the drafter ensures that features are correctly located on the finished product.

As the completed example in Figure 4.14 shows, the following features are regarded as essential:

1. The axis of the bored holes is vertical, centrally located and is a toleranced distance from the back surface, which is machined.
2. The top surface of the boss must be correctly located in relation to the three 16 mm diameter fixing holes.
3. The bore of the boss is a toleranced size.

The above features are functional and must be dimensioned accordingly. Hence the centre line of the bored hole is dimensioned directly off the back surface. The axis of the boss is located centrally between the fixing holes, and the top of the boss is spotfaced and located 10 mm above the horizontal centre line of the two top fixing holes. The bottom fixing hole is dimensioned from the centre line of the top holes as well.

It is necessary when dimensioning a drawing to decide on one or more base or datum lines from which functional dimensions are taken. The datum lines for the above example are the back surface lines on the top and side views, the vertical centre line on the front view, and the horizontal centre line through the two fixing holes on the front view. The overall height (140 mm) and the width (105 mm) are given as auxiliary dimensions. The dimensioning of the three views (Fig. 4.14) follows the above rules as well as those given in Chapter 1.

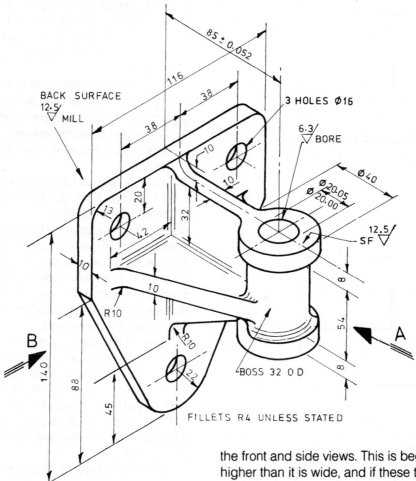

Fig. 4.12 *Wall bracket to be drawn*

Dimensions and labels shown on figure:
85 ± 0.052
116
38
38
BACK SURFACE
12.5 / MILL
3 HOLES Ø16
6.3 / BORE
10
10
Ø40
Ø 20.05 / 20.00
12.5 / SF
20
13
12
32
8
10
10
R 10
54
B
140
88
R 10
45
22
8
BOSS 32 0 D
A
FILLETS R4 UNLESS STATED

Drawing of title block, parts list and revisions table

A suitable layout for these three items is given in Figure 1.6, and a general description on page 8. For this exercise a title block only is required, and it is inserted in the bottom right-hand corner of the sheet, as shown in Figures 4.9 and 4.10.

The completed orthogonal projection

Figures 4.12, 4.13 and 4.14 demonstrate the drawing of a simple mechanical component in third-angle orthogonal projection.

Figure 4.12 shows an isometric view of a cast-steel wall bracket. The following views in third-angle orthogonal projection are to be drawn:
1. a front view in direction A
2. a side view in direction B
3. a top view

The drawing is to be fully dimensioned and supplied with a suitable title block. The scale is full size.

Figure 4.13 shows the rough sketch for the calculation of the positions of the three views on the drawing sheet. Notice the space between the top and front views is 40 mm compared with 75 mm between

the front and side views. This is because the bracket is higher than it is wide, and if these two spaces were made the same, the drawing would appear cramped on the paper.

The completed orthogonal projection is shown in Figure 4.14. This should be studied carefully to ensure full understanding of the relationship that exists between the detail on the views. Attempt to do the drawing within an A2 size drawing frame using the measurements given in Figures 4.12 and 4.13 and without referring to Figure 4.14.

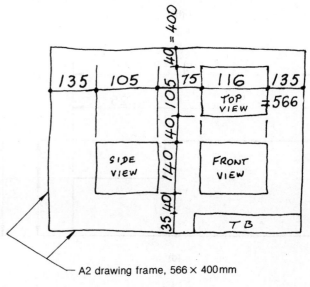

135 | 105 | 75 | 116 | 135
400 = 400
105
140
35 40
=566
TOP VIEW
SIDE VIEW
FRONT VIEW
T B

A2 drawing frame, 566 × 400 mm

Fig. 4.13 *Calculations for view positions*

118

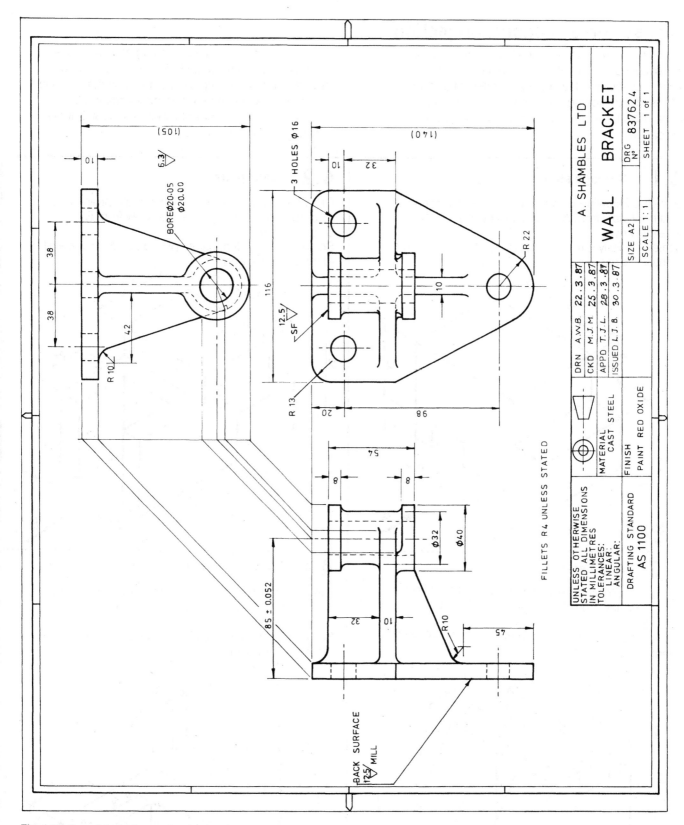

Fig. 4.14 *Completed orthogonal projection*

The drawing contains the following labelled information:

- (105)
- 10
- 6.3 ▽
- BORE Ø20.05 Ø20.00
- 38
- 38
- 42
- R 10
- 3 HOLES Ø16
- (140)
- 10
- 32
- 116
- R 22
- 10
- 12.5 ▽ SF
- R 13
- 20
- 98
- 54
- 8
- 8
- Ø32
- Ø40
- 85 ± 0.052
- 32
- 10
- R 10
- 45
- BACK SURFACE
- 12.5 ▽ MILL
- FILLETS R4 UNLESS STATED

Title block:

- UNLESS OTHERWISE STATED ALL DIMENSIONS IN MILLIMETRES
- TOLERANCES: LINEAR: ANGULAR:
- DRAFTING STANDARD AS 1100
- MATERIAL CAST STEEL
- FINISH PAINT RED OXIDE
- DRN A.W.B. 22.3.87
- CKD M.J.M. 25.3.87
- APPD T.J.L. 28.3.87
- ISSUED L.J.B. 30.3.87
- A. SHAMBLES LTD
- WALL BRACKET
- SIZE A2
- DRG Nº 837624
- SHEET 1 of 1
- SCALE 1:1

Exercises on orthogonal projection

The following exercises are graded in approximate order of difficulty. Start at 4.1 and work through, referring, as the need arises, to the relevant text on sections, dimensioning and so on. The exercises may also be used for technical sketching on squared or plain paper.

As a general rule, dimensioning of drawings is carried out with a full knowledge of the functional requirements of a component and those dimensions that are critical are inserted. In dimensioning the following exercises students should accept that the dimensions given are critical.

4.1 Carriage stop

Draw the following views in third-angle projection:
1. a front view from A; 2. a side view from B;
3. a top view.
Scale 1:1. Fully dimension, provide a title block and identify the drawing.

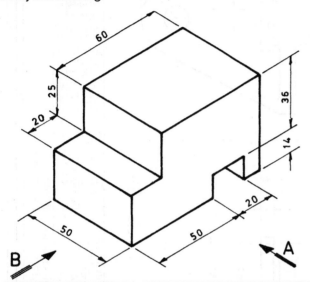

4.2 Lock plate

Draw the following views in third-angle projection:
1. a front view from A; 2. a side view from B;
3. a top view.
Scale 1:1. Fully dimension, provide a title block and identify the drawing.

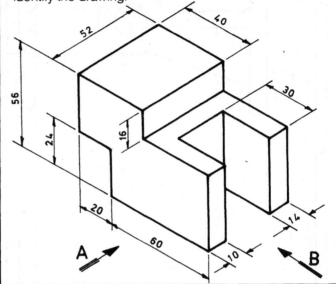

4.3 Angle stop

Draw the following views in third-angle projection:
1. a front view from A; 2. a side view from B;
3. a top view.
Scale 1:1. Fully dimension, provide a title block and identify the drawing.

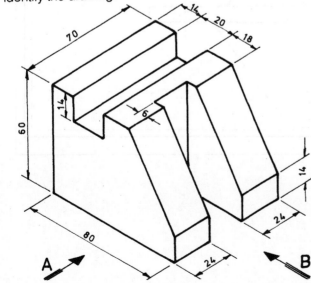

4.4 Dovetail template

Draw the following views in third-angle projection:
1. a front view from A; 2. a side view from B;
3. a top view.
Scale 1:1. Fully dimension, provide a title block and identify the drawing.

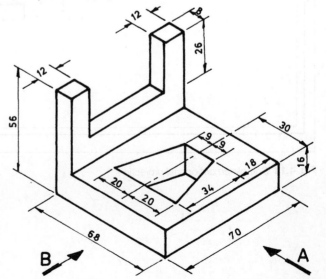

4.5 Angle guide

Draw the following views in third-angle projection:
1. a front view from A
2. a side view
3. a top view
Scale 1:1
Fully dimension, provide a title block and identify the drawing.

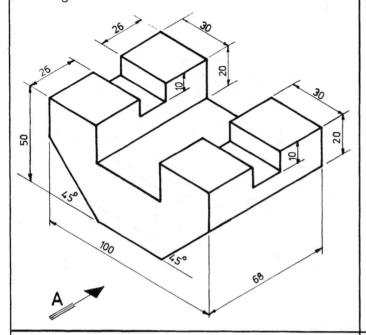

4.6 MS gauge

Draw the following views in third-angle projection:
1. a front view from A
2. a side view
3. a top view
Scale 1:1
Fully dimension, provide a title block and identify the drawing.

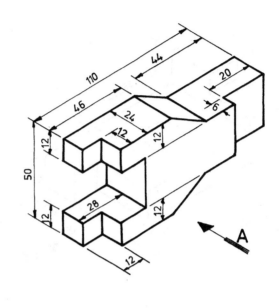

4.7 Angle plate

Draw the following views in third-angle projection:
1. a front view from A
2. a side view from B
3. a top view
Scale 1:1
Fully dimension, provide a title block and identify the drawing.

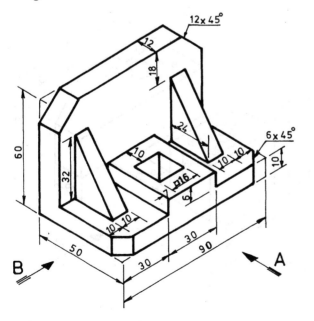

4.8 Shaft support

Draw the following views in third-angle projection:
1. a front view from A
2. a side view from B
3. a top view
Scale 1:1
Fully dimension, provide a title block and identify the drawing.

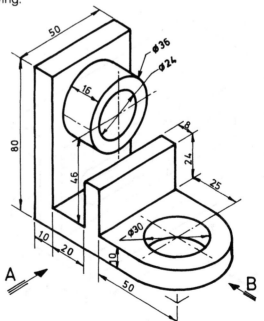

4.9 Bearing plate

Draw the following views in third-angle projection:
1. a front view from A
2. a side view from B
3. a top view
Scale 1:1
Fully dimension, provide a title block and identify the drawing.

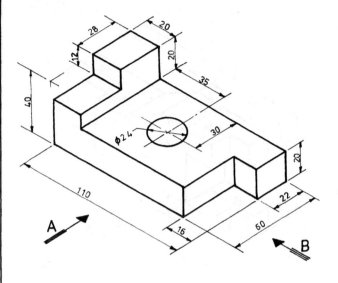

4.10 Cross base

Draw the following views in third-angle projection:
1. a front view from A
2. a side view from B
3. a top view
Scale 1:1
Fully dimension, provide a title block and identify the drawing.

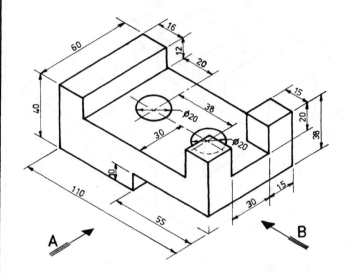

4.11 Bed plate

Draw the following views in third-angle projection:
1. a front view from A
2. a side view from B
3. a top view
Scale 1:1
Fully dimension, provide a title block and identify the drawing.

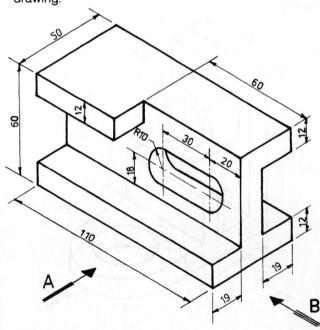

4.12 Swing bracket

Draw the following views in third-angle projection:
1. a front view from A
2. a side view from B
3. a top view
Scale 1:1
Fully dimension, provide a title block and identify the drawing.

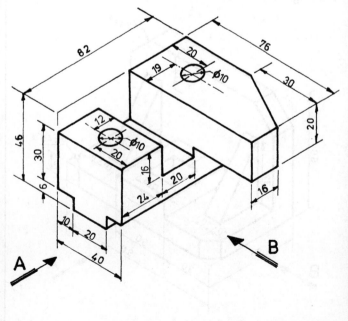

4.13   CS bracket

Draw the following views in third-angle projection:
1. a front view from A
2. a side view
3. a top view

Scale 1:1

Fully dimension, provide a title block and identify the drawing.

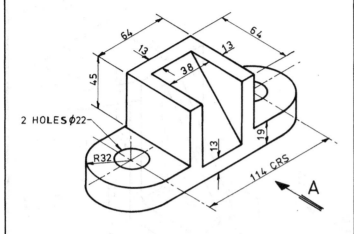

4.14   Brass step

Draw the following views in third-angle projection:
1. a front view from A
2. a side view
3. a top view

Scale 1:1

Fully dimension, provide a title block and identify the drawing.

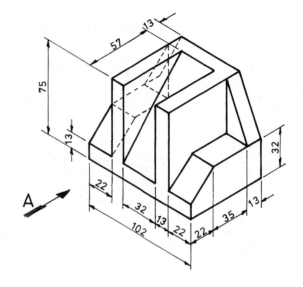

4.15   CI bench block

Draw the following views in third-angle projection:
1. a front view from A
2. a side view from B
3. a top view

Scale 1:1

Fully dimension, provide a title block and identify the drawing.

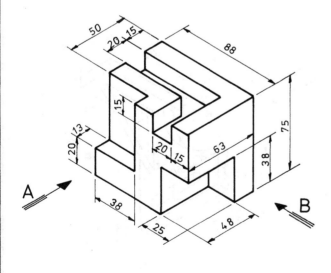

4.16   CI bracket

Draw the following views in third-angle projection:
1. a front view from A
2. a side view from B
3. a top view

Scale 1:1

Fully dimension, provide a title block and identify the drawing.

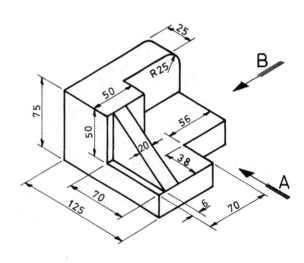

4.17 MS bracket

Draw the following views in third-angle projection:
1. a front view from A
2. a side view from B
3. a top view
Scale 1:1
Fully dimension, provide a title block and identify the drawing.

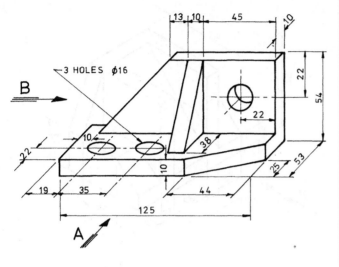

4.18 CI support

Draw the following views in third-angle projection:
1. a front view from A
2. a side view from B
3. a top view
Scale 1:1
Fully dimension, provide a title block and identify the drawing.

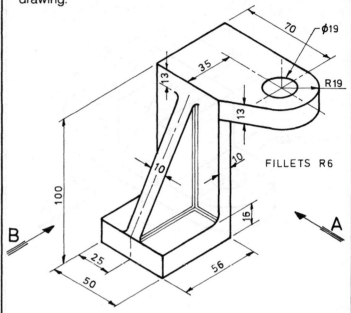

4.19 Jig frame

Draw the following views in third-angle projection:
1. a front view from A
2. a side view from B
3. a top view
Scale 1:1
Fully dimension, provide a title block and identify the drawing.

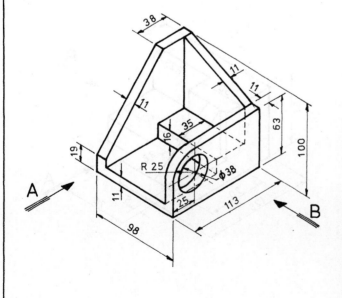

4.20 Grinder rest

Draw the following views in third-angle projection:
1. a front view from A
2. a side view from B
3. a top view
Scale 1:1
Fully dimension, provide a title block and identify the drawing.

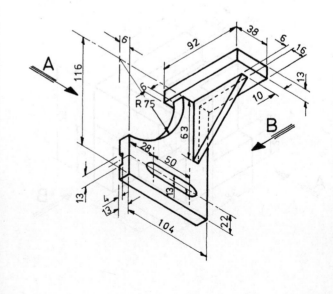

4.21 MS rod bracket

Draw the following views in third-angle projection:
1. a front view from A
2. a side view from B
3. a top view
Scale 1:1
Fully dimension, provide a title block and identify the drawing.

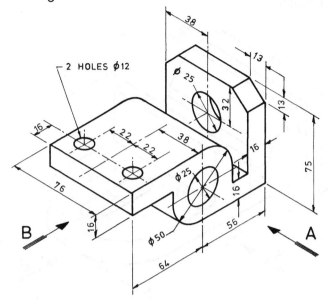

4.22 Safety bracket

Draw the following views in third-angle projection:
1. a front view from A
2. a side view from B
3. a top view
Scale 1:1
Fully dimension, provide a title block and identify the drawing.

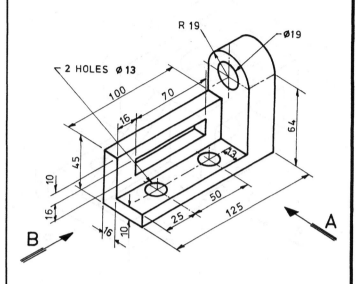

4.23 CI bracket

Draw the following views in third-angle projection:
1. a front view from A
2. a side view from B
3. a top view
Scale 1:1
Fully dimension, provide a title block and identify the drawing.

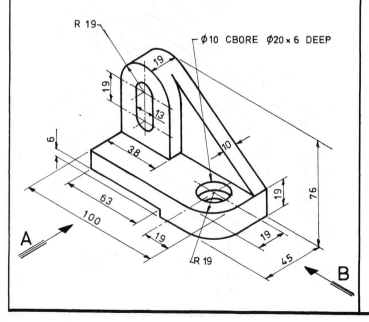

4.24 Support bracket

Draw the following views in third-angle projection:
1. a front view from A
2. a side view from B
3. a top view
Scale 1:1
Fully dimension, provide a title block and identify the drawing.

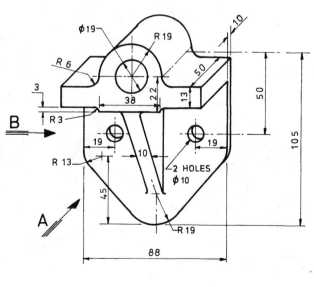

4.25 Adjustable holder

Draw the following views in third-angle projection:
1. a front view from A
2. a side view from B
3. a top view
Scale 1:1
Fully dimension, provide a title block and identify the drawing.

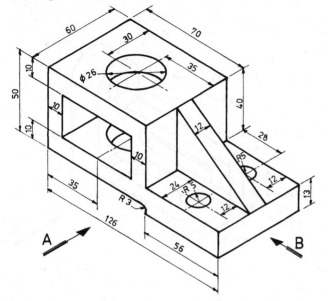

4.26 Vee block

Draw the following views in third-angle projection:
1. a front view from A
2. a side view from B
3. a top view
Scale 1:1
Fully dimension, provide a title block and identify the drawing.

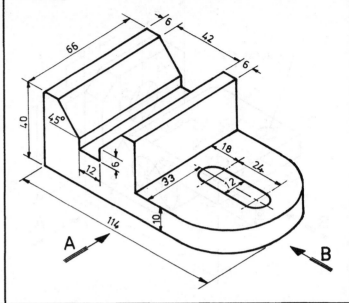

4.27 Dovetail slide

Draw the following views in third-angle projection:
1. a front view from A
2. a side view from B
3. a top view
Scale 1:1
Fully dimension, provide a title block and identify the drawing.

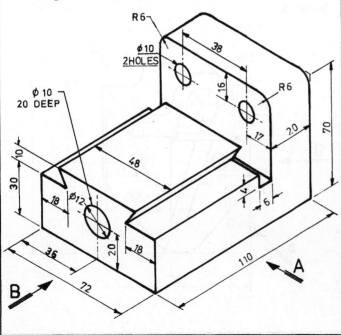

4.28 Chuck jaw

Draw the following views in third-angle projection:
1. a front view from A
2. a side view from B
3. a top view
Scale 1:1
Fully dimension, provide a title block and identify the drawing.

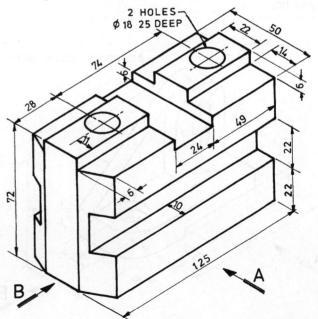

4.29 CS socket

Draw the following views in third-angle projection:
1. a sectional front view from A–A
2. a side view from B
3. a top view
Scale 1:1
Fully dimension, provide a title block and identify the drawing.

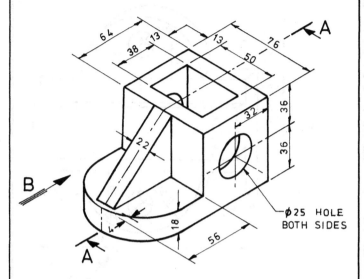

4.30 MS clamp

Draw the following views in third-angle projection:
1. a front view from A
2. a sectional side view on B–B
3. a top view
Scale 1:1
Fully dimension, provide a title block and identify the drawing.

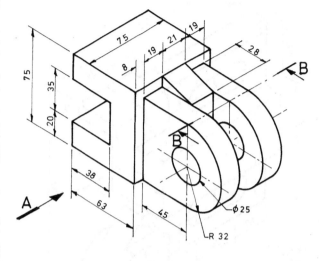

4.31 CS bracket

Draw the following views in third-angle projection:
1. a front view from A
2. a side view from B
3. a top view
Scale 1:1
Fully dimension, provide a title block and identify the drawing.

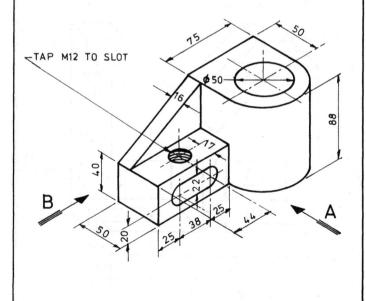

4.32 MS sleeve bracket

Draw the following views in third-angle projection:
1. a front view from A
2. a sectional side view on B–B
3. a top view
Scale 1:1
Fully dimension, provide a title block and identify the drawing.

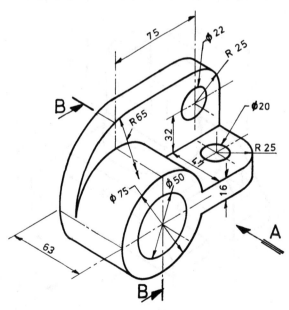

4.33 Adjustable vee block

Draw the following views in third-angle projection:
1. a front view from A
2. a side view from B
3. a top view
Scale 1:1
Fully dimension, provide a title block and identify the drawing.

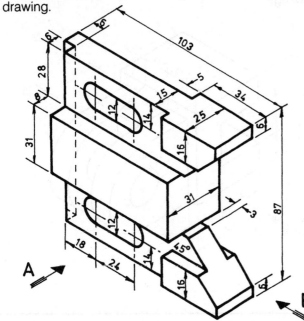

4.34 Cross slide

Draw the following views in third-angle projection:
1. a front view from A
2. a side view from B
3. a top view
Scale 1:1
Fully dimension, provide a title block and identify the drawing.

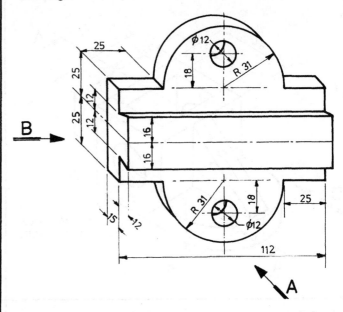

4.35 Sliding gauge

Draw the following views in third-angle projection:
1. a front view from A
2. a side view from B
3. a top view
Scale 1:1
Fully dimension, provide a title block and identify the drawing.

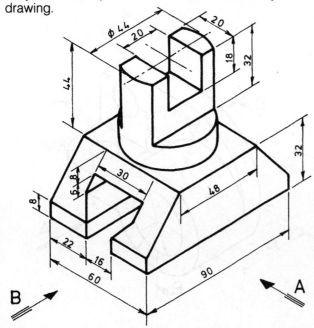

4.36 Vee block

Draw the following views in third-angle projection:
1. a front view from A
2. a side view from B
3. a top view
Scale 1:1
Fully dimension, provide a title block and identify the drawing.

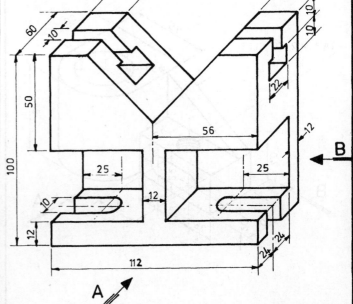

4.37 CI jaw support

Draw the following views in third-angle projection:
1. a sectional front view on A–A
2. a side view from B
3. a top view
Scale 1:1
Fully dimension, provide a title block and identify the drawing.

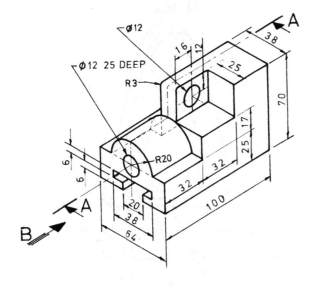

4.38 Offset crank

Draw the following views in third-angle projection:
1. a front view from A
2. a sectional side view from B–B
3. a top view
Scale 1:1
Fully dimension, provide a title block and identify the drawing.

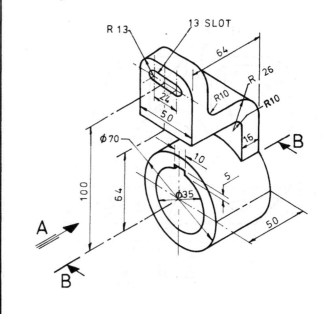

4.39 Main bearing cap

Draw the following views in third-angle projection:
1. a sectional front view on a vertical plane through A–A
2. a side view
3. a top view
Scale 1:1
Fully dimension, provide a title block and identify the drawing.

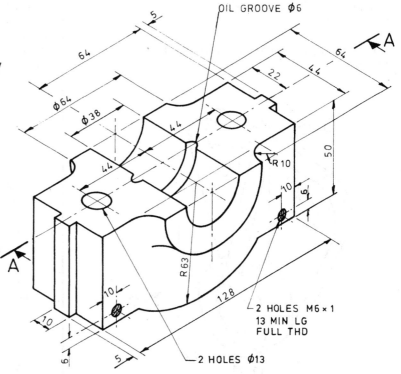

129

4.40 Bearing blank

Draw the following views in third-angle projection:
1. a front view from A
2. a sectional side view on B–B
3. a top view
Scale 1:1
Fully dimension, provide a title block and identify the drawing.

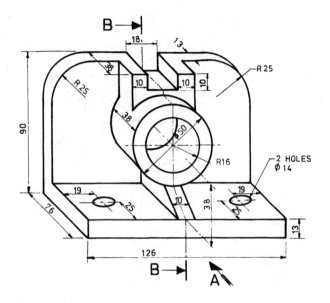

4.41 Shaft support

Draw the following views in third-angle projection:
1. a front view from A
2. a sectional side view on B–B
3. a top view
Scale 1:1
Fully dimension, provide a title block and identify the drawing.

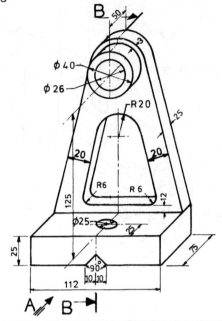

4.42 Rod guide

Draw the following views in third-angle projection:
1. a front view from A
2. a sectional side view on B–B
3. a top view
Scale 1:1
Fully dimension, provide a title block and identify the drawing.

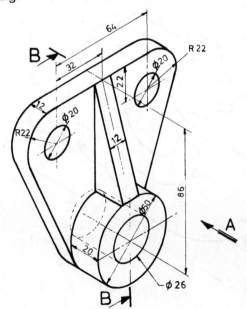

4.43 Bearing bracket

Draw the following views in third-angle projection:
1. a front view from A
2. a sectional side view on B–B
3. a top view
Scale 1:1
Fully dimension, provide a title block and identify the drawing.

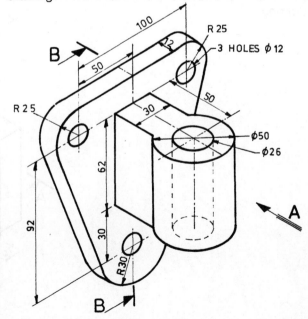

130

4.44 CI bearing bracket

Draw the following views in third-angle projection:
1. a front view from A
2. a sectional side view on the centre line, from B
3. a top view

Scale 1:1

Fully dimension, provide a title block and identify the drawing.

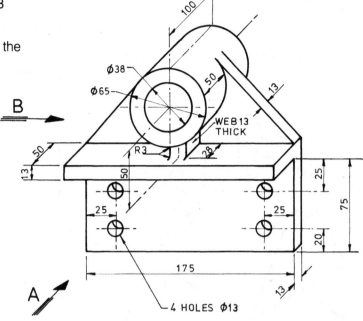

4.45 Bearing retainer

Draw the following views in third-angle projection:
1. a sectional front view on the centre line viewed from A
2. a side view from B
3. a top view

Scale 1:1

Fully dimension, provide a title block and identify the drawing.

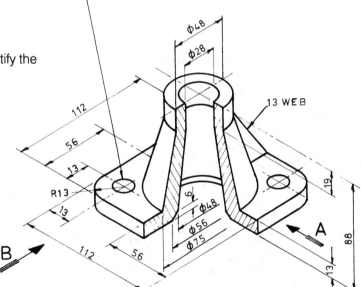

4.46 Profile guide

Draw the following views in third-angle projection:
1. a front view from A
2. a side view from B
3. a top view
Scale 1:1
Fully dimension, provide a title block and identify the drawing.

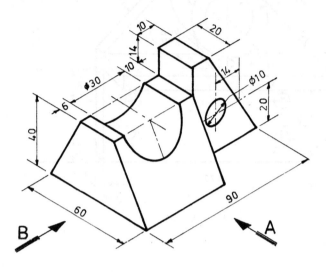

4.47 Control block

Draw the following views in third-angle projection:
1. a front view from A
2. a side view from B
3. a top view
Scale 1:1
Fully dimension, provide a title block and identify the drawing.

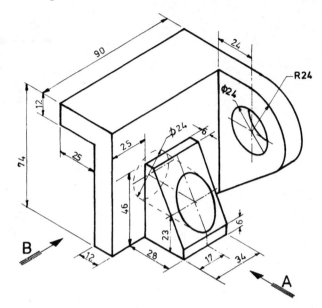

4.48 Support block

Draw the following views in third-angle projection:
1. a front view from A
2. a side view from B
3. a top view
Scale 1:1
Fully dimension, provide a title block and identify the drawing.

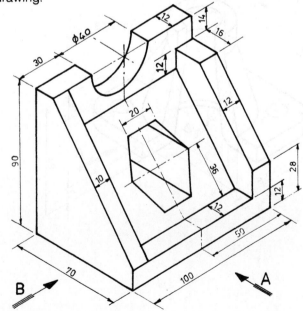

4.49 Drill guide

Draw the following views in third-angle projection:
1. a front view from A
2. a side view from B
3. a top view
Scale 1:1
Fully dimension, provide a title block and identify the drawing.

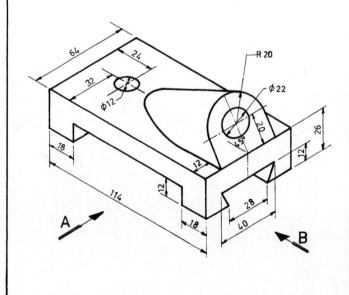

4.50   Adjustable bearing

Draw the following views in third-angle projection:
1. a sectional top view on A–A
2. a front view from B
3. a side view from C
Scale 1:1
Fully dimension, provide a title block and identify the drawing.

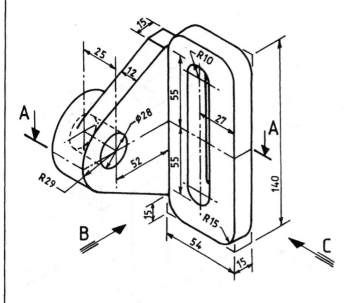

4.51   Tool post

Draw the following views in third-angle projection:
1. a front view from A
2. a side view from B
3. a top view
Scale 1:1
Fully dimension, provide a title block and identify the drawing.

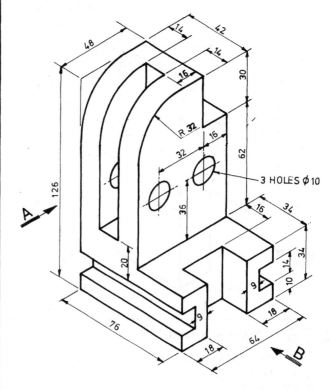

4.52   Pivot head

Draw the following views in third-angle projection:
1. a front view from A
2. a side view from B
3. a top view
Scale 1:1
Fully dimension, provide a title block and identify the drawing.

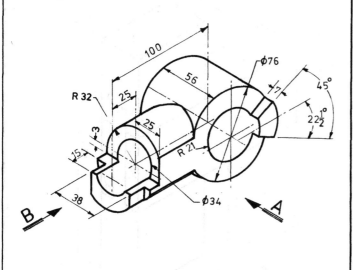

4.53   Locking block

Draw the following views in third-angle projection:
1. a front view from A
2. a sectional side view on B–B
3. a top view
Scale 1:1
Fully dimension, provide a title block and identify the drawing.

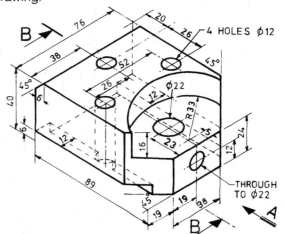

4.54 End half check

Draw the following views in third-angle projection:
1. a front view from A
2. a side view from B
3. a top view
Scale 1:1
Fully dimension, provide a title block and identify the drawing.

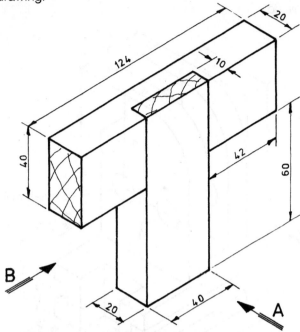

4.55 Mortice and tenon

Draw the following views in third-angle projection:
1. a front view from A
2. a side view from B
3. a top view
Scale 1:1
Fully dimension, provide a title block and identify the drawing.

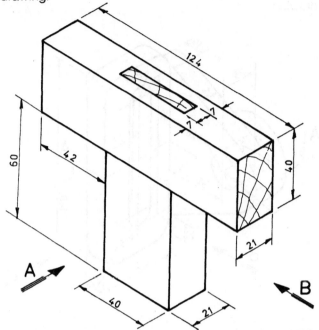

4.56 Dovetail halving

Draw the following views in third-angle projection:
1. a front view from A
2. a side view from B
3. a top view
Scale 1:1
Fully dimension, provide a title block and identify the drawing.

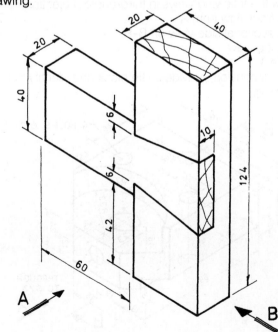

4.57 Bridle joint

Draw the following views in third-angle projection:
1. a front view from A
2. a side view from B
3. a top view
Scale 1:1
Fully dimension, provide a title block and identify the drawing.

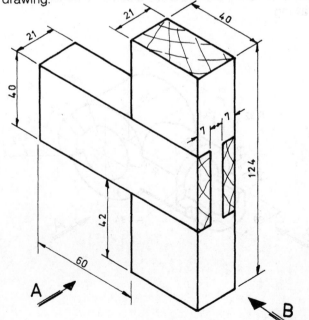

4.58 Sawing exercise

Draw the following views in third-angle projection:
1. a front view from A
2. a side view from B
3. a top view
Scale 1:1
Fully dimension, provide a title block and identify the drawing.

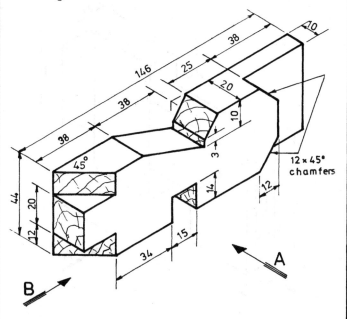

4.59 Grooving exercise

Draw the following views in third-angle projection:
1. a front view from A
2. a side view from B
3. a top view
Scale 1:1
Fully dimension, provide a title block and identify the drawing.

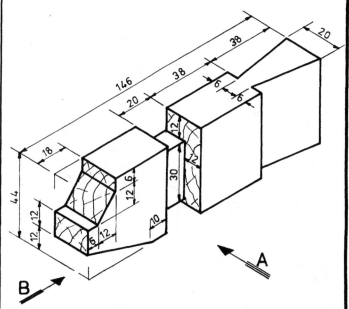

4.60 Oblique grooving

Draw the following views in third-angle projection:
1. a front view from A
2. a side view from B
3. a top view
Scale 1:1
Fully dimension, provide a title block and identify the drawing.

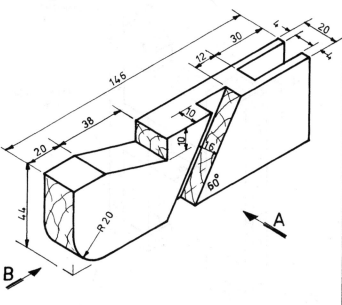

4.61 Sawing and grooving

Draw the following views in third-angle projection:
1. a front view from A
2. a side view from B
3. a top view
Scale 1:1
Fully dimension, provide a title block and identify the drawing.

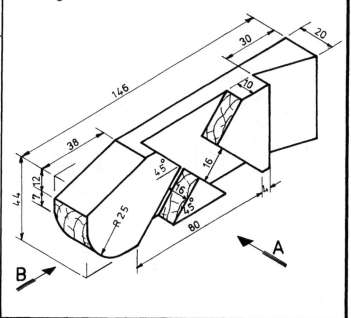

135

4.62 Jointing exercise

Draw the following views in third-angle projection:
1. a front view from A
2. a side view from B
3. a top view
Scale 1:1
Fully dimension, provide a title block and identify the drawing.

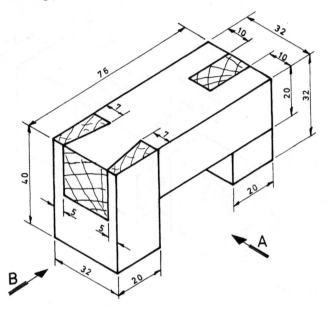

4.63 Woodwork test

Draw the following views in third-angle projection:
1. a front view from A
2. a side view from B
3. a top view
Scale 1:1
Fully dimension, provide a title block and identify the drawing.

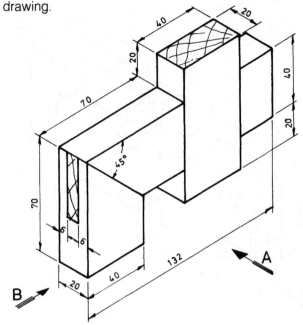

4.64 Test exercise

Draw the following views in third-angle projection:
1. a front view from A
2. a side view from B
3. a top view
Scale 1:1
Fully dimension, provide a title block and identify the drawing.

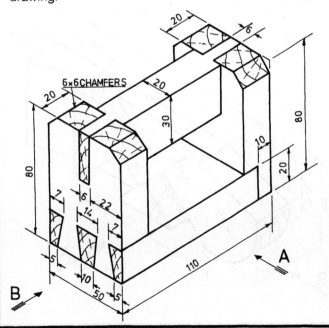

4.65 Jointed frame

Draw the following views in third-angle projection:
1. a front view from A
2. a side view from B
3. a top view
Scale 1:1
Fully dimension, provide a title block and identify the drawing.

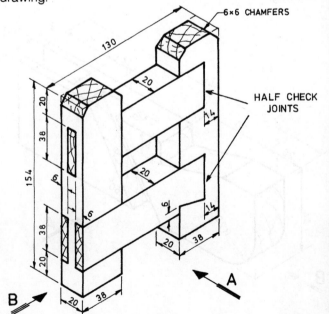

4.66 Forked bracket

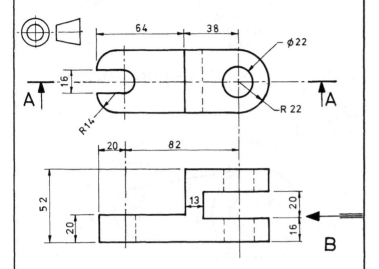

Draw the following views in third-angle projection:
1. a sectional front view on A–A
2. a side view from B
3. the given top view
Scale 1:1
Fully dimension, provide a title block and identify the drawing.

4.67 MS slotted link

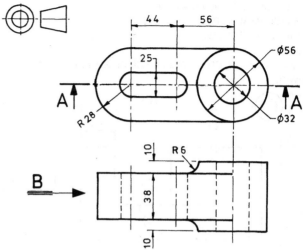

Draw the following views in third-angle projection:
1. a sectional front view on A–A
2. a side view from B
3. the given top view
Scale 1:1
Fully dimension, provide a title block and identify the drawing.

4.68 MS half clip

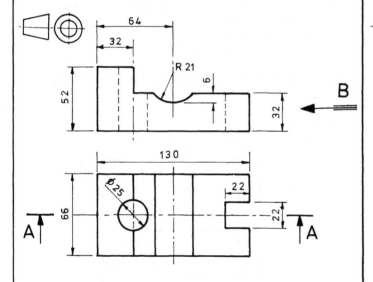

Draw the following views in third-angle projection:
1. a sectional front view on A–A
2. a side view from B
3. the given top view
Scale 1:1
Fully dimension, provide a title block and identify the drawing.

4.69 CI forked end

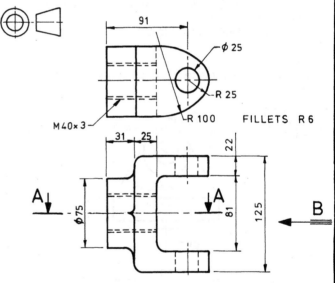

Draw the following views in third-angle projection:
1. the front view as given
2. a side view from B
3. a sectional top view on A–A
Scale 1:1
Fully dimension, provide a title block and identify the drawing.

4.70 Shaft bracket

Draw the following views in third-angle projection:
1. a sectional front view on A–A
2. a side view from B
3. the given top view
Scale 1:1
Fully dimension, provide a title block and identify the drawing.

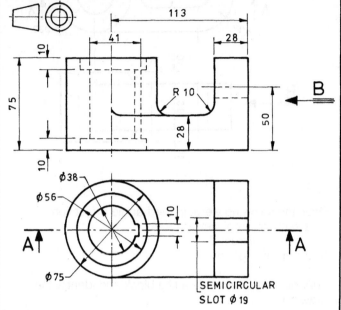

4.71 CI crank arm

Draw the following views in third-angle projection:
1. a sectional front view on A–A
2. a side view from B
3. the given top view
Scale 1:1
Fully dimension, provide a title block and identify the drawing.

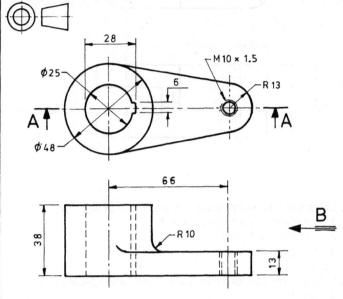

4.72 Sliding support

Draw the following views in third-angle projection:
1. the given front view
2. a sectional side view on B–B
3. a bottom view
Scale 1:1
Fully dimension, provide a title block and identify the drawing.

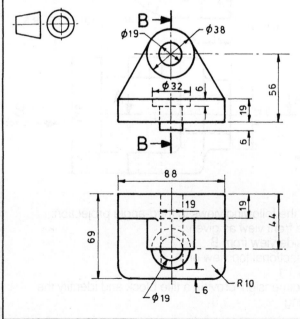

4.73 Guide bracket

Draw the following views in third-angle projection:
1. the front view as given
2. a side view from B
3. a sectional top view on A–A
Scale 1:1
Fully dimension, provide a title block and identify the drawing.

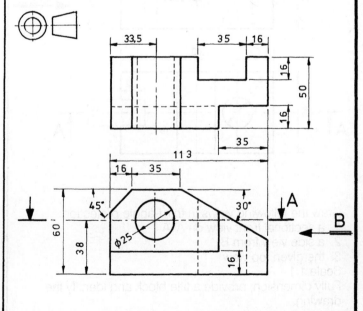

138

4.74   Vee stop

Draw the following views in third-angle projection:
1. a sectional front view on A–A
2. a side view from B
3. a top view
Scale 1:1
Fully dimension, provide a title block and identify the drawing.

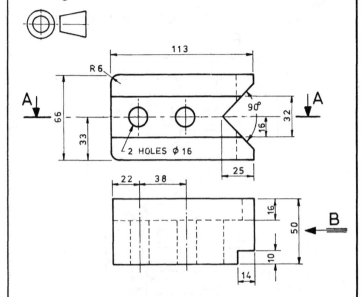

4.75   CI support

Draw the following views in third-angle projection:
1. the front view as given
2. a sectional side view on A–A
3. the given top view
Scale 1:1
Fully dimension, provide a title block and identify the drawing.

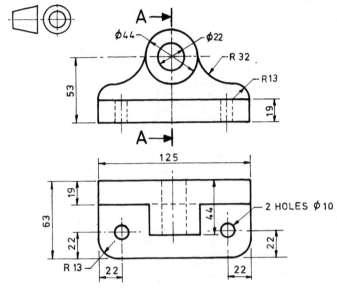

4.76   CI guide block

Draw the following views in third-angle projection:
1. a sectional front view on A–A
2. a side view from B
3. a sectional top view on C–C
Scale 1:1
Fully dimension, provide a title block and identify the drawing.

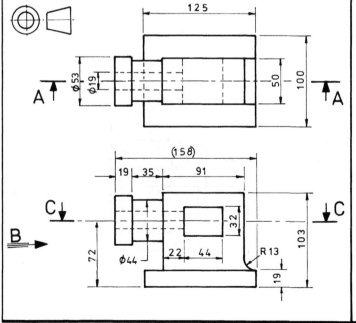

4.77   Brass support

Draw the following views in third-angle projection:
1. a sectional front view on A–A
2. a sectional side view on B–B
3. the given top view
Scale 1:1
Fully dimension, provide a title block and identify the drawing.

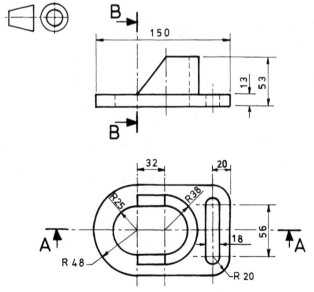

4.78 CI jig

Draw the following views in third-angle projection:
1. the given front view
2. a sectional side view on A-A
3. the given top view
Scale 1:1
Fully dimension, provide a title block and identify the drawing.

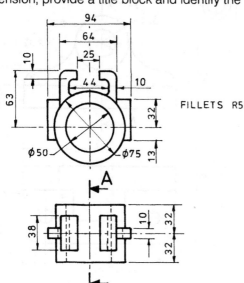

FILLETS R5

4.79 CI rod guide

Draw the following views in third-angle projection:
1. a rear view
2. an aligned sectional side view on B-B
3. a bottom view
Scale 1:1
Fully dimension, provide a title block and identify the drawing.

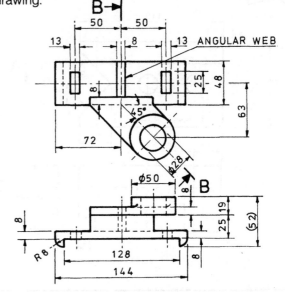

ANGULAR WEB

4.80 Truss bearing

Draw the following views in third-angle projection:
1. the front view as given
2. a half-sectional side view from B
3. an offset section on A-A
Scale 1:1
Fully dimension, provide a title block and identify the drawing.

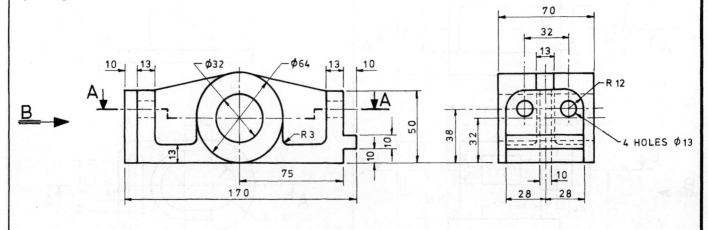

4 HOLES ⌀13

4.81   Lever hub

Draw the following views in third-angle projection:
1. a front view from A
2. a side view from B
3. a sectional top view on the centre line
Scale 1:1
Fully dimension, provide a title block and identify the drawing.

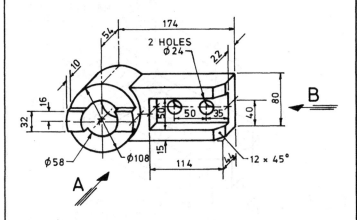

4.82   Roller stud

Draw the following views in third-angle projection:
1. a front view from A
2. a side view from B
3. a top view
Scale 1:1
Fully dimension, provide a title block and identify the drawing.

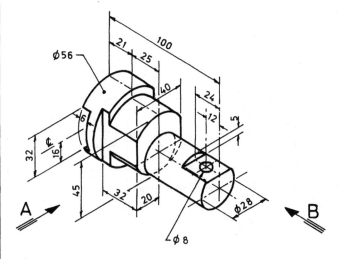

4.83   Index slide

Draw the following views in third-angle projection:
1. a front view from A
2. a side view from B
3. a sectional top view on the centre line
Scale 1:1
Fully dimension, provide a title block and identify the drawing.

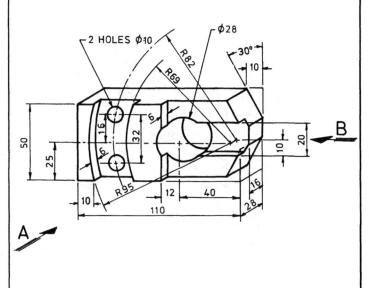

4.84   Toggle lever

Draw the following views in third-angle projection:
1. a front view from A
2. a side view from B
3. a top view
Scale 1:1
Fully dimension, provide a title block and identify the drawing.

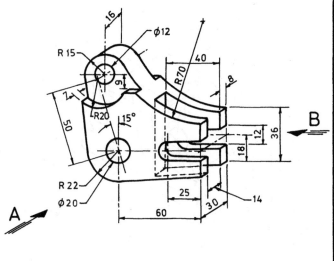

141

4.85 Hinge base

Draw the following views in third-angle projection:
1. a front view from A
2. a side view from the left-hand side of A
3. a top view
Scale 1:1
Fully dimension, provide a title block and identify the drawing.

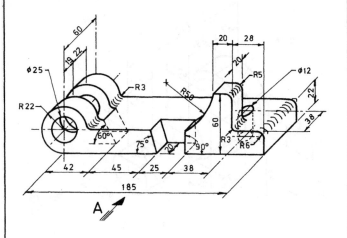

4.86 Chuck jaw

Draw the following views in third-angle projection:
1. a sectional front view from A
2. a side view from B
3. a bottom view
Scale 1:1
Fully dimension, provide a title block and identify the drawing.

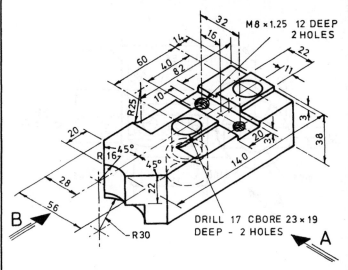

4.87 Tool holder

Draw the following views in third-angle projection:
1. a front view from A
2. a side view from B
3. a top view
Scale 1:1
Fully dimension, provide a title block and identify the drawing.

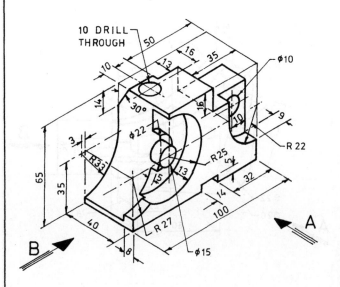

4.88 Locating block

Draw the following views in third-angle projection:
1. a front view from A
2. right and left side views
3. a top view
Scale 1:1
Fully dimension, provide a title block and identify the drawing.

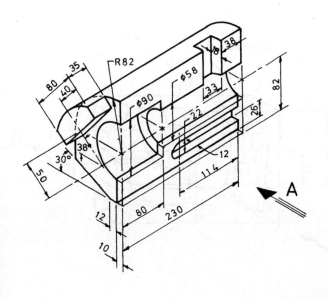

4.89   Cover plate

Draw the following views in third-angle projection:
1. a front view from A
2. a side view from B
3. a half-sectional top view
Scale 1:1
Fully dimension, provide a title block and identify the drawing.

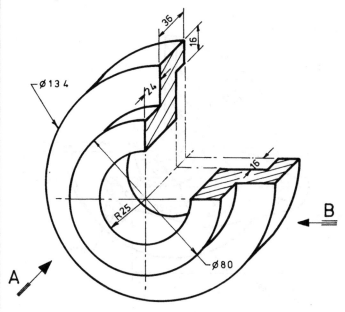

4.90   Oil seal cap

Draw the following views in third-angle projection:
1. a front view from A
2. a half-sectional side view from B
3. a top view
Scale 1:1
Fully dimension, provide a title block and identify the drawing.

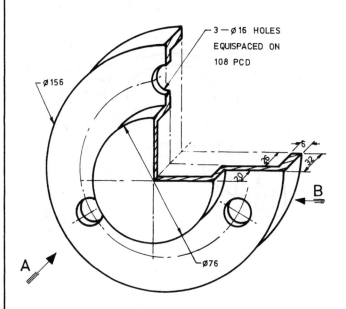

4.91   Face plate blank

Draw the following views in third-angle projection:
1. a front view from A
2. a side view from B
3. a half-sectional top view
Scale 1:1
Fully dimension, provide a title block and identify the drawing.

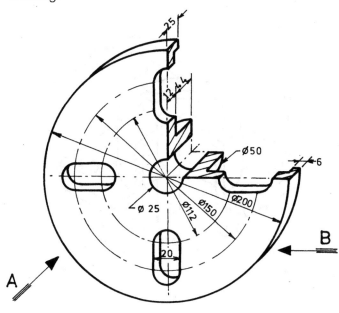

4.92   CI retainer

Draw the following views in third-angle projection:
1. a front view from A
2. a half-sectional side view from B
3. a top view
Scale 1:1
Fully dimension, provide a title block and identify the drawing.

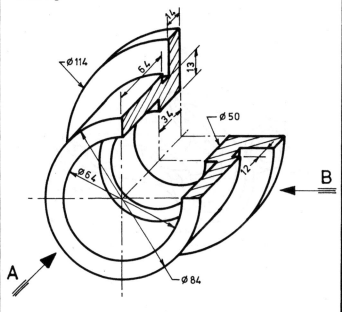

4.93 Packing gland

Draw the following views in third-angle projection:
1. a half-sectional front view from A
2. a side view from B
3. a top view
Scale 1:1
Fully dimension, provide a title block and identify the drawing.

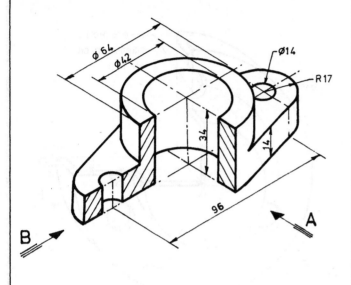

4.94 Stuffing box

Draw the following views in third-angle projection:
1. a front view from A
2. a side view from B
3. a half-sectional top view
Scale 1:1
Fully dimension, provide a title block and identify the drawing.

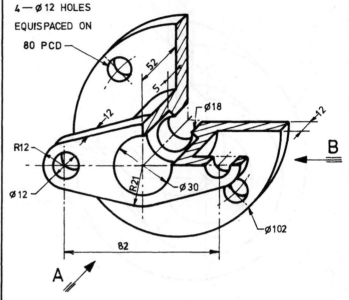

4.95 CI bearing

Draw the following views in third-angle projection:
1. a front view from A
2. a side view from B
3. a half-sectional top view
Scale 1:1
Fully dimension, provide a title block and identify the drawing.

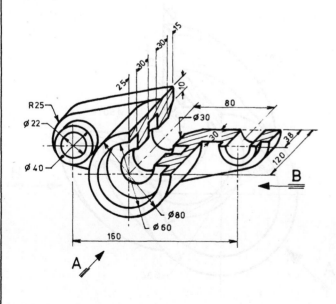

4.96 Flanged cover

Draw the following views in third-angle projection:
1. a half-sectional front view from A
2. a side view from B
3. a top view
Scale 1:1
Fully dimension, provide a title block and identify the drawing.

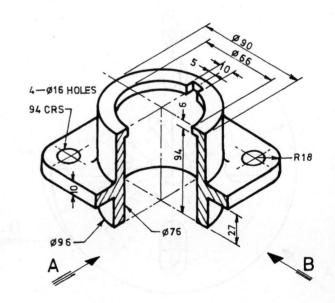

Drawing analysis

5

The ability to analyse a drawing and hence to be able to 'take in' all the information contained on it is a skill not easily obtained. Tradespeople, engineers and drafters must have this ability if they are to communicate with the originator of the drawing.

Many symbols, abbreviations and conventions have been universally agreed upon to condense information so that it may be put on a drawing without congestion. This chapter describes the basic vocabulary used on a mechanical drawing (see Fig. 5.1). It gives a series of exercises that will provide a sound knowledge of orthogonal projection and questions that will promote understanding of the drawing content.

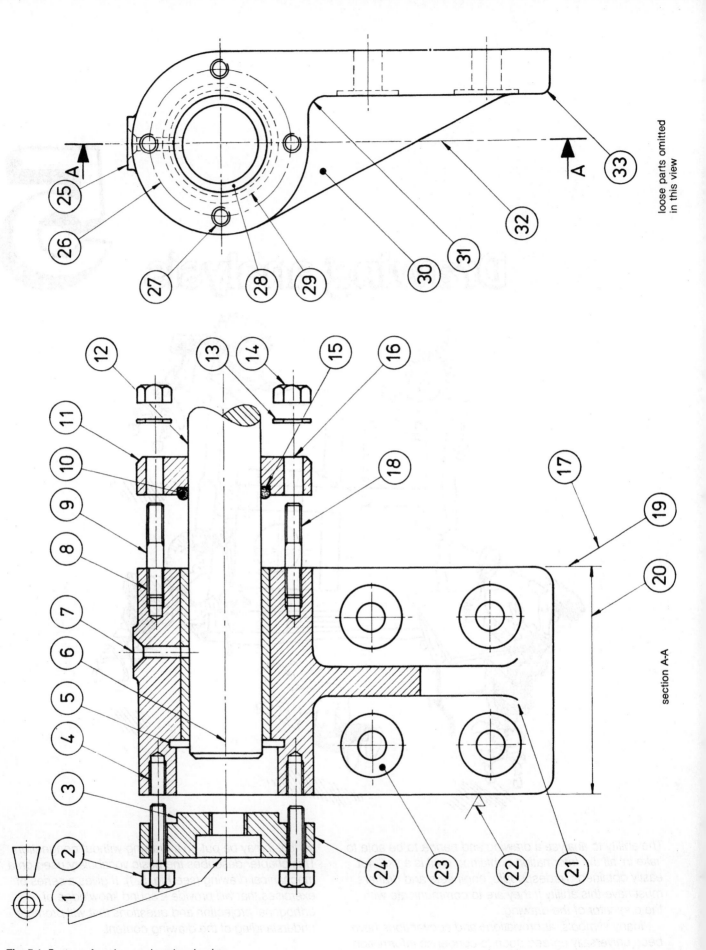

loose parts omitted
in this view

section A-A

Fig. 5.1 *Features found on engineering drawings*

Sample analysis

The following features are found on engineering drawings (see Fig. 5.1).

1. A *counterbored hole* is used to house a screw or bolt head so that it does not project from the surface. It also provides a surface, square to the hole axis, for bolt head seating.

2. A *bolt* is designated by the material, head shape, ISO metric thread diameter (mm) and the length (mm) of its shank.

3. A *spigot* is a piece of material (usually circular) that projects from the face of a member. It is used to locate a member precisely when assembling it with another member. It may also be used to carry any shear load applied to bolts holding the two members together.

4. As this is a sectional view, the cross-hatch lines pass over the *internal thread* section.

5. A *recess* allows a member to engage right to the bottom of a hole without interference from a rounded corner. A recess can also be used externally, for example when turning a thread up to a shoulder.

6. A *centre line* is a light, long-short dash line (type G) used to indicate axes of holes and the centres of part and full circles.

7. A *countersunk hole* in this case is used as an oil hole but generally would be used to house the countersunk head of a screw.

8. Note that the cross-hatch lines do *not* pass over the *assembled threads*, but where the thread stands alone, item 4 above applies.

9. A *stud* is a member threaded both ends and screwed firmly into the main part. Studs are used to attach coverplates and housings, as shown.

10. A *seal* is generally a plastic ring seal which, when compressed against the main housing, squeezes against the rotating shaft and prevents entry of dust and grit into the main bearing. It also prevents lubricant from leaking out.

11. A *chamfer* is generally 45°, its purpose being to eliminate the sharp edge.

12. A *shaft* is a rotating member used to transmit torque. Note the chamfer on the end and the method of showing a break in the shaft, that is, the shaft actually extends beyond the length shown.

13.–14 A *washer* (13) is used with a *nut* (14) on the stud. It prevents scoring of the plate when the nut is tightened.

15. A *housing* is a general term used to describe the location of items such as seals, bearings and gears. A *seal housing* is shown here.

16. A *clearance hole* is a hole just a little larger than the diameter of the stud, so that assembly is made easy. Recommended diameters of clearance holes for various sizes of metric thread diameter are given in Table 1.5.

17. *Leaders* are used to indicate where dimensions or notes are intended to apply. They are thin full lines that terminate in arrowheads or dots. Arrowheads terminate on a line, dots should fall within the outline of the object, as shown by items 30, 28, 23 and 10.

18. An *external* or *male thread* is the representation of the outside view of a threaded member.

19. A *projection line* is a thin full line (type B) extending from the outline, but not touching it. These lines denote the extremities of a dimension and should extend a little beyond the dimension line.

20. A *dimension line* is a thin full line (type B) extending between projection lines with arrowheads on either end to indicate the length of the dimension, which is placed above the dimension line near the centre.

21. A *runout* is used to indicate the intersection of two surfaces that do not meet at a sharp corner.

22. A *surface finish symbol* indicates the finish of the surface to which it is applied. See page 32.

23. A *spotface* is an area around a hole that is machined perpendicular to the hole axis. It provides a flat true seating for the head of a nut or bolt.

24. *Flange* is a term used to describe a section of a member that carries holes through which bolts or screws pass to fasten the member.

25. A *boss* is a raised or extra portion of metal machined on top to support the screw head. The term boss can be applied to extra projections of metal that provide additional support or an extension of the function, for example shaft bosses provide extra bearing length, screw or bolt bosses provide for adequate thread length.

26. *Pitch circle diameter (PCD)* is a light, long-short dash circle that passes through the centres of a series of holes. The holes are generally pitched evenly around its circumference.

27. The end view of an *internal thread* has a full circle on the inside and a broken circle on the outside, as opposed to the end view of an external thread. See page 21 for more details.

28. *Bush* is a term used to describe a plain bearing for a shaft. It is a sleeve, usually made of bronze material and fitting tightly into the housing.

29. A thin short-dash line (type E) is used to indicate *hidden detail*, such as corners or edges that cannot be seen from the outside.

30. A *web* is a strengthening or stiffening member.

31. All castings have *fillets* on internal corners to prevent the formation of stress fatigue cracks, which originate in sharp corners.

32. The *course of a section plane* is indicated by a chain line (type H), thick at the ends and where it changes direction, but thin elsewhere. The view in Figure 5.1 (section A-A) reveals detail seen at the level of this plane in the direction of the arrows A-A.

33. A *round* is similar to a fillet but is found on external corners of a casting.

Exercises on drawing analysis

5.1

The surfaces of the pictorial view have been identified by a letter. The surfaces and edges of surfaces in the orthogonal projection have been identified by a number.

Complete the table with the numbers that correspond to the surfaces lettered in the pictorial view.

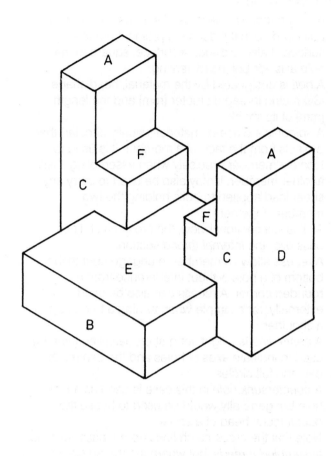

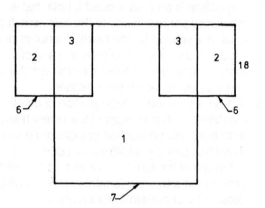

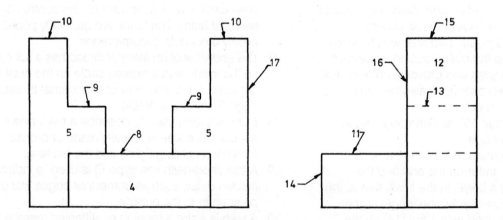

	A	B	C	D	E	F
Top	2					
Front	10					
Right side	15					

5.2

The surfaces of the pictorial view have been identified by a letter. The surfaces and edges of surfaces on the orthogonal views have been identified by a number.

Fill in the table with the numbers that correspond to the surfaces lettered in the pictorial view.

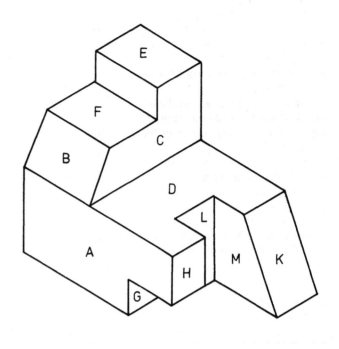

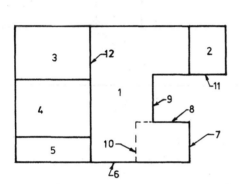

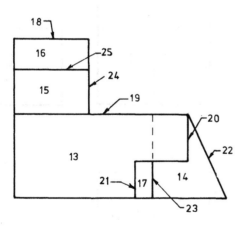

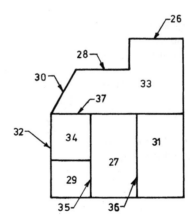

	A	B	C	D	E	F	G	H	K	L	M
Top											
Front											
Right side											

5.3

1. Examine the drawing and answer the following questions:
   (a) Name the angle of projection.
   (b) Is surface D above or below surface H?
   (c) Which two surfaces are on the same horizontal plane?
   (d) Is surface J in front of or behind surface G?
   (e) Which is the highest surface?
   (f) Which is the nearest surface?
   (g) How many plane surfaces make up the whole block?
2. Complete the following views:
   (a) the other side view
   (b) the bottom view
   (c) the rear view
3. Sketch an isometric view with corner Y nearest the viewer using the axes indicated. Print the correct letter on each face.

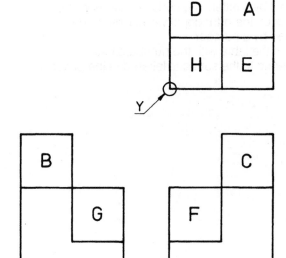

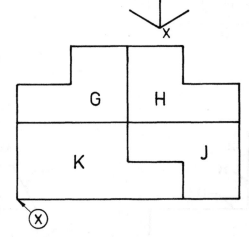

5.4

1. Examine the drawing and answer the following questions:
   (a) What is the angle of projection?
   (b) Is surface H in front of surface G?
   (c) Which surface is nearer the viewer, K or J?
   (d) Is surface E below surface F?
   (e) Which of the surfaces G, H, K or J are in the same plane?
   (f) Which surfaces are the highest?
   (g) What surface is level with surface D?
   (h) What surfaces are shown in hidden detail on the side view?
   (i) What surfaces are shown in hidden detail on the top view?
   (j) Which of the surfaces M, N, L, O and P are in the same plane?
2. Make an isometric sketch of the block shown using the axes indicated. Place the correct letter on each surface.

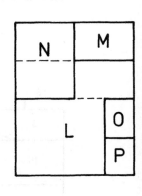

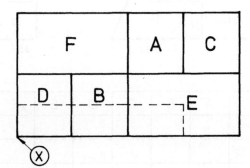

150

5.5　Mounting bracket

Study the drawing carefully and answer the questions below.

1. How many holes require drilling?
2. What is the overall height of the part?
3. What is dimension A in millimetres?
4. What is the tapping size of the M8 holes?
5. What is dimension B in millimetres?
6. What general tolerance is specified?
7. What is dimension C in millimetres?
8. What is the thread series for the 16 mm threaded hole?
9. What is dimension E in millimetres?
10. What is the specified diameter of hole D?
11. What is dimension F in millimetres?
12. What is the tolerance of the large hole?
13. What does PCD on the top view mean?
14. What surface finish is required?
15. What is the radius of fillets and rounds?
16. What is dimension H in millimetres?
17. How many threaded holes are on the component?
18. What is dimension J in millimetres?
19. How many surfaces are machine finished?
20. What dihedral angle is used for projection of views?

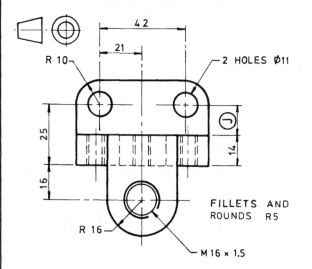

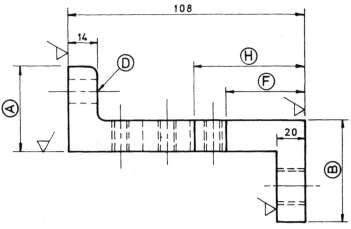

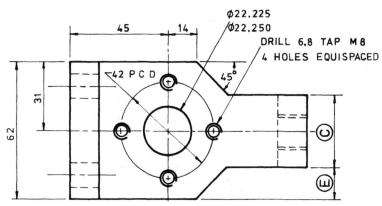

						TOLERANCES	MATERIAL	DRN.	A SHAMBLES PTY LTD	
						± 0.5 EXCEPT	C I	CKD	**MOUNTING BRACKET**	
						AS STATED	FINISH	APPD		
2	THREAD	WAS M 16	AwB				▽ = ▽MILL			
1	THREAD	WAS M8 x 1	AwB				HEAT TREAT	ISSUED	SCALE 1 : 2	DRAWING NO. A−161	A1
CHANGE NO	ITEM	CHANGE	CKD	OK	DATE	DATE: 12.03:85				

151

5.6 Slide block

1. Answer the questions below:
   (a) What is the centre-to-centre distance of the 6 mm tapped holes?
   (b) What does the dotted line S represent?
   (c) What tap drill is used for the holes marked Z?
   (d) How many surfaces are machine finished?
   (e) Is the true shape of face R shown on the drawing?
   (f) Which view shows the true shape of surface X?
   (g) What are the dimensions of the face marked R?
   (h) What are the dimensions of surface Y?
   (i) In what dihedral angle is the drawing made?
   (j) What are the dimensions of face M?
   (k) Would the bare-faced dovetail slot be turned, shaped, milled or ground?
   (l) In the surface finish symbol, we have 3.2 units. What are they?
2. Find the appropriate dimension for each letter down the left-hand side of the diagram.
3. Draw the following views in third-angle projection:
   (a) a sectional front view on A–A
   (b) a side view from B
   (c) a top view
   Scale: 1:1
   Fully dimension and subtitle your drawing.

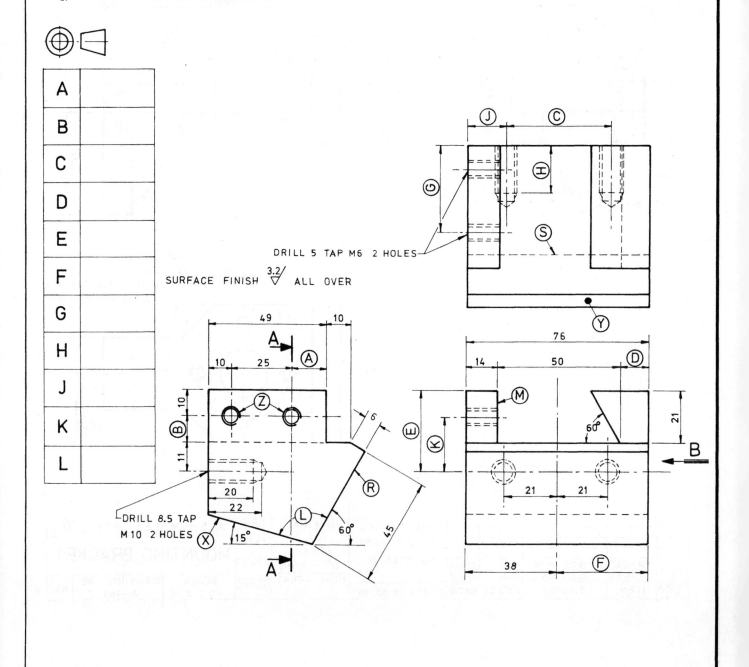

A	
B	
C	
D	
E	
F	
G	
H	
J	
K	
L	

DRILL 5 TAP M6 2 HOLES

SURFACE FINISH 3.2/ ▽ ALL OVER

DRILL 8.5 TAP M10 2 HOLES

Primary auxiliary views

Sometimes it is desirable to show the true shape and dimensions of an irregular surface inclined to one or more of the principal planes of projection. In this case a view must be projected onto a plane parallel to the surface in question. This plane is called an auxiliary plane and the view projected onto this plane is called an auxiliary view.

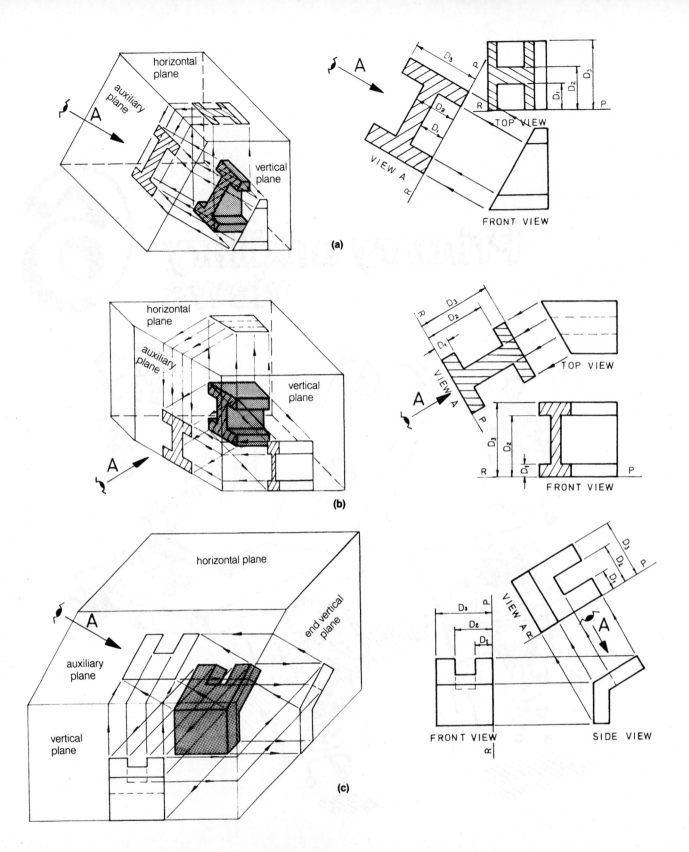

Fig. 6.1 *Types of primary auxiliary views in third-angle projection*

Auxiliary orthogonal views

Sometimes an object has an irregular face inclined to the normal planes of projection. In this case a true shape view of the irregular face can only be obtained by projecting it onto an auxiliary plane parallel to the face. A view so obtained is called an *auxiliary view*. There are two kinds of auxiliary views: *primary* and *secondary*.

Primary auxiliary views

When the auxiliary plane is inclined to four of the six principal planes of projection and is square with the other two, the view so obtained is termed a *primary auxiliary view*. Such is the case illustrated by the pictorial views in Figure 6.1.

Types of primary auxiliary views

An auxiliary view must be projected from the principal view which provides the edge view of the irregular face. Hence there are basically three types of auxiliary views:

1. the view obtained when the edge view of the irregular face is shown on the front (or rear) view (Fig. 6.1(a))
2. the view obtained when the edge view of the irregular face is shown on the top (or bottom) view (Fig. 6.1(b))
3. the view obtained when the edge view of the irregular face is shown on the side (right or left) view (Fig. 6.1(c))

Each of the three figures shows that dimensions in one direction are projected from the edge view of the irregular face. Dimensions required to complete the auxiliary view in the other direction are transferred from the other principal view and are normally measured from a reference plane (RP). On Figure 6.1 they are marked as D_1, D_2 and D_3.

The position of the reference plane can be established by considering its edge representation on the orthogonal views and then relating it to the pictorial view in each case.

Partial auxiliary views

If a component has an irregular face, which must be detailed, it is often only necessary to draw an auxiliary view of the irregular face and not an auxiliary view of the complete object. In many cases a complete auxiliary view distorts another part of the view, making it of little value.

The use of partial views saves drawing time, simplifies the drawing and makes it easier to read. Figure 6.2(a) illustrates a complete auxiliary view, part of which is completely distorted because of the angle of viewing so that it is of little descriptive value. Figure 6.2(b) shows the partial auxiliary view, omitting the distorted portion.

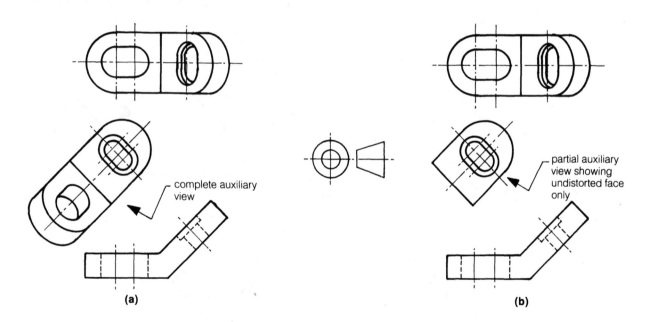

Fig. 6.2 *Comparison of full and partial auxiliary views*

155

Orientation of auxiliary views

Auxiliary views should always be drawn using *third-angle projection*, irrespective of whether the method of projection used on the other views is first-angle or third-angle. The type of auxiliary view most commonly used is the normal view obtained by looking perpendicularly at the inclined face and projecting the true shape onto an auxiliary plane perpendicular to the line of viewing (Fig. 6.3(a)).

The principle of auxiliary projection may also be used when drawing removed views (Fig. 6.3(b)). A full or partial auxiliary view may be removed from its normal position, without changing its orientation, if it is more convenient or allows greater clarity, for example for dimensioning purposes or to achieve a better layout on a drawing sheet. The word 'view' followed by a direction indicator, for example 'A', should be used to identify the view, and the direction of viewing should be indicated by an arrow, together with indicator 'A'. If the removed view needs to be re-orientated as well, the number of degrees of rotation and the direction must be stated (Fig. 6.3(c)).

Removed views drawn to a larger scale are labelled with the word 'detail' followed by a letter, as well as an indication of the scale used (Fig. 6.4). The portion of the actual view removed is enclosed in a circle or a rectangle drawn with a thin type B line. If the removed view is close to the detail on the actual view, the circle or rectangle may be joined to 'detail' by a leader.

Example of a primary auxiliary view

Figure 6.5 shows the construction of a primary auxiliary view from a front and a side view. The front view shows the edge of the inclined 30° face. The right side view is included to enable an understanding of how the widths of this normal view are applicable to the auxiliary view. Note the aid for constructing the elliptical shape of the top.

A reference line (RP) has been drawn parallel to the edge view of the inclined face. The RP may be selected in any position; an alternative is shown by the dashed line R'P'. The centre line position was chosen, as in a symmetrical view of this type most detail is located on the centre line and the rest is equally spaced on either side of it. The lengths of the auxiliary view have been projected from the front view onto the RP and beyond where necessary.

Widths of detail on the auxiliary view, as well as arcs and circles, were then drawn to complete the view. In this example widths could be measured but the widths of detail on the auxiliary view are always identical to widths of the same detail on the right side view.

Hidden detail is not shown on this auxiliary view as little is to be gained by including it. However, where it would allow a better shape description, hidden detail should be included. Auxiliary views should be dimensioned, but in this case dimensions were omitted in favour of showing projection lines for a better understanding of the method. Projection lines are always removed from the final drawing in practice.

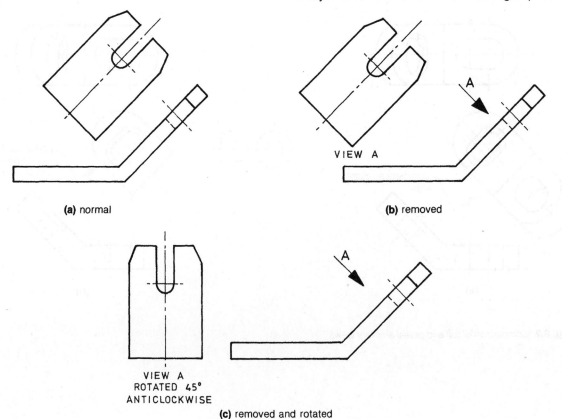

(a) normal

(b) removed

VIEW A
ROTATED 45°
ANTICLOCKWISE

(c) removed and rotated

Fig. 6.3 *Orientation of auxiliary views*

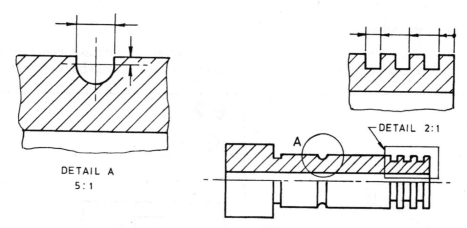

Fig. 6.4 *Removed and enlarged views*

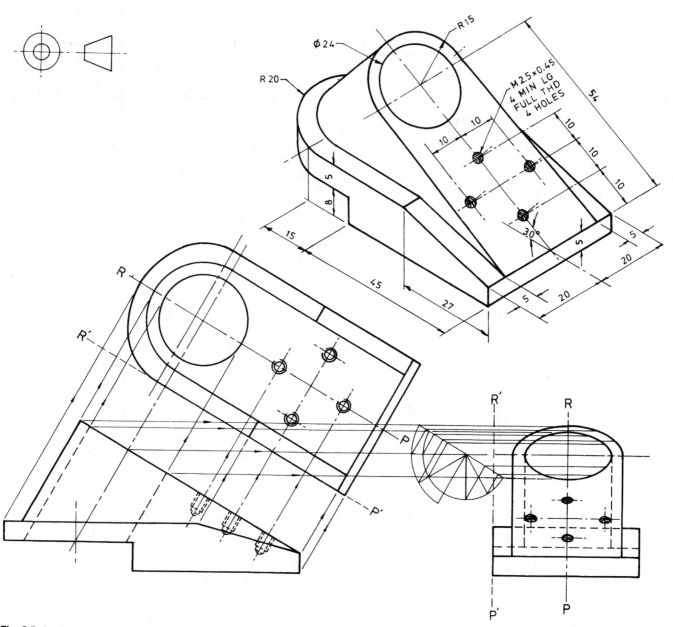

Fig. 6.5 *A primary auxiliary view*

Exercises on auxiliary views

From a practical point of view, most of the exercises in this section would be best drawn as 'partial' auxiliary views, but, in order to make them more challenging, 'complete' views should be constructed.

6.1

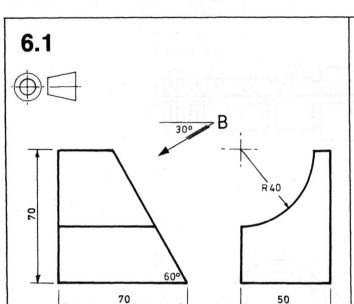

Draw the following views in third-angle projection:
1. the front and right side views
2. a complete auxiliary view from B
Scale 1:1
Fully dimension and show hidden detail.

6.2

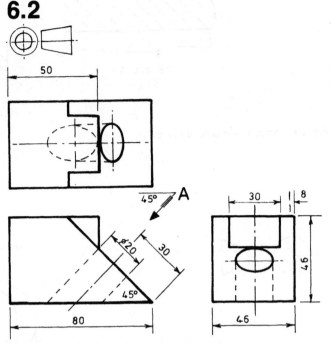

Draw the following views in third-angle projection:
1. the front and top views
2. a complete auxiliary view from A
Scale 1:1
Fully dimension and show hidden detail.

6.3

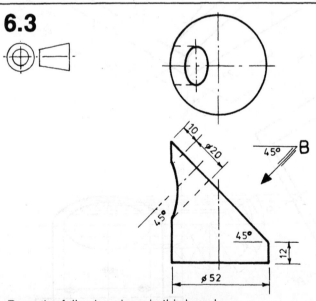

Draw the following views in third-angle projection:
1. the front and top views
2. an auxiliary view from B showing the true shape
Scale 1:1
Fully dimension each view.

6.4

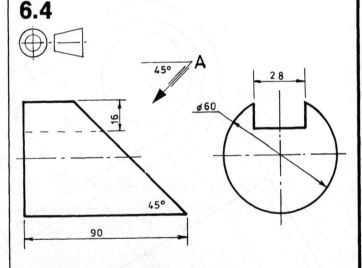

Draw the following views in third-angle projection:
1. the front and right side views
2. a top view
3. an auxiliary view from A showing the true shape
Scale 1:1
Fully dimension and show hidden detail.

6.5

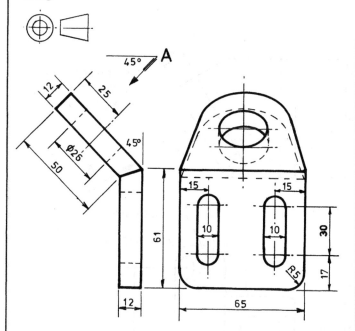

Draw the following views in third-angle projection:
1. the given front and right side views
2. a partial auxiliary top view from A

Scale 1:1

Fully dimension and show hidden detail.

6.6

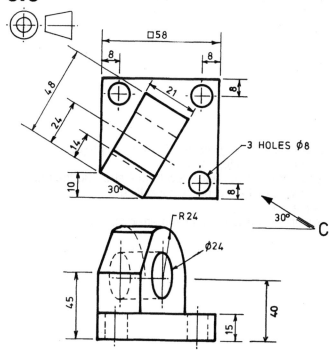

Draw the following views in third-angle projection:
1. the given front and top views
2. a complete auxiliary view from C

Scale 1:1

Fully dimension and show hidden detail.

6.7

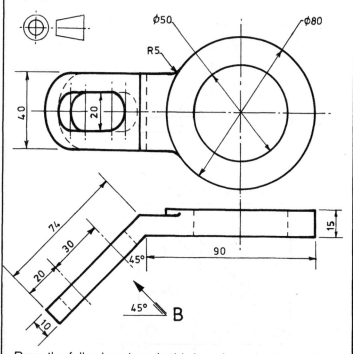

Draw the following views in third-angle projection:
1. the given top and front views
2. a partial auxiliary view from B

Scale 1:1

Fully dimension and show hidden detail.

6.8

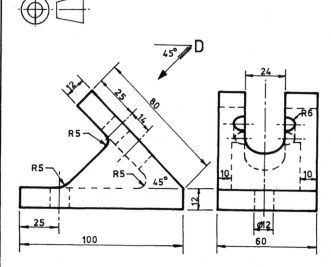

Draw the following views in third-angle projection:
1. the given front and right side views
2. a complete auxiliary view from D

Scale 1:1

Fully dimension and show hidden detail.

6.9 Drill press bracket

Draw the following views in third-angle projection:
1. the front and top views
2. a complete auxiliary view from A showing the true shape of the inclined vertical face

Scale 1:1
Subtitle all views but omit dimensions.

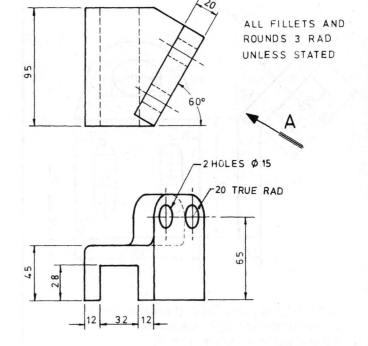

ALL FILLETS AND
ROUNDS 3 RAD
UNLESS STATED

20

95

60°

A

2 HOLES ⌀15

20 TRUE RAD

65

45

28

12 32 12

6.10 Angle support

Draw the following views in third-angle projection:
1. the given top and a complete front view showing all hidden detail
2. a complete auxiliary view showing the true shape of the inclined vertical face

Also show welded joint symbols as follows:
1. 6 mm fillet all round
2. 6 mm fillet both sides

Scale 1:1
Fully dimension and subtitle the views.

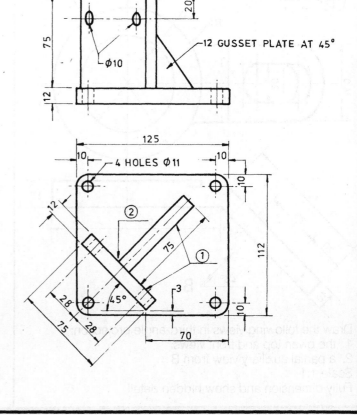

75

20

112

⌀10

12 GUSSET PLATE AT 45°

125

10 4 HOLES ⌀11 10

10

112

2

75

1

12

28

45°

3

28

75

70

10

6.11 Box spanner

Draw the following views in third-angle projection:
1. the given front and top views
2. a complete auxiliary view projected in the direction of arrow B

Scale 1:1
Show all hidden detail but omit dimensions.

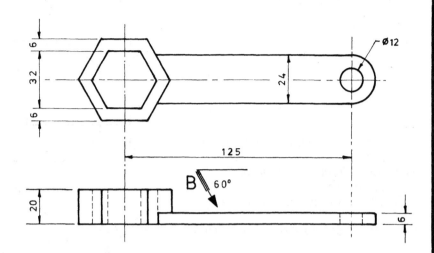

6.12 Bevel washer

Draw the following views in third-angle projection:
1. a sectional view on A–A
2. a top view
3. a complete auxiliary view from B

Scale 1:1
Fully dimension and show hidden detail.

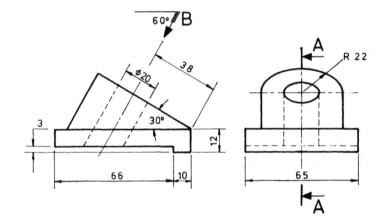

6.13 Angle bracket

Draw the following views in third-angle projection:
1. the front view as given
2. a complete auxiliary view from A

Scale 1:1
Fully dimension and show hidden detail.

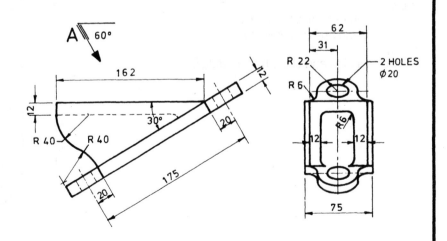

6.14 MS bracket

Draw the following views in third-angle projection:
1. the given front and top views
2. a left side view
3. a partial auxiliary top view from A
Scale 1:1
Fully dimension and show hidden detail.

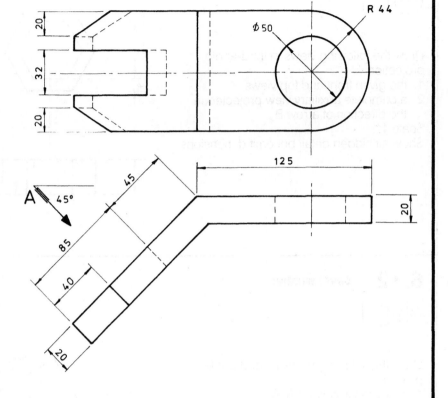

6.15 Tie-rod support

Draw the following views in third-angle projection:
1. the given front and top views
2. a complete auxiliary view from A
Scale 1:1
Fully dimension and show hidden detail.

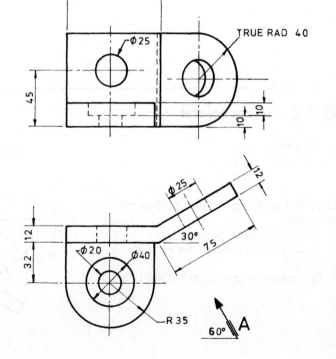

Pictorial drawing

Isometric and oblique parallel projection
Planometric and dimetric views
Mechanical perspective

Orthogonal views are two-dimensional and two or more views can convey an idea of shape and form to people who are familiar with this type of drawing. The need frequently arises, however, to convey the idea of shape and form to persons untrained in the use of orthogonal projection. In these cases pictorial views are used because of their three-dimensional aspect, which conveys a full shape description to the viewer.

Introduction

Pictorial views are not intended to transmit dimensions, and so they are not normally dimensioned. Sometimes, however, an engineer may wish to give a drafter a pictorial sketch of a design in mind and in that case may add dimensions that are applicable or considered necessary.

There are three general classifications of pictorial drawings:
1. axonometric projection
2. oblique projection
3. perspective projection

Perspective views are more complicated to produce than the first two but are more realistic and are used mainly by architects. Engineers prefer either axonometric or oblique views.

Axonometric projection

This involves turning the object so that three principal faces can be seen from one viewing position. There are an infinite number of views possible, and they all result in shortening of the edges by varying degrees, depending on the angles involved.

Accordingly, certain positions have been classified as *isometric*, *dimetric* and *trimetric*, and one of these is used when an axonometric projection is required. The most commonly used is the isometric; it will be described in detail. The other two are described in *AS 1100* Part 101.

Isometric projection

The word 'isometric' means 'equal measure', and to produce an isometric projection it is necessary to view an object so that its principal edges are equally inclined to the viewer and so are foreshortened equally. This is best illustrated by considering orthogonal views of an inclined cube. Figure 7.1 shows the front, top and side views of a cube resting on one of its faces, with its horizontal edges inclined at 45° to the vertical plane. The front view shows that all horizontal edges are equally foreshortened, but the vertical edges are not altered.

Now consider the cube to be pivoted on corner A and tilted forward until all the edges are equally inclined to the vertical plane. This is shown in Figure 7.2. (The angle between the edges and the vertical plane is approximately 35° 15′.) In this position, edges OA, OB and OC make 120° with each other in the front view and are called the *isometric axes*. The axis OA is vertical, OB is inclined at 30° to the right, and OC at 30° to the left. Any line parallel to one of these axes is called an *isometric line* and can be drawn with a 30° set square. All other lines are called *non-isometric lines* and have to be plotted.

Isometric scale

For correct isometric projection, a scale that allows for the foreshortening of isometric lines is used. The construction of such a scale is shown in Figure 7.3. When an isometric view is drawn using an isometric scale, it is termed an *isometric projection*.

Isometric drawing

An isometric drawing differs from an isometric projection in that it is prepared without shortening measurements. An isometric drawing gives about a 22.5% larger view than the isometric projection, but the pictorial value of each view is the same. Hence for most purposes the isometric drawing is used.

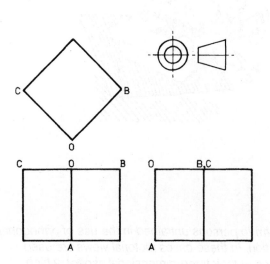

Fig. 7.1 *Foreshortening of cube edges*

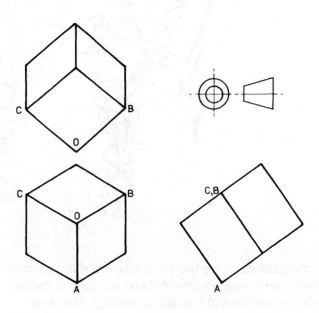

Fig. 7.2 *Concept of isometric axes*

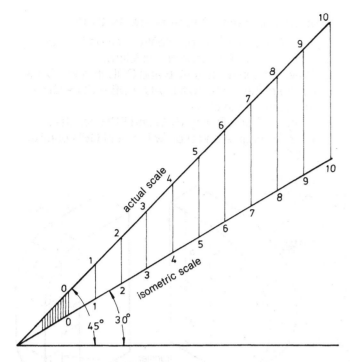

Fig. 7.3 *Making an isometric scale*

Selection of isometric axes

The main purpose of an isometric view is to provide a pictorial view that reveals as much detail as possible, and this fact should be remembered when selecting the principal edges for the isometric axes.

Figure 7.4(a)–(h) shows eight isometric views of the same block with the isometric axes intersecting at the circled point in each view. View (a) is preferred as it reveals more detail than the others.

The isometric axes can be rotated to make one axis horizontal, as shown in Figure 7.4(i) and (j). This is sometimes preferred for long narrow objects, where the long axis can be placed horizontally for best effect.

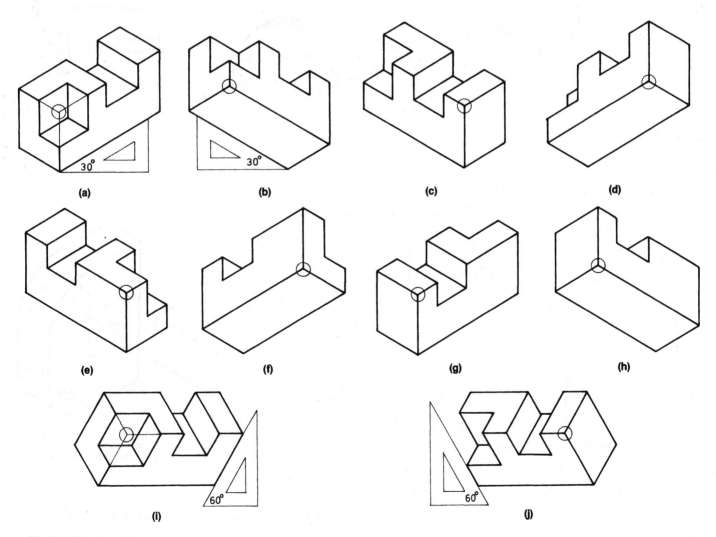

Fig. 7.4 *Selection of isometric axes*

Isometric circles—ordinate method

Circles may be drawn whole or in part in isometric view by the use of ordinates constructed on an orthogonal view and then transferred to the isometric view, as shown in Figure 7.5. A smooth curve is drawn either freehand or with a French curve through the ends of the ordinates to give the isometric circle or curve.

Isometric circles—four-centre method

Figure 7.6(a) shows the four-centre method for drawing isometric circles. The steps are as follows:
1. Draw the centre lines AOB and COD through O, the centre of the circle, so that AO = OB = CO = OD = the radius of the circle.
2. Through C and D draw FCG and EDH parallel to AOB. Through A and B draw FAE and GBH parallel to COD.

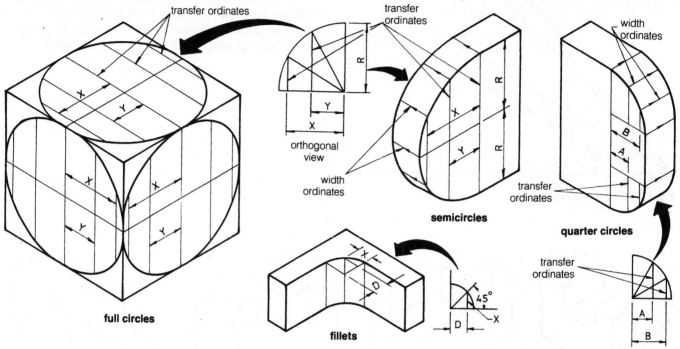

Fig. 7.5 *Isometric circles—ordinate method*

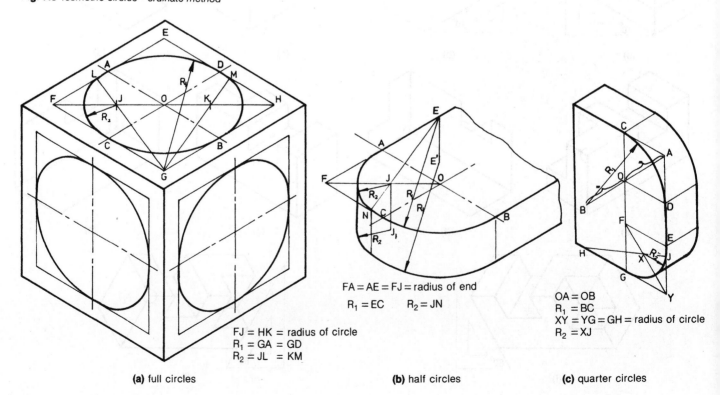

(a) full circles

FJ = HK = radius of circle
R_1 = GA = GD
R_2 = JL = KM

(b) half circles

FA = AE = FJ = radius of end
R_1 = EC R_2 = JN

(c) quarter circles

OA = OB
R_1 = BC
XY = YG = GH = radius of circle
R_2 = XJ

Fig. 7.6 *Isometric circles—four-centre method*

166

3. Draw the long diagonal FOH, and locate points J and K on it such that FJ = HK = the radius of the circle.
4. With centre G and radius R_1 = GA, draw an arc between GJ, produced at L, and GK, produced at M. Follow the same procedure with centre E.
5. With centres J and K and radius R_2 = JL = KM, complete the figure.

Half and quarter circles may also be drawn by this method, as shown in Figure 7.6(b) and (c) respectively, using part of the construction method outlined above.

Isometric curves

Points on these curves are plotted by the method of ordinates taken from an orthogonal view, as shown in Figure 7.7.

A smooth curve is drawn through the plotted points, which are obtained by transferring lengths from the orthogonal view to the other by means of dividers.

Isometric angles and non-isometric lines

These have to be plotted by the use of horizontal and vertical measurements, as shown in Figure 7.8.

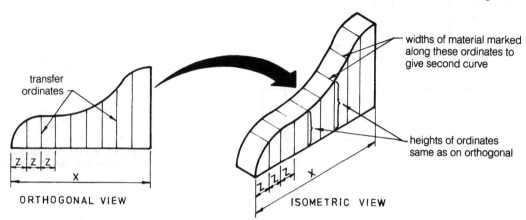

Fig. 7.7 *Isometric curves*

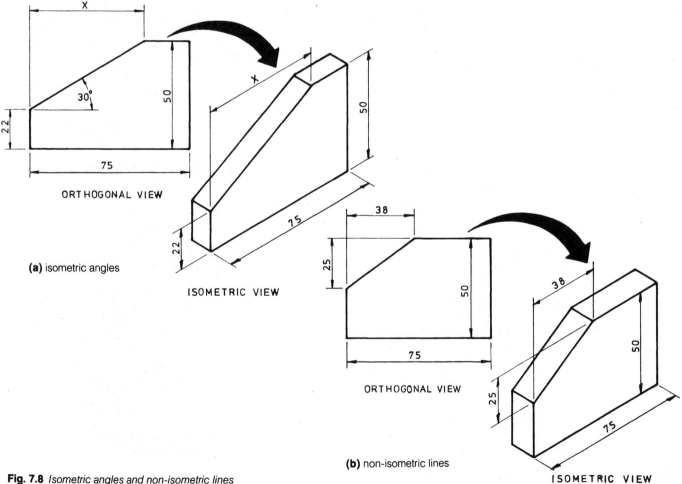

(a) isometric angles

(b) non-isometric lines

Fig. 7.8 *Isometric angles and non-isometric lines*

167

Making an isometric drawing

The five views in Figure 7.9 show step-by-step production of a simple isometric drawing.

The isometric axes meet at the circled point in Figure 7.9(a). This point is carefully chosen so that the view will reveal as much detail as possible.

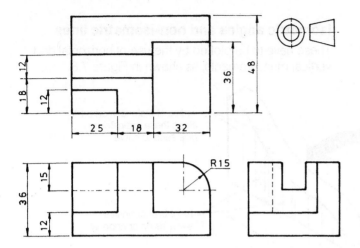

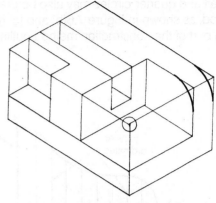

(a) Draw in light construction lines (circles and curves full thickness)

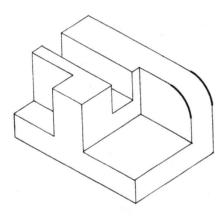

(b) Remove excess lines (simplified if construction lines lightly drawn)

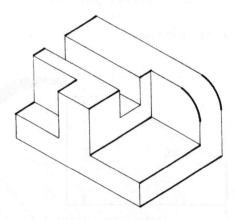

(c) Line in 30° right lines

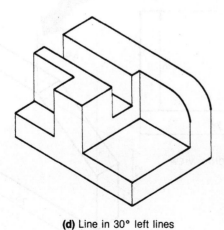

(d) Line in 30° left lines

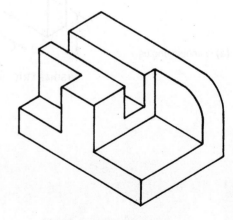

(e) Line in vertical lines to complete view

Fig. 7.9 *Making an isometric drawing*

Representation of details common to pictorial drawings

Fillets and rounds

Filleted corners and rounded edges may be represented by either straight or curved lines, as shown in Figure 7.10 using a type B (thin) line.

Threads

Threads may be represented by a series of ellipses or circles (depending on the type of drawing) evenly spaced along the centre line of the threaded section using a type B (thin) line (Fig. 7.11).

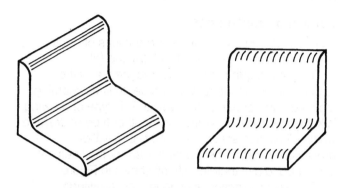

Fig. 7.10 *Pictorial representation of fillets and rounds*

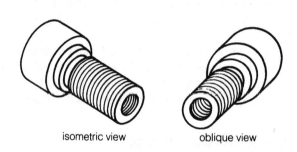

isometric view   oblique view

Fig. 7.11 *Pictorial representation of threads*

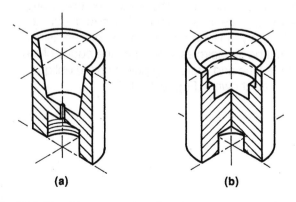

(a)   (b)

Fig. 7.12 *Pictorial representation of sections*

Sectioning

Pictorial drawings should be sectioned along centre lines, the sectioning plane cutting parallel to one of the principal viewing planes of the object (Fig. 7.12(a)).

Hatching on half-sections should be drawn in the opposite direction on the adjacent cut faces coinciding at the axis (Fig. 7.12(b)).

Dimensioning

Dimensioning on pictorial views may sometimes be required and should follow the same general rules as for orthogonal views, that is, the dimension line, projection lines and the dimension itself should lie in the same plane.

One of the following two methods should be used:
1. *unidirectional*—where all dimensions are read from the bottom of the drawing (Fig. 7.13(a))
2. *principal plane dimensioning*—where dimensions lie in one or more of the three principal planes (Fig. 7.13(b))

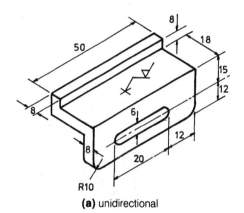

(a) unidirectional

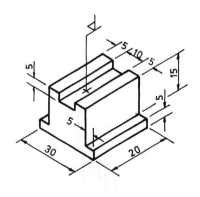

(b) principal plane

Fig. 7.13 *Dimensioning pictorial views*

169

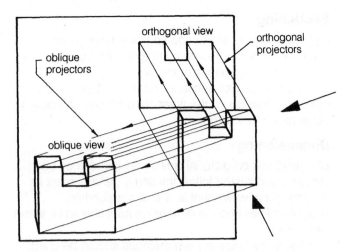

Fig. 7.14 *Oblique parallel projection*

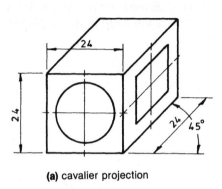

(a) cavalier projection

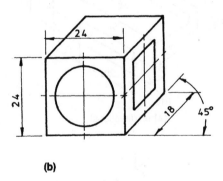

(b)

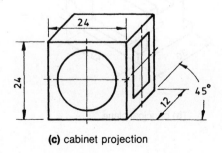

(c) cabinet projection

Fig. 7.15 *Lengths of depth lines in oblique parallel projection*

Oblique parallel projection

With this type of projection the object is viewed from an oblique angle so that the resulting view is three-dimensional. The view is produced by drawing parallel projectors from the object to the picture plane, as shown in Figure 7.14. As the object is placed so that its front face is parallel to the picture plane, the oblique projectors will produce this face on the picture plane. Depth lines will also be reproduced, and their lengths will vary with the viewing angle. Depth lines are usually taken as receding at angles of 45°, 30° or 60° as these angles are easily drawn with set squares. However, any angle that shows the detail to the best advantage may be used.

Length of depth lines

A cube is drawn using various proportions of depth lines, as shown in Figure 7.15(a), (b) and (c).

In (a) the depth lines are not reduced, and the appearance is unnatural with the depth lines seeming too long and appearing to diverge. This type of drawing is known as *cavalier projection*. Another type of drawing, which eliminates some of the faults of cavalier projection, is *cabinet projection*. Here depth lines are shortened to half their length, as shown in Figure 7.15(c). This projection is used in most drawings.

Three rules are worth remembering when making an oblique drawing.
1. Place the object so that the irregular face is parallel to the picture plane. This is illustrated in Figure 7.16.
2. Place the object so that the longest dimension is parallel to the picture plane, as shown in Figure 7.17.
3. When rules 1 and 2 conflict, rule 1 has preference as the advantage gained by having the irregular face without distortion is greater than that gained by observing rule 2. This rule is illustrated in Figure 7.18.

Circles on the oblique face

These circles are plotted using a plotting view, which consists of a true size quadrant of the circle, together with a half size quadrant on the same view (Fig. 7.19). The circles are plotted in a similar manner to isometric circles, except that measurements along the 45° axis are taken from the half size quadrant. Alternatively, oblique circles may be plotted using true shape semicircles located on the edges of the oblique face and projecting points on the oblique circles as shown in Figure 7.19.

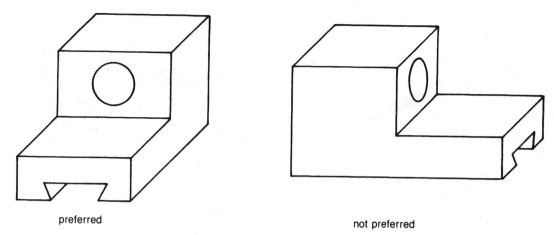

preferred

not preferred

Fig. 7.16 *Irregular face parallel to picture plane*

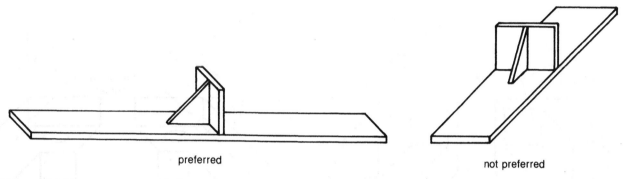

preferred

not preferred

Fig. 7.17 *Longest dimension parallel to picture plane*

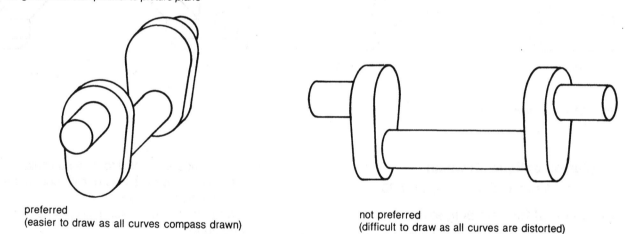

preferred
(easier to draw as all curves compass drawn)

not preferred
(difficult to draw as all curves are distorted)

Fig. 7.18 *Preferred pictorial view*

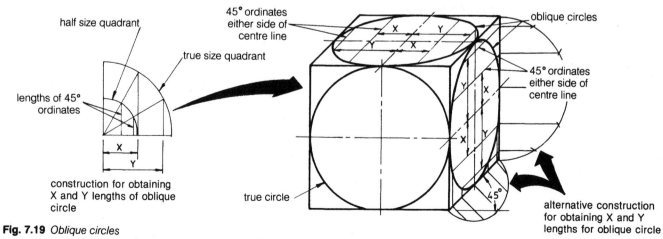

half size quadrant

true size quadrant

lengths of 45°
ordinates

45° ordinates
either side of
centre line

oblique circles

45° ordinates
either side of
centre line

X

Y

construction for obtaining
X and Y lengths of oblique
circle

true circle

45°

alternative construction
for obtaining X and Y
lengths for oblique circle

Fig. 7.19 *Oblique circles*

171

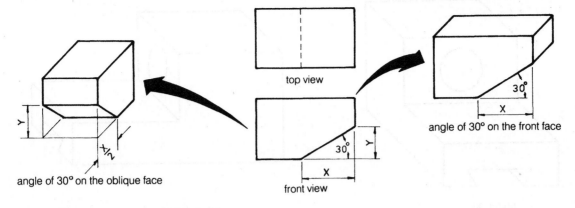

top view

front view

angle of 30° on the front face

angle of 30° on the oblique face

Fig. 7.20 *Oblique angles*

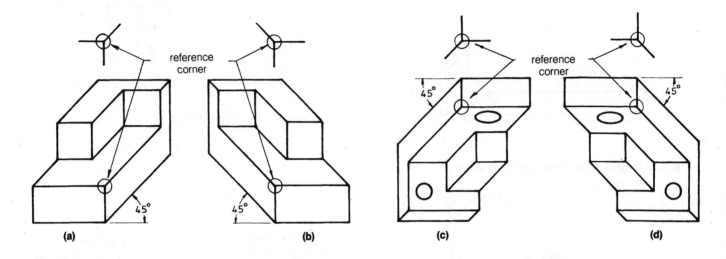

reference corner

reference corner

(a)   (b)   (c)   (d)

Fig. 7.21 *Selection of oblique axes*

Angles on oblique drawings

These are drawn as shown in Figure 7.20.

Selection of the receding axis

A number of views that can be obtained by varying the angle of the receding axis are shown in Figure 7.21. Each view is chosen because it reveals the maximum amount of detail for that particular orientation of the object, taking into account rules 1, 2 and 3 mentioned above.

The reference corner for each view is circled and outlined above.

Exercises on isometric and oblique parallel view

Draw an isometric or an oblique parallel view of each of the following objects.

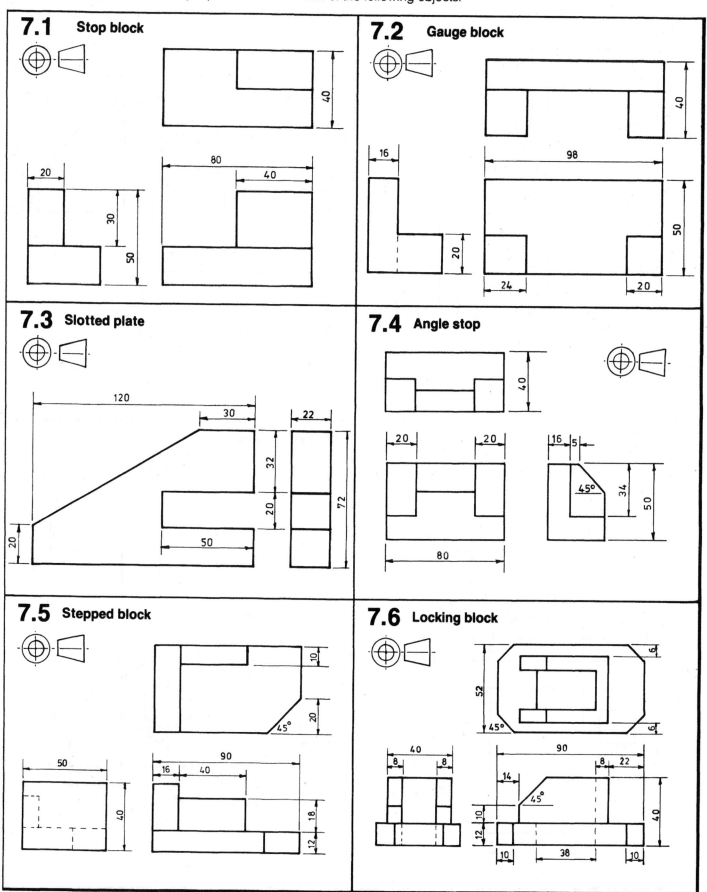

7.1 Stop block

7.2 Gauge block

7.3 Slotted plate

7.4 Angle stop

7.5 Stepped block

7.6 Locking block

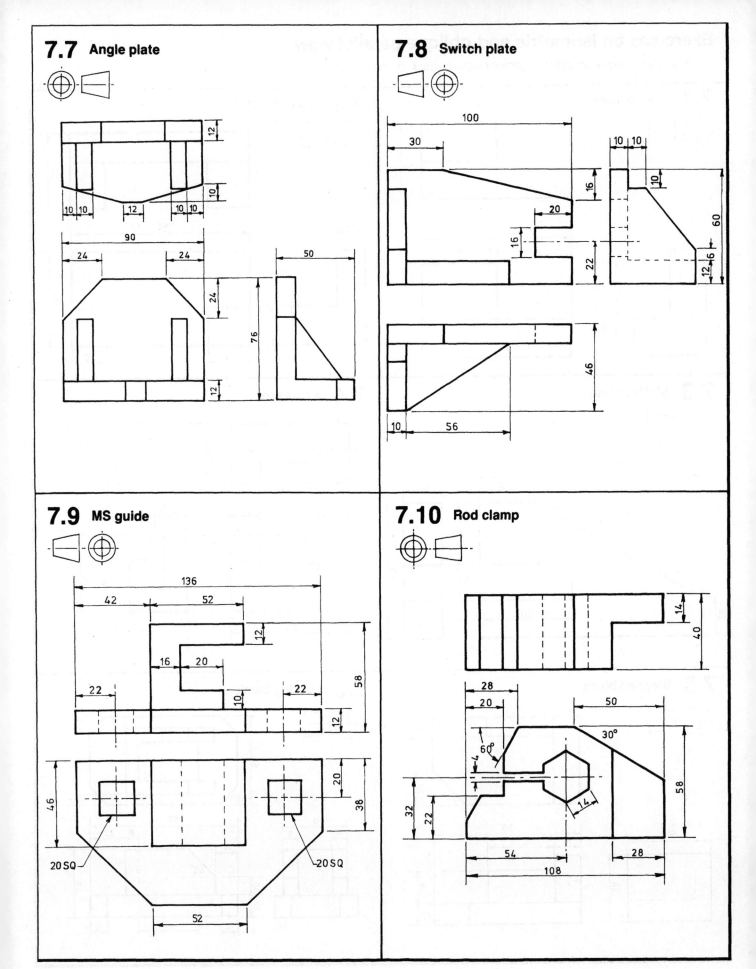

7.7 Angle plate

7.8 Switch plate

7.9 MS guide

7.10 Rod clamp

174

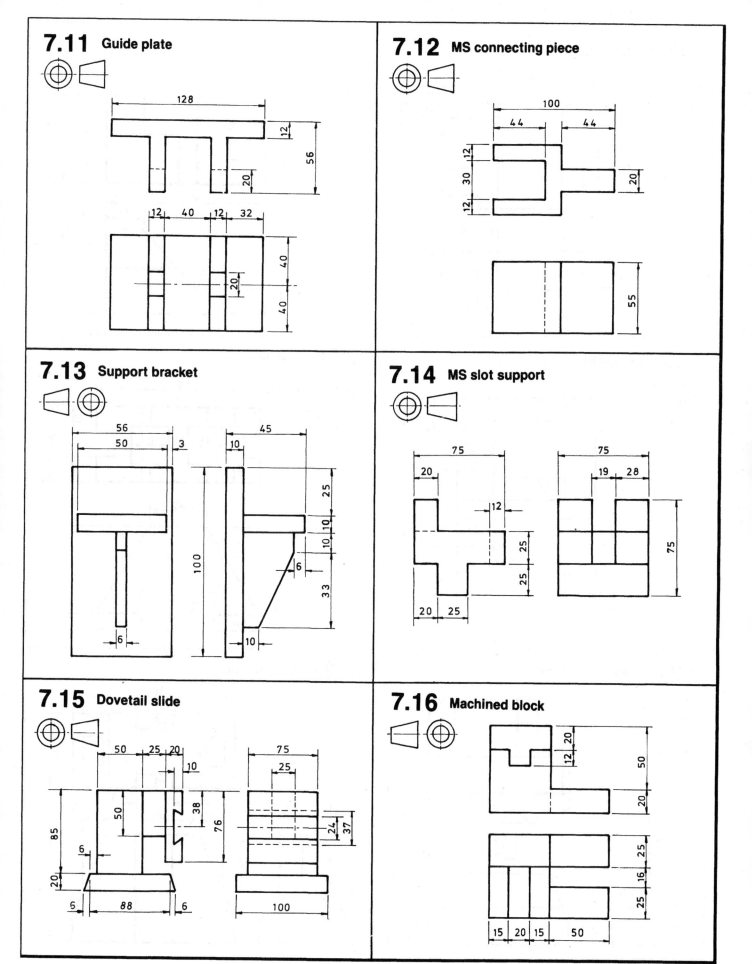

7.11 Guide plate

128 · 56 · 12 · 20 · 12 · 40 · 12 · 32 · 40 · 20 · 40

7.12 MS connecting piece

100 · 44 · 44 · 12 · 30 · 12 · 20 · 55

7.13 Support bracket

56 · 50 · 3 · 45 · 10 · 25 · 10 · 10 · 100 · 6 · 33 · 6 · 10

7.14 MS slot support

75 · 20 · 12 · 25 · 25 · 20 · 25 · 75 · 19 · 28 · 75

7.15 Dovetail slide

50 · 25 · 20 · 10 · 50 · 38 · 76 · 85 · 6 · 24 · 37 · 20 · 6 · 88 · 6 · 75 · 25 · 100

7.16 Machined block

20 · 12 · 50 · 20 · 25 · 16 · 25 · 15 · 20 · 15 · 50

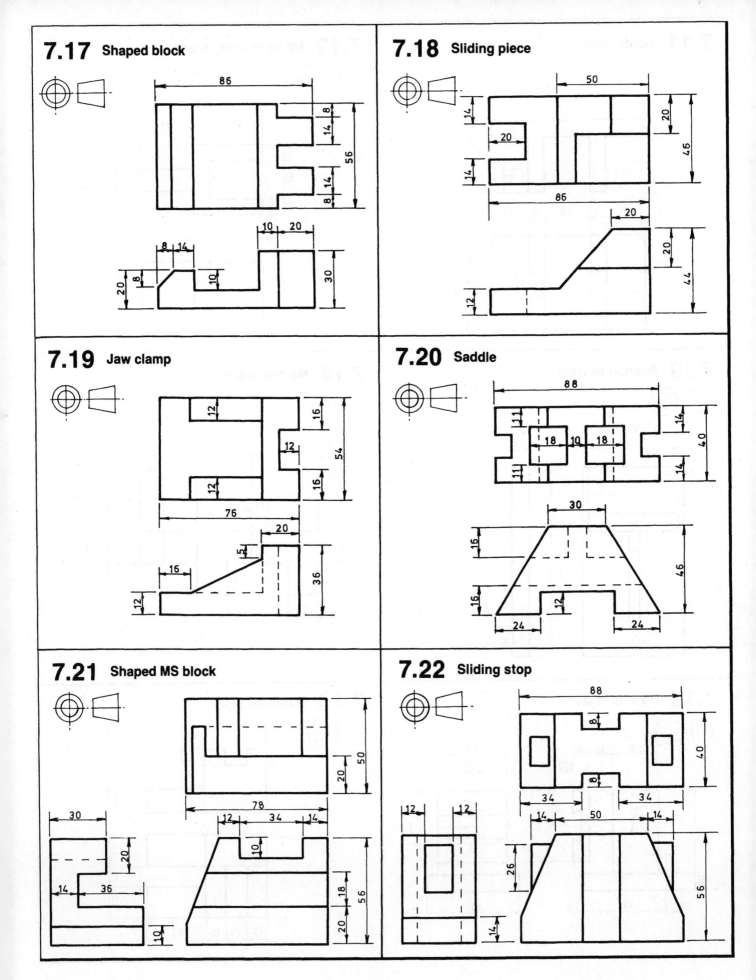

7.17 Shaped block

7.18 Sliding piece

7.19 Jaw clamp

7.20 Saddle

7.21 Shaped MS block

7.22 Sliding stop

176

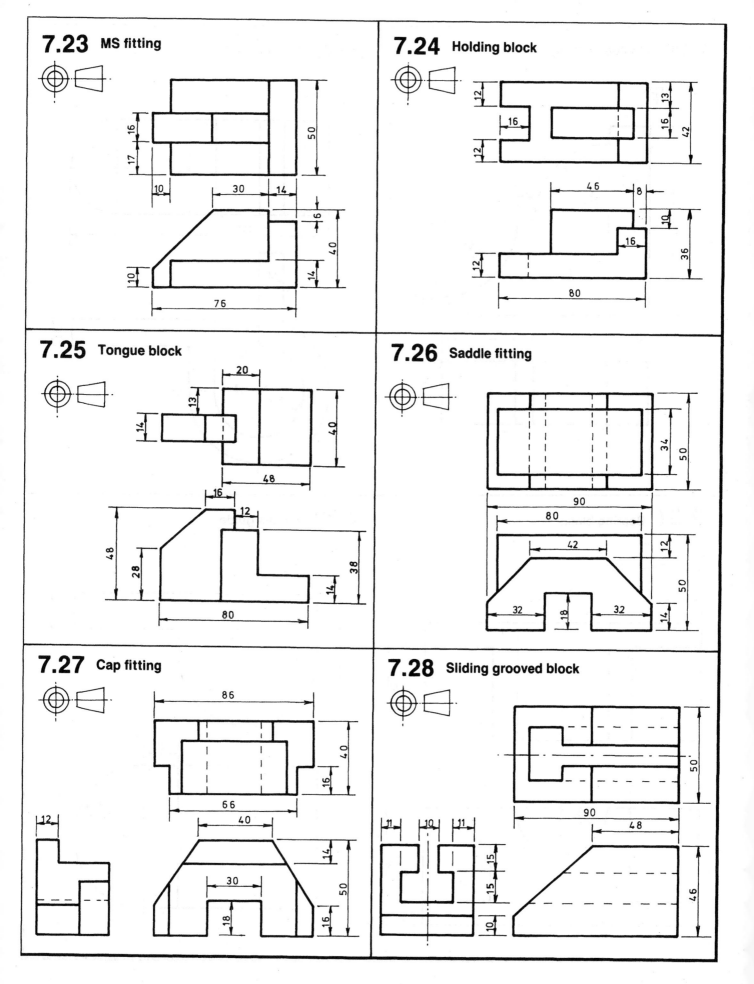

7.23 MS fitting

7.24 Holding block

7.25 Tongue block

7.26 Saddle fitting

7.27 Cap fitting

7.28 Sliding grooved block

177

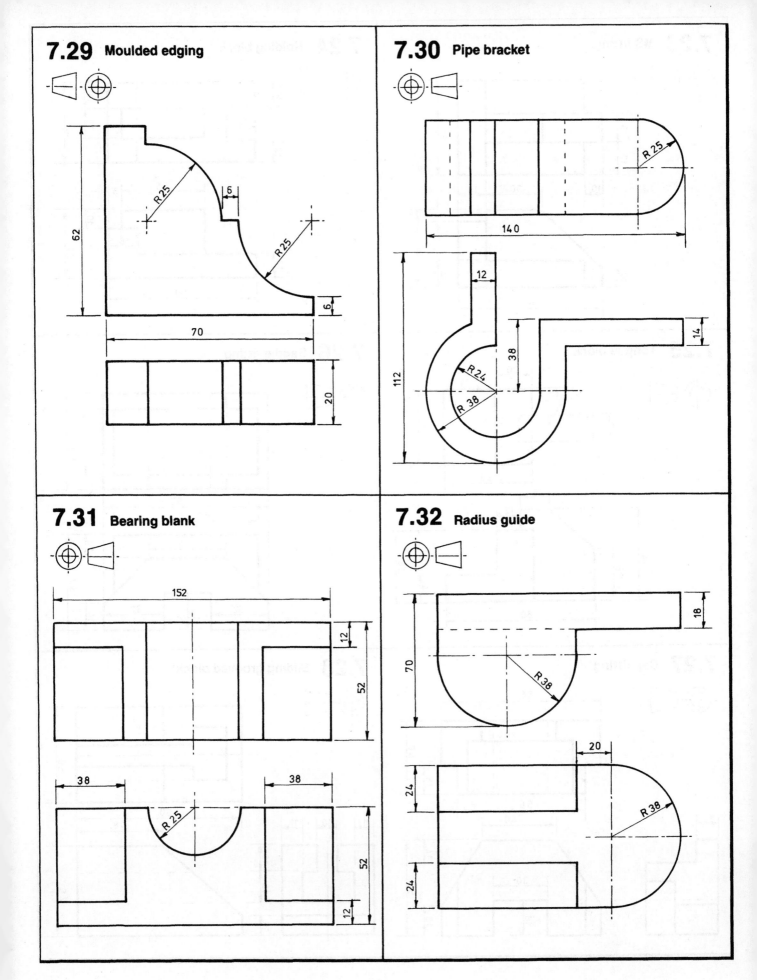

7.29 Moulded edging

7.30 Pipe bracket

7.31 Bearing blank

7.32 Radius guide

178

7.33 Bearing standard

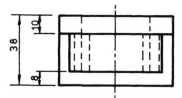

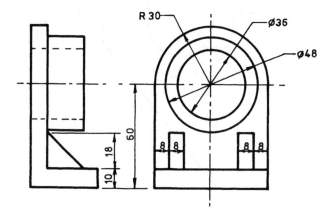

R 30 Ø36 Ø48

7.34 Tracking block

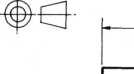

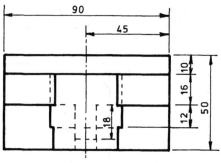

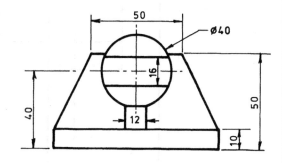

90 45 10 16 12 50 18

50 Ø40 16 50 40 12 10

7.35 Shaft support

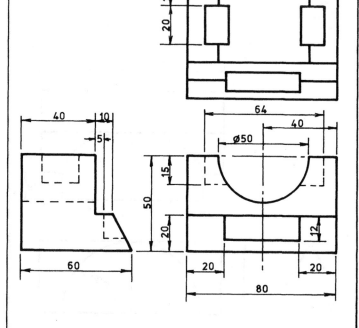

10 20 40 10 5 60 64 40 Ø50 15 50 20 12 20 80 20

7.36 Self-oiling bearing

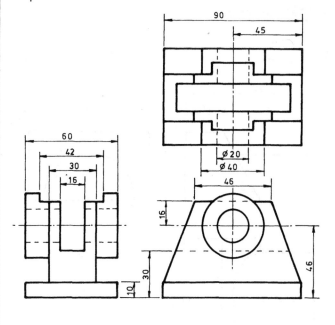

90 45 Ø20 Ø40 60 42 30 16 46 16 46 30 10

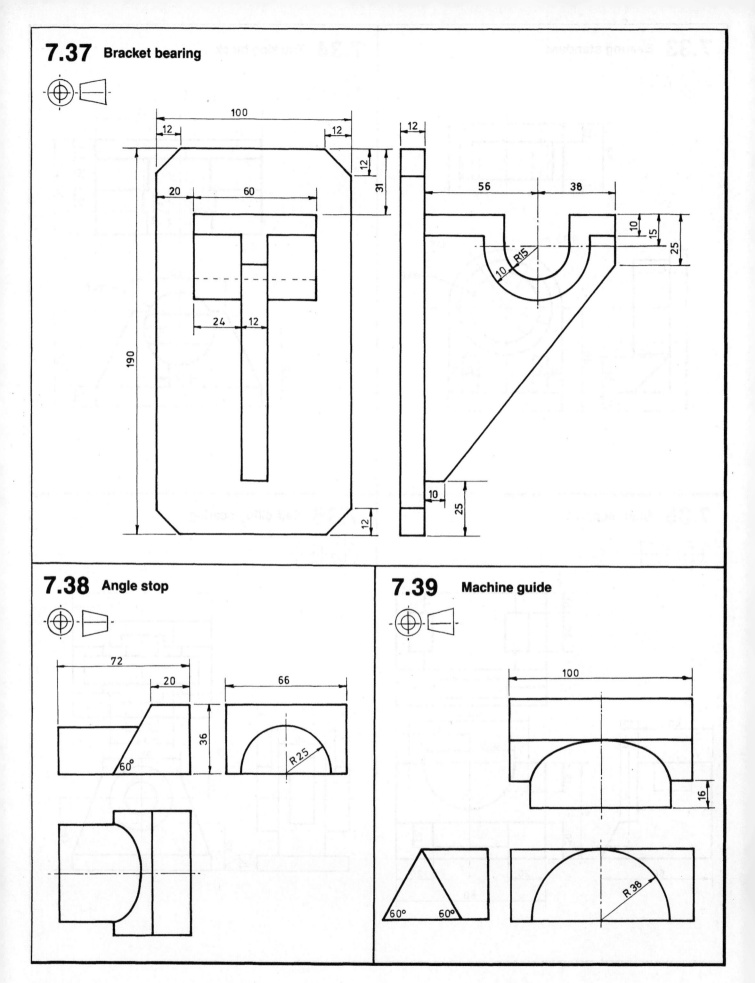

7.37 Bracket bearing

7.38 Angle stop

7.39 Machine guide

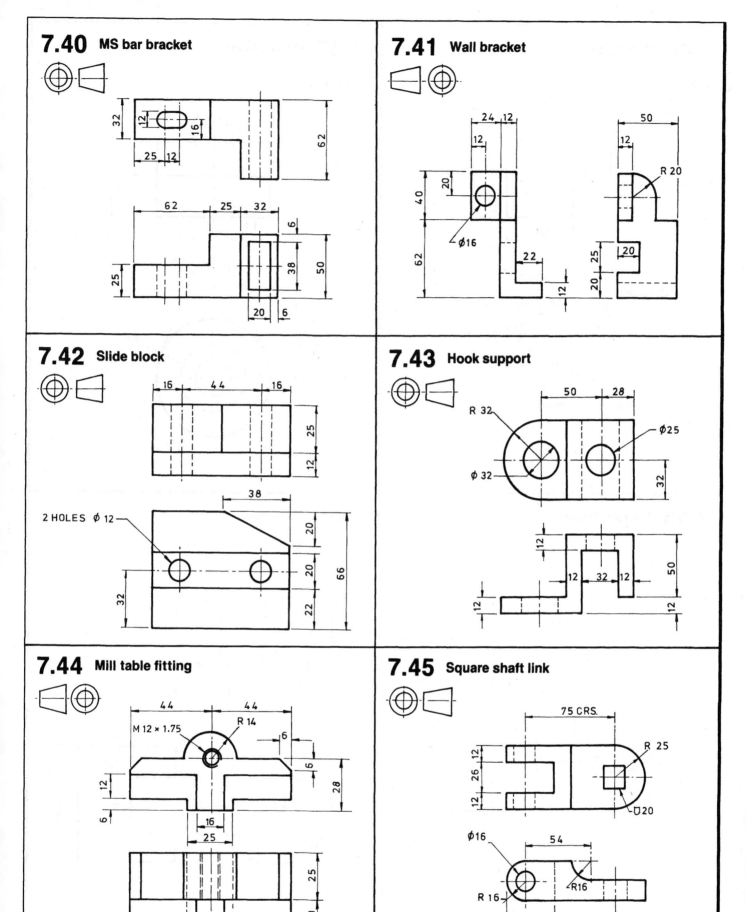

7.40 MS bar bracket

7.41 Wall bracket

7.42 Slide block

2 HOLES ⌀ 12

7.43 Hook support

R 32
⌀ 32
⌀25

7.44 Mill table fitting

M 12 × 1.75   R 14

7.45 Square shaft link

75 CRS.
R 25
⌴20
⌀16   54
R 16   R16
40

181

7.46 Driving dog

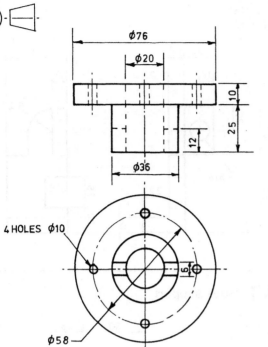

φ76

φ20

10

25

12

φ36

4 HOLES φ10

6

φ58

7.47 Machine cam

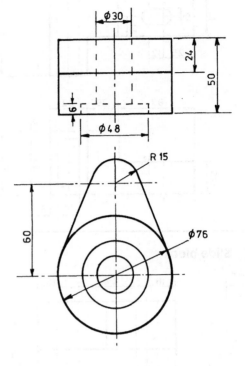

φ30

24

50

6

φ48

R 15

60

φ76

7.48 Locking collar

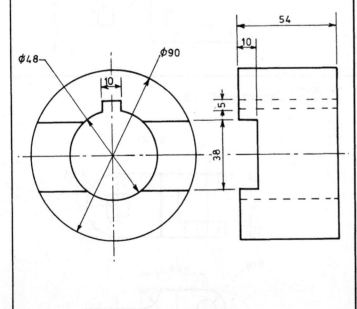

φ48

φ90

10

54

10

5

38

7.49 Cam guide

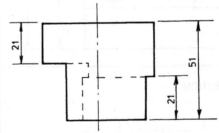

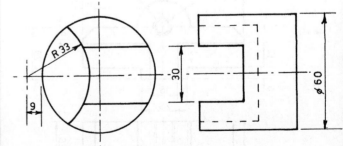

21

51

21

R 33

30

9

φ 60

7.50 Coolant cam

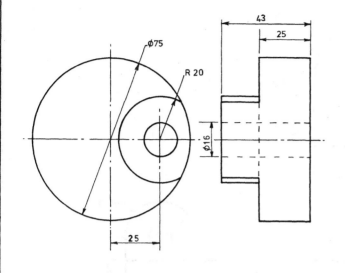

7.51 Guide block

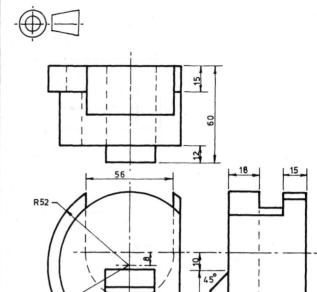

7.52 Shaft cap

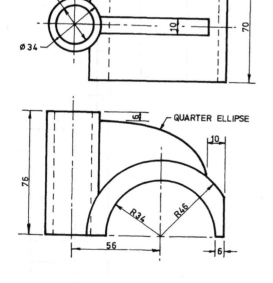

QUARTER ELLIPSE

7.53 Bearing

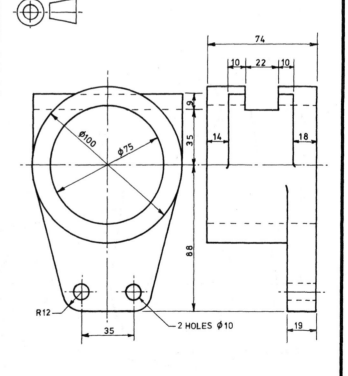

2 HOLES ⌀10

183

7.54 Angle bracket

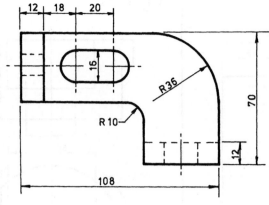

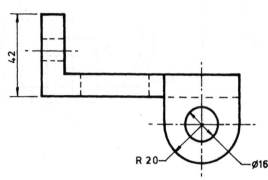

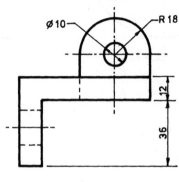

7.55 Rocker block

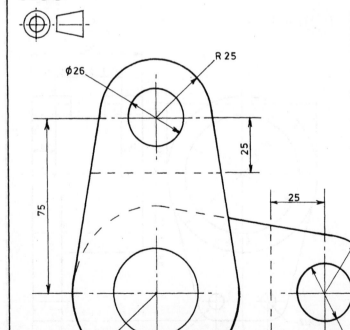

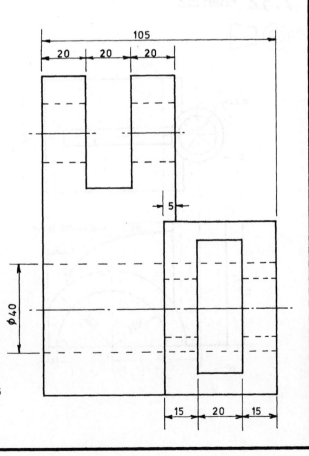

7.56 Table dog

Draw an isometric or an oblique parallel view of
the table dog.
Do not use an isometric scale.
Scale 1:1

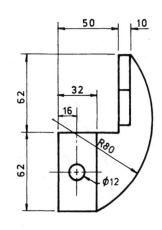

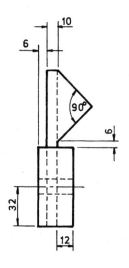

7.57 Fabricated bracket

Draw the given side view of
the fabricated bracket
and using this view as an aid
make an isometric
drawing of the bracket.
Scale 1:2

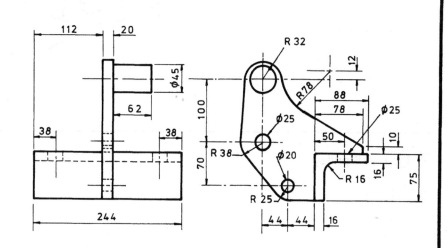

7.58 Mounting bracket

Draw, full size, an isometric or
an oblique parallel view of the
mounting bracket.

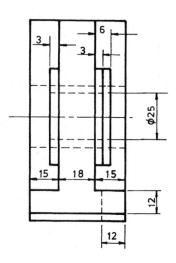

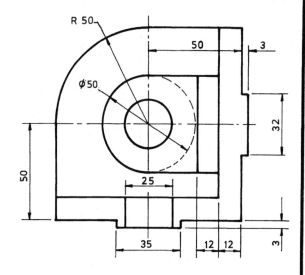

185

7.59 Grinder guide

Make an oblique parallel drawing of the grinder guide using a scale of 1:2.

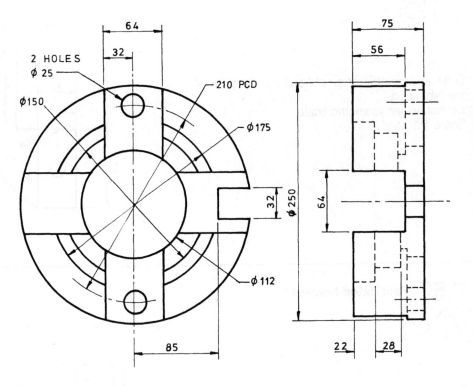

7.60 MS pulley

Make an oblique parallel drawing of the pulley.
Scale 1:1

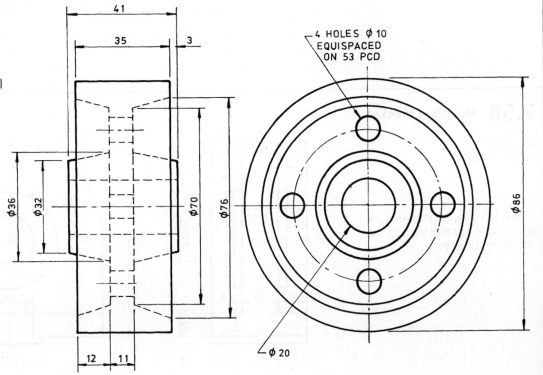

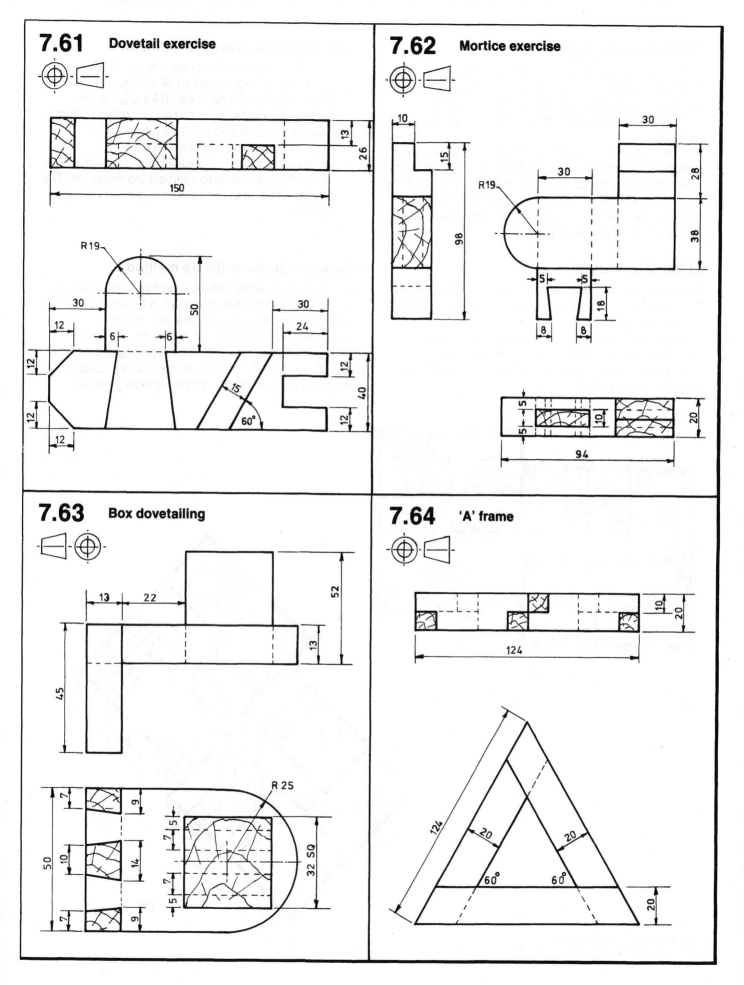

7.61 Dovetail exercise

7.62 Mortice exercise

7.63 Box dovetailing

7.64 'A' frame

187

Planometric views

A planometric view is another form of pictorial drawing, which enables the drafter to show the interior of a building. It is based on the top view or plan (Fig. 7.22(a)), which shows the true shape and size of a dwelling.

The planometric view (Fig. 7.22(b)), has been drawn with the long sides at 45° to the horizontal. The verticals have been reduced to two-thirds of the original height to allow more of the interior detail to be shown. It is also permissible to remove vertical surfaces to reveal detail that would normally be hidden.

Dimetric views

The word 'dimetric' implies two scales of measurement. Two axes are foreshortened an equal amount, and the third foreshortened a different amount.

Figure 7.23 illustrates approximate scales and angles of axes used to construct a dimetric drawing. The resulting drawings will be sufficiently accurate for all practical purposes.

Dimetric circle—four-centre method

This method of construction can only be used on surfaces such as those marked 'A' in Figure 7.23.

Figure 7.24 shows the method for a full circle:
1. Draw in the centre lines and an enclosing rhombus equal to the diameter of the required circle.
2. Draw the diagonals of the rhombus.
3. From points a and b draw lines at right angles to the sides of the rhombus to intersect the diagonals. The centres of the arcs are located at points, 1, 2, 3 and 4.

Dimetric circle—ordinate method

Circles may be drawn whole or in part in dimetric view by the use of ordinates, as shown in Figure 7.24. In this case two aids are required: one for full-size measurements, the other for the fractional scale.

Figure 7.25 shows a dimetric drawing of a machine part. The circles on both faces were constructed using the four-centre method and the ordinate method.

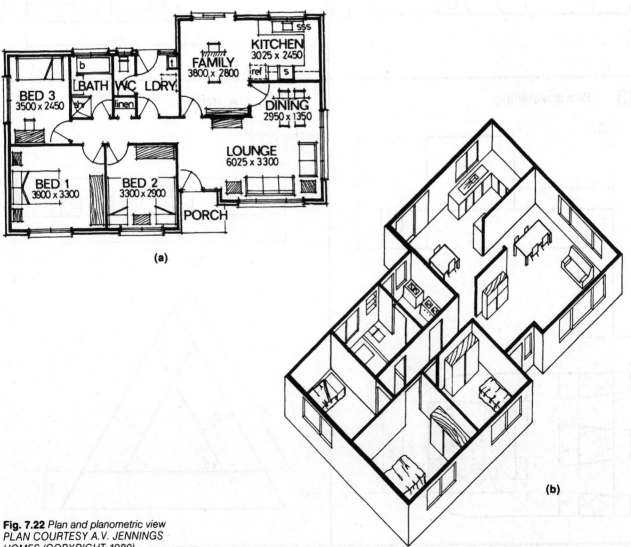

(a)

(b)

Fig. 7.22 *Plan and planometric view*
PLAN COURTESY A.V. JENNINGS HOMES (COPYRIGHT 1980)

188

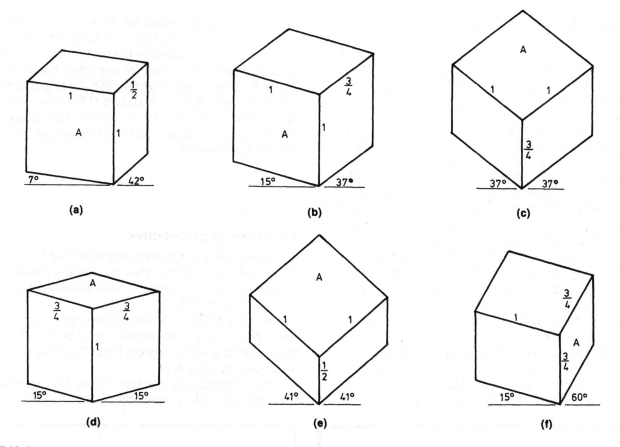

Fig. 7.23 *Dimetric scales and angles (full-size axes marked '1', half-size ones '½' and so on)*

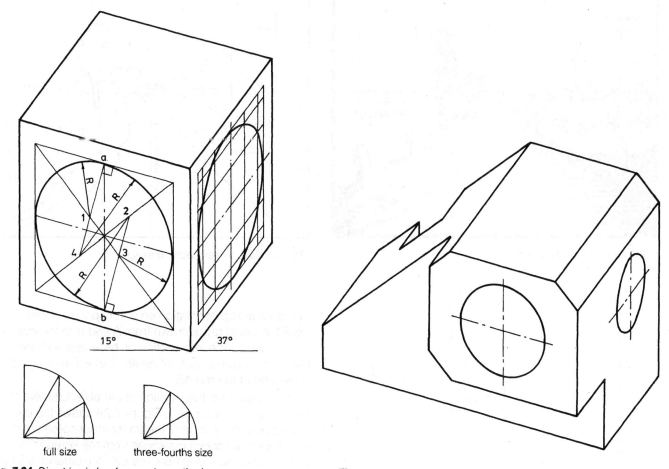

Fig. 7.24 *Dimetric circle—four-centre method*

Fig. 7.25 *Dimetric drawing of a machine part*

full size   three-fourths size

Perspective drawing

In isometric and oblique parallel projections lines representing corners and edges are drawn in a given direction, as parallel lines. This creates a slightly false impression of the object for the viewer. When the eye sees an object the edges and corners will tend to converge towards a distant point.

This fact can be illustrated by referring to Figures 7.26 and 7.27. Figure 7.26 is a drawing of a roadside disappearing into the distance. A closer inspection of the drawing will show that the sides of the road, the tops and bottoms of the fence posts and the telegraph poles all appear to converge to a single point on the horizon. This is shown more clearly in Figure 7.27, where the convergent lines are drawn and only the outline of detail is present. The viewer is standing in the middle of the road, which is going away towards the horizon in a right-angled direction, as are the fence and telegraph posts. The convergent point for these features is on the horizon and is known as the centre of vision (CV).

If the road had been going away in an angled direction, then the convergent point would be on the horizon line to the right or left of the CV, depending on the direction in which the road was heading.

The purpose of perspective drawing is to present a pictorial view of an object as it would actually appear to the eye. This allows a viewer to constructively criticise an object before it is made, for example a house can be drawn by an architect in perspective drawing and the client can see a true-to-life picture of the house before it is built. If any of the features are not satisfactory, then the architect can be advised to make changes before the plans are finalised.

Principles of perspective

The principles of perspective projection can be best understood by students if they can imagine themselves viewing an object through a vertical transparent screen, such as a pane of glass.

Rays of light are reflected from the object, pass through the transparent vertical screen and enter the eye of the spectator in straight lines. These rays of light are known as visual rays.

The accompanying diagram, Figure 7.28, illustrates the relationship of a simple object, a rod AB, to its

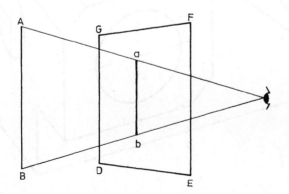

Fig. 7.26 Perspective drawing

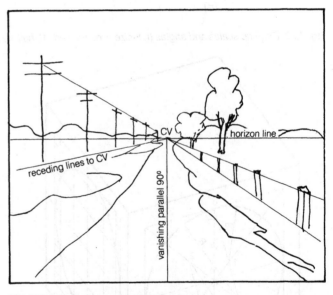

Fig. 7.27 Basis for perspective drawing

perspective representation on a transparent screen DEFG. A visual ray is drawn from A and B to the eye of the spectator. The intersection of these rays with the transparent screen determine ab. This is the perspective projection of the rod AB.

In perspective, the position of the object relative to the eye is most important. Figure 7.29 shows the eye in a fixed position in front of the transparent screen and a line placed at varying distances from the screen. The further away the line is from the eye, the smaller will be its perspective representation.

Fig. 7.28 Relationship of AB to its perspective representation

However, as demonstrated in Figure 7.30, if the position of the object relative to the screen is fixed and the position of eye varies, then it can be shown that as the eye approaches the transparent screen, the perspective representation will become smaller.

For the purpose of perspective projection the transparent screen is termed the picture plane (PP).

Figure 7.31 pictorially illustrates perspective theory and defines the points, lines and planes used. It shows a square, ABCD, placed on the ground plane, with the side AB touching the picture plane. An observer has been positioned at a convenient distance from the picture plane. A central visual ray, or vanishing parallel of 90° from the observer's eye to the picture plane, determines the centre of vision (CV). A line drawn through this point, parallel to the ground line, determines the horizon line and represents the eye level of the observer.

To obtain the perspective representation of the square ABCD on the picture plane, visual rays are drawn from the eye to points D and C. Points A and B are joined to the centre of vision and these converging lines will intersect the visual rays at c and d. The figure abcd is the perspective projection of the square ABCD.

Note: The perspective points a and b coincide with A and B because the square ABCD touches the picture plane along AB.

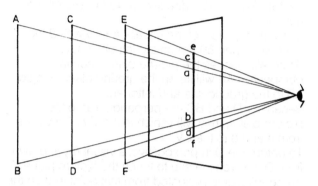

Fig. 7.29 *Position of object relative to eye*

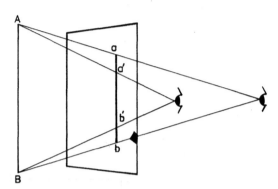

Fig. 7.30 *Position of object relative to eye*

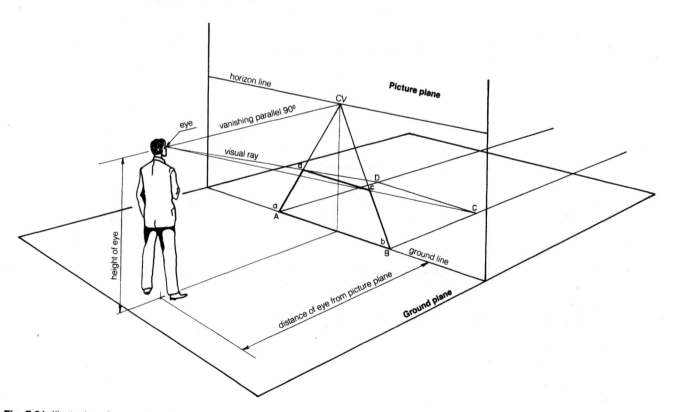

Fig. 7.31 *Illustration of perspective theory*

191

Exercises on perspective drawing

Exercises 7.65–7.80 have solutions provided for the students to follow. No solutions are provided for 7.81–7.94 and these can be used to test students' knowledge of perspective drawing.

In Exercises 7.67–7.82 the picture plane (PP) is assumed to be the horizon line (HL). This method saves time, projection lines are shorter and the drawing occupies less space.

7.65

Represent in perspective a square, side 120 cm, lying on the ground plane. One side touches the picture plane, and the nearest point of the square is 60 cm on the left of the spectator.
Distance of eye from the PP = 280 cm
Height of eye = 160 cm
Scale 1:20

Solution
1. Draw a horizontal line to represent the ground line. Mark the position of the eye.
2. Position the picture plane 280 cm above the ground line.
3. At a point 160 cm above the ground line, draw a horizontal line to represent the horizon line. Mark the centre of vision, CV, directly above the eye on the horizon line.
4. Draw the top view of the square ABCD with B the nearest corner, 60 cm on the left of the CV.
5. Since AB touches the picture plane, then in perspective AB will lie on the ground plane. Project AB to the ground line and letter ab.
6. Because AD and BC are perpendicular to the picture plane, they will vanish to the CV. Draw lines from a and b to the CV.
7. To determine D in perspective, a visual ray is drawn from D towards the eye to touch the picture plane. A perpendicular is projected from this point to d on the line from a to the CV.
8. Draw dc parallel to the ground line to meet the line from b to the CV to complete the perspective representation of the square.

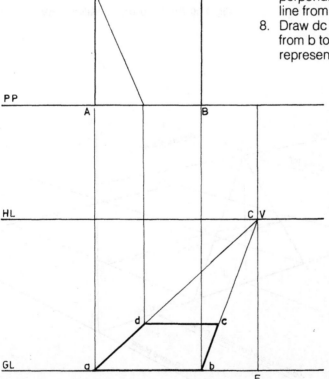

7.66

Represent in perspective a square, side 120 cm, when it lies on the ground with one side parallel to and 20 cm from the picture plane. The nearest corner of the square is 80 cm on the right of the spectator.

Distance of eye from the PP = 280 cm
Height of eye = 160 cm
Scale 1:20

Solution

1. Position the PP, HL and GL, and mark the CV.
2. Draw the top of the square ABCD with AB 20 cm from the PP and point A 80 cm to the right of the CV.
3. Produce DA and CB to meet the ground line at x and y.
4. Since DA and CB are perpendicular to the picture plane, they will vanish to the CV. Draw lines from x and y to the CV.
5. To determine B and C in perspective, visual rays are drawn from these points towards the eye to touch the PP. Perpendiculars are projected from these points to b and c on the line from y to the CV.
6. Draw ab and dc parallel to the GL to complete the perspective representation of the square.

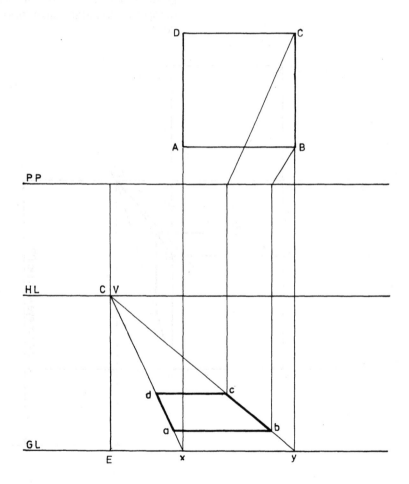

7.67

Represent in perspective a cube, edge 120 cm, standing on one of its faces, another touching the picture plane, with its nearer vertical edge 60 cm on the left of the spectator.
Distance of eye from the PP = 280 cm
Height of eye = 200 cm
Scale 1:20

Solution
1. Position the PP and GL and mark the CV and eye.
2. Draw the top view of the cube with the edge AB touching the PP and point B 60 cm on the left of the CV.
3. Project A and B to the GL and letter a and b. Draw lines from a and b to the CV.
4. To obtain D in perspective, a visual ray is drawn from D towards the eye to touch the PP. A perpendicular is drawn from this point to d on the line from a to the CV.
5. Draw dc parallel to ab.
6. Measure off aa' 60 cm high. Draw a'b' parallel to and equal to ab.
7. Since the top of the cube is horizontal, its receding edges will vanish towards the CV.
8. Draw vertical lines from d to d' and c to c' to complete the perspective representation of the cube.

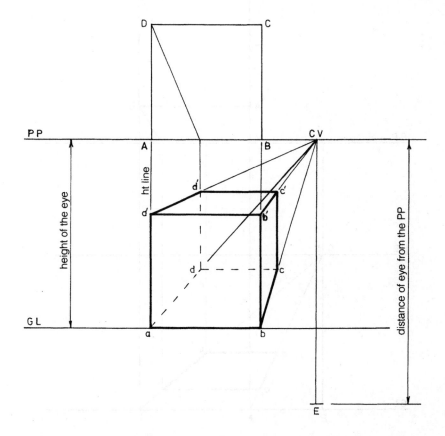

7.68

Represent in perspective a square prism, edges of ends 120 cm, axis 180 cm, when it rests on one of its rectangular faces on the ground plane. Its nearer vertical rectangular surface is 40 cm from and parallel to the picture plane. The nearest corner is 60 cm on the right of the spectator.

Distance of eye from the PP = 260 cm
Height of eye = 180 cm
Scale 1:20

Solution

1. Position the PP and GL and mark the CV and eye.
2. Draw the top view of the prism ABCD with AB 40 cm from the PP and A 60 cm on the right of the CV.
3. Draw the rectangle in perspective. Letter points a, b, c and d.
4. Measure off b' 120 cm high. Draw a'b' parallel to ab.
5. Draw lines from a' and b' to the CV.
6. Draw vertical lines from d to d' and c to c' to complete the perspective representation of the prism.

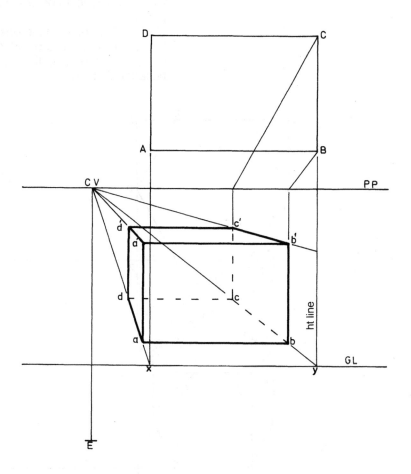

7.69

Obtain the perspective projection of a right hexagonal prism, edges of ends 80 cm, height 80 cm, when it stands on one of its ends on the ground plane with one rectangular face touching the picture plane and its axis opposite the spectator.

Distance of eye from the PP = 280 cm
Height of eye = 220 cm
Scale 1:20

Solution

1. Position the PP and GL and mark the CV and eye.
2. Draw the top view of the hexagonal prism, ABCDEF, and enclose it in a rectangle 1234.
3. Draw the rectangle in perspective. Number points 1', 2', 3' and 4'.
4. Project B and C to b and c on the GL and join to the CV. Letter points e and f.
5. To obtain A in perspective, draw a visual ray from A towards the eye to touch the PP. Draw a vertical line from this point to meet the line from 1' to the CV at point a.
6. Draw ad parallel to bc; abcdef is the end of the prism.
7. Mark off the height of the prism. Draw a horizontal line and project the top surface of the prism to complete the perspective representation of the right hexagonal prism.

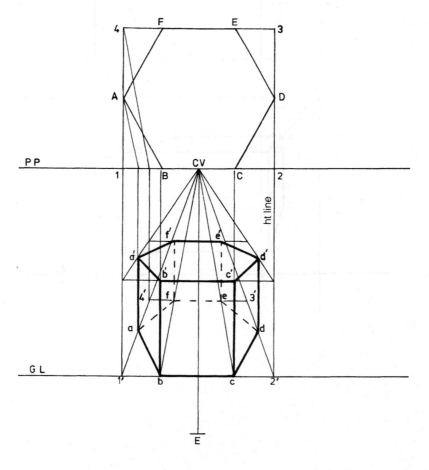

7.70

Represent in perspective a right square pyramid, edges of base 120 cm, height 200 cm, when it stands on its base on the ground plane with one edge parallel to and touching the picture plane. The nearest corner is 60 cm on the right of the spectator.

Distance of eye from the PP = 280 cm
Height of eye = 220 cm
Scale 1:20

Solution

1. Position the PP and GL and mark the CV and eye.
2. Draw the top view of the pyramid ABCD with the edge AB touching the PP and A 60 cm on the right of the CV.
3. Draw the square in perspective. Letter points a, b, c and d.
4. To obtain the height of the pyramid, project a line from O to touch the GL. Measure off 200 cm and join to the CV.
5. Draw a vertical line from the intersection of the diagonals to point o, the apex of the pyramid.
6. Join all points to complete the perspective representation of the pyramid.

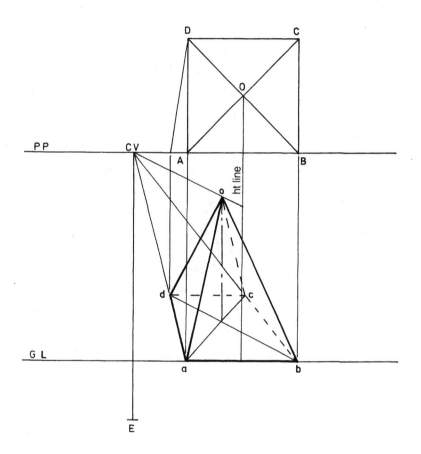

197

7.71

Represent in perspective a right hexagonal pyramid, edges of base 60 cm, height 200 cm, when it stands on its base on the ground plane with one diagonal perpendicular to the picture plane. One corner of the base touches the picture plane and is 140 cm to the left of the spectator.

Distance of eye from the PP = 280 cm
Height of eye = 220 cm
Scale 1:20

Solution
1. Position the PP and GL and mark the CV and eye.
2. Draw the top view of the pyramid ABCDEF with point C touching the PP and 140 cm to the left of the CV.
3. Draw the hexagon in perspective. Letter points a, b, c, d, e and f.
4. Measure off on the height line 200 cm, the height of the pyramid, and join to the CV.
5. Draw a vertical line from the intersection of the diagonals to point o, the apex of the pyramid.
6. Join all points to complete the perspective representation of the pyramid.

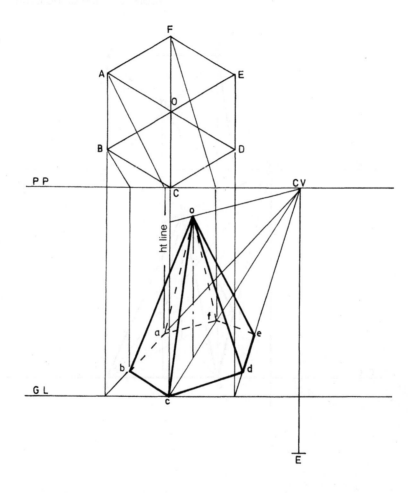

7.72

Represent in perspective a right hexagonal pyramid 160 cm high, having one diagonal of the base 144 cm long and parallel to the picture plane. The centre of the base is 152 cm to the right of the spectator and 100 cm from the picture plane.

Distance of eye from the PP = 280 cm
Height of eye = 220 cm
Scale 1:20

Solution
1. Position the PP and GL and mark the CV and eye.
2. Draw the top view of the pyramid ABCDEF with centre O 152 cm on the right and 100 cm from the PP.
3. Enclose the top view in a rectangle.
4. Draw the hexagon in perspective. Letter points a, b, c, d, e and f.
5. Measure off on the height line 160 cm, the height of the pyramid, and join to the CV.
6. Draw a vertical line from the centre of the base to point o, the apex of the pyramid.
7. Join all points to complete the perspective representation of the pyramid.

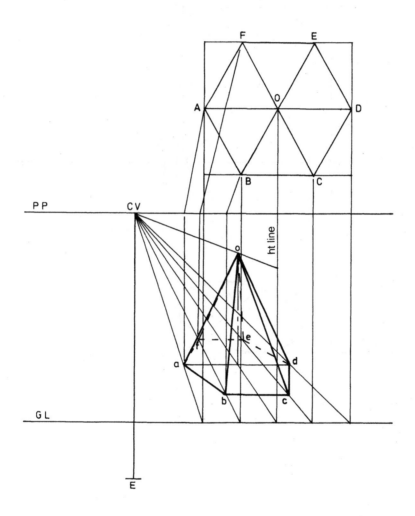

7.73

Represent in perspective a circle, diameter 120 cm,
lying on the ground plane with its curved surface
touching the picture plane. The centre of the circle is
100 cm to the left of the spectator.
Distance of eye from the PP = 280 cm
Height of eye = 200 cm
Scale 1:20

Solution
1. Position the PP and GL and mark the CV and eye.
2. Draw the top view of the circle and enclose it within
   the square 1234.
3. Draw the diagonals 1–3 and 2–4.
4. Draw two diameters to the circle CG and AE.
5. Draw the perspective representation of the square
   and draw the diagonals.
6. Produce C to the GL and draw to the CV to obtain c
   and g.
7. Draw a smooth freehand curve through points
   a, b, c, d, e, f, g and h to obtain the perspective
   representation of the circle.

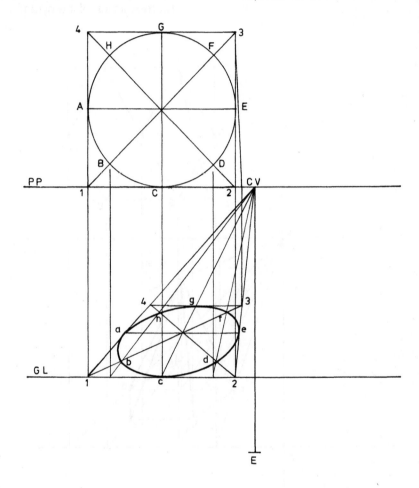

7.74

Represent in perspective a square slab 50 mm × 50 mm × 10 mm, surmounted by a right square pyramid, side 34 mm, axis 45 mm. The pyramid is centrally positioned on the slab. The nearest vertical surface of the slab is 10 mm behind the picture plane and the nearest point is 20 mm on the right of the spectator.
Distance of eye from the PP = 70 mm
Height of eye = 60 mm
Scale 1:1

Solution
1. Position the PP and GL and mark the CV and eye.
2. Draw the top view of the solids with the AB 10 mm from the PP and A 20 mm on the right of the CV.
3. Draw the square in perspective. Letter points a, b, c and d.
4. Measure off b' 10 mm high. Draw a'b' parallel to ab.
5. Draw lines from a' and b' to the CV.
6. Draw vertical lines from d to d' and c to c' to complete the slab.
7. Draw the base of the pyramid. Letter points e, f, g and h.
8. Measure off 55 mm along the height line and join to the CV.
9. Join e, f, g and h to o' to complete the pyramid.

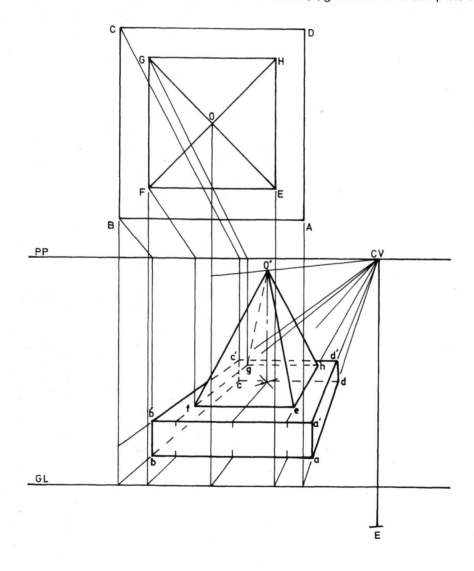

201

7.75

Represent in perspective a hexagon, side 30 mm, when it rests on the ground plane on a side and is contained in a vertical plane that recedes to the left at 30° from the picture plane. The nearest point is 17 mm behind the picture plane and 30 mm on the left of the spectator.
Distance of eye from the PP = 80 mm
Height of eye = 60 mm
Scale 1:1

Solution
1. Position the PP and GL and mark the CV and the eye.
2. Draw the top view of the hexagon with point D 10 mm from the PP and 10 mm on the left of the CV.
3. Find VP 60° and VP 30° by drawing vanishing parallels from the eye to touch the PP.
4. Produce AD to the PP and project to the GL at point o.
5. Draw a line from o to VP 30°.
6. Mark off on the height line the height of the hexagon and draw lines to VP 30°.
7. To find points a, b, c, d, e and f, visual rays are drawn from A, BF, CE and D to the PP and projected down to meet the lines receding to VP 30°.
8. Complete the perspective representation of the hexagon by joining all points.

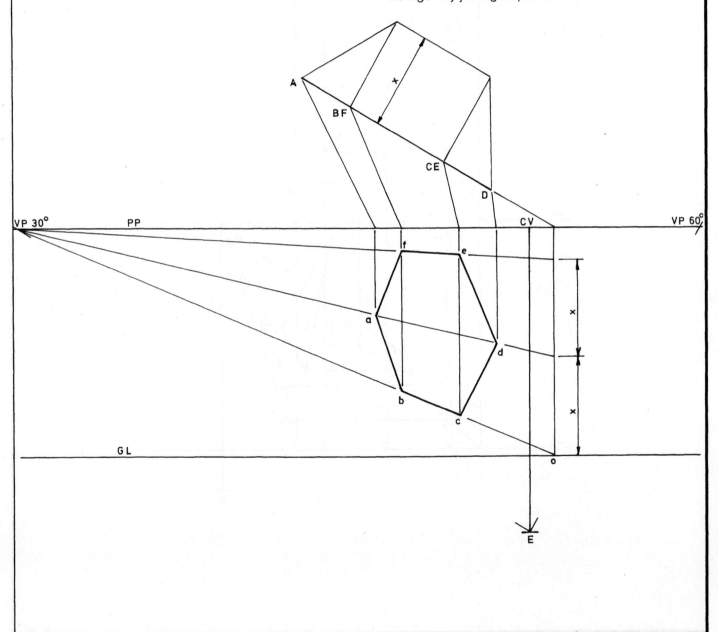

7.76

Represent in perspective a circle, diameter 50 mm, when it is contained in a vertical plane that recedes to the right at 45°. The centre of the circle is 25 mm behind the picture plane and opposite the spectator.
Distance of eye from the PP = 80 mm
Height of eye = 50 mm
Scale 1:1

Solution
1. Position the PP and GL and mark the CV and the eye.
2. Draw the top view of the circle with the centre 25 mm behind the PP and in line with the CV.
3. Obtain the vanishing points of 45° by drawing vanishing parallels from the eye to touch the PP.
4. Produce 0/6 to the PP and project to the GL.
5. Construct an aid and divide it into six equal parts.
6. Project these points onto the height line and draw lines to VP 45°.
7. To find points 0 to 6, visual rays are drawn to the PP and projected down to meet the lines receding to VP 45°.
8. Complete the perspective representation of the circle by drawing a freehand curve through all points.

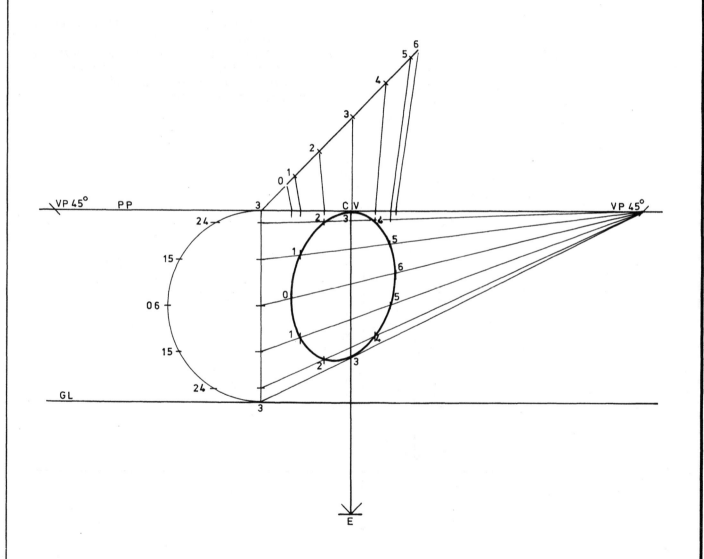

7.77

Represent in perspective a cube, edges 120 cm, when it stands on the ground plane with its nearest corner 60 cm on the left of the spectator and 20 cm from the picture plane. The sides recede at 45° from the picture plane.

Distance of eye from the PP = 260 cm
Height of eye = 200 cm
Scale 1:20

Solution

1. Position the PP and GL and mark the CV and eye.
2. Draw the top view of the cube ABCD at 45° to the PP with B 60 cm on the left and 20 cm from the PP.
3. Obtain the vanishing points of 45° by drawing vanishing parallels of 45° from the eye to touch the PP.
4. Produce BC to the PP and project to the GL at point x.
5. Since the line BC recedes to the right at 45°, point x will vanish to VP 45° on the right of the eye.
6. To determine b and c, visual rays are drawn from B and C to the PP and projected down to meet the line from x to VP 45° right.
7. From b and c draw lines to VP 45° left and the visual ray from A will determine a on the line from b.
8. Complete the rectangle by drawing a line from a to VP 45° right to cut the line from c to VP 45° left at d.
9. Mark off on the height line 120 cm, the height of the cube, and join to VP 45° right.
10. Complete the rectangle b'c'd'a' in the same way in which the rectangle bcda on the ground was obtained to complete the perspective representation of the cube.

7.78

Obtain the perspective representation of a right square pyramid, side of base 120 cm, height 160 cm, when it stands on the ground plane with its nearest corner 80 cm on the right of the spectator and 40 cm behind the picture plane. The side BA recedes from the picture plane at 60° to the left.

Distance of eye from the PP = 240 cm
Height of eye = 180 cm
Scale 1:20

Solution

1. Position the PP and GL and mark the CV and eye.
2. Draw the top view of the pyramid ABCD with B 80 cm on the right and 40 cm from the PP, and BA inclined to the right at 60°.
3. Find VP 60° and VP 30° by drawing vanishing parallels from the eye to touch the PP.
4. Produce AB to the PP and project to the GL at point x.
5. Draw a line from x to VP 60°.
6. To determine a and b, visual rays are drawn from A and B to the PP and projected down to meet the line from x to VP 60°.
7. From a and b, draw lines to VP 30° and a visual ray from C will determine c on the line from b.
8. Complete the square abcd by drawing a line from c to VP 60° to cut the line from a to VP 30° at d.
9. The height line from the apex is obtained by drawing a line parallel to BC through the apex O to meet the PP. Project this point down to meet the GL at right angles.
10. Mark off on this line 160 cm, the height of the pyramid, and draw to VP 30°.
11. From the intersection of the diagonals, draw a vertical line to point o. Join all points to complete the pyramid.

7.79

Given the side view of three steps 160 cm long, obtain their perspective representation when they are standing on the ground plane with their long edges inclined at 30° to the right of the picture plane. The corner of the lowest step is opposite the spectator and 40 cm from the picture plane.
Distance of eye from the PP = 240 cm
Height of eye = 160 cm
Scale 1:20

Solution
1. Position the PP and GL and mark the CV and eye.
2. Draw the top view of the steps with their long edges at 30° to the PP and the corner of the lowest step opposite the spectator and 40 cm from the PP.
3. Find VP 30° and VP 60°.
4. Produce the short side of the top view to meet the PP and project it to the GL at point x.
5. Mark off on the height line points a, b and c from the side view and draw to VP 60°
6. Draw visual rays to the PP and project them down to give the perspective representation of the side view of the steps.
7. Draw vertical and receding lines to both VPs to complete the perspective representation of the steps.

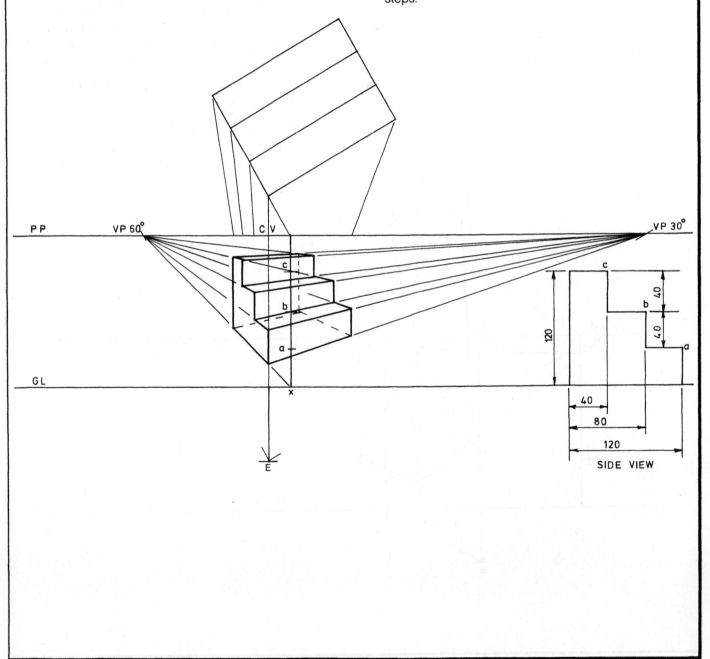

SIDE VIEW

7.80

Represent in perspective a rectangular prism 160 cm × 132 cm × 40 cm lying on one of its 160 cm × 132 cm surfaces with its short edges making 45° with the picture plane. Point A touches the picture plane opposite the spectator. On it, place another rectangular prism with its long edges at 30° to the picture plane towards the right, with point E located at the mid-point of AD. A hexagonal hole, side 52 cm, is centrally positioned in the prism. Height of the prism is 128 cm, thickness 60 cm.

Distance of eye from the PP = 280 cm
Height of eye = 200 cm
Scale 1:20

Solution

1. Position the PP, HL and GL and mark the CV.
2. Draw the top view of both prisms with point A opposite the spectator and touching the PP.
3. Find VP 45° left and VP 45° right..
4. Project A to the GL and draw to VP 45° left and VP 45° right. Visual rays and projectors from B and D will complete the bottom rectangle abcd.
5. Mark off the height of 40 cm and complete the perspective representation of the bottom prism.
6. Find VP 30° right and VP 60° left.
7. Visual rays and projectors will locate e and f in perspective. Draw e and f to VP 60° to obtain g and h.
8. Mark off on the height line the heights for the top prism and hexagonal opening and draw to VP 30°.
9. Draw visual rays and projectors from 1′, 2′, 3′, 4′, 5′ and 6′, in perspective.
10. Draw lines from these points to VP 60°.
11. Visual rays and projectors will determine the depth of the hexagonal hole in perspective to complete the perspective representation of the prisms.

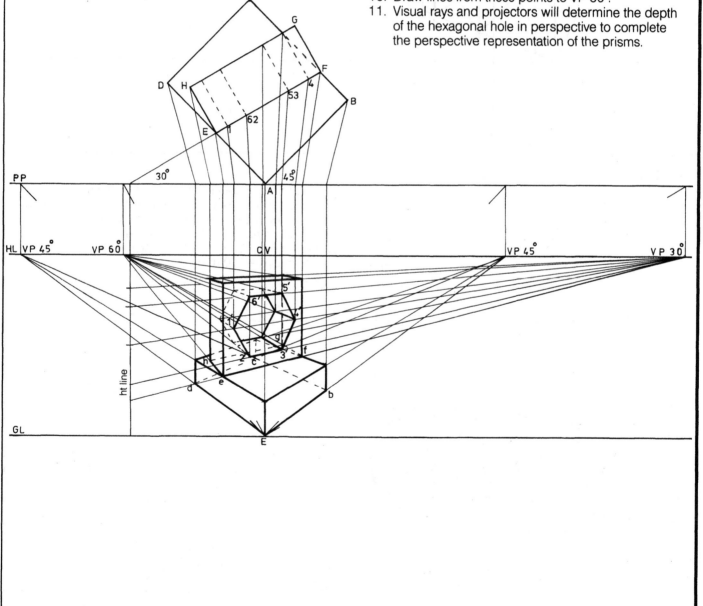

207

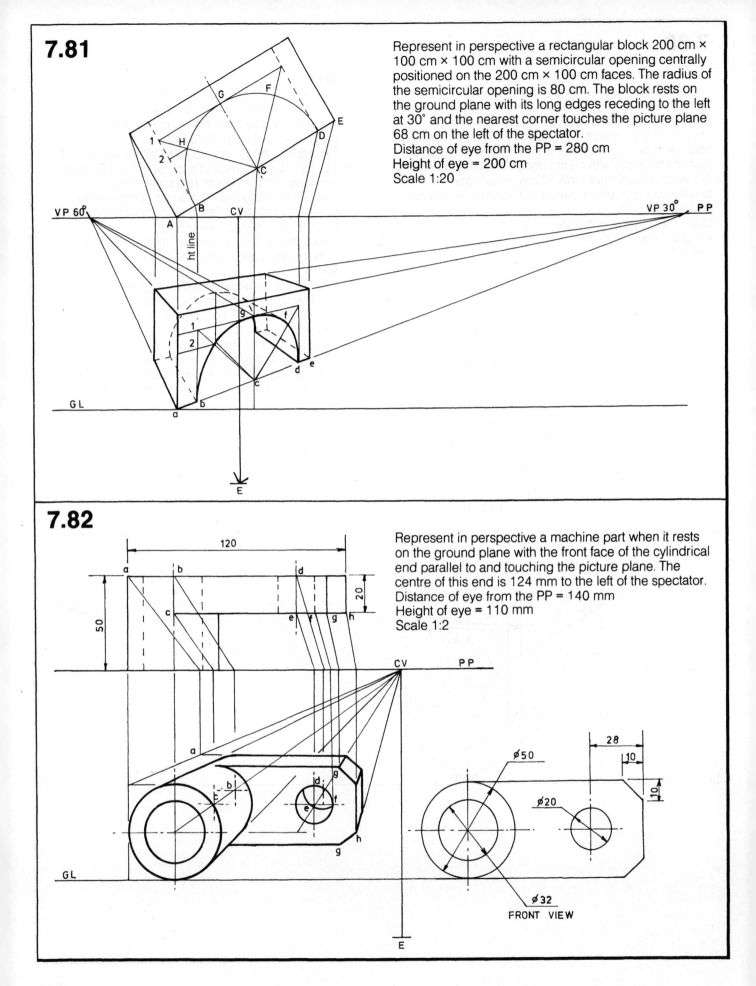

7.81

Represent in perspective a rectangular block 200 cm × 100 cm × 100 cm with a semicircular opening centrally positioned on the 200 cm × 100 cm faces. The radius of the semicircular opening is 80 cm. The block rests on the ground plane with its long edges receding to the left at 30° and the nearest corner touches the picture plane 68 cm on the left of the spectator.
Distance of eye from the PP = 280 cm
Height of eye = 200 cm
Scale 1:20

7.82

Represent in perspective a machine part when it rests on the ground plane with the front face of the cylindrical end parallel to and touching the picture plane. The centre of this end is 124 mm to the left of the spectator.
Distance of eye from the PP = 140 mm
Height of eye = 110 mm
Scale 1:2

7.83

Represent in perspective a rectangle 150 cm × 90 cm, lying on the ground plane with one of its long edges touching the picture plane and with the nearest corner 20 cm to the right of the eye.
Distance of eye from the PP = 200 cm
Height of eye = 140 cm
Scale 1:20

7.84

Represent in perspective a right square prism, edges of ends 90 cm, axis 120 cm, when it rests on one of its ends on the ground plane. The nearest vertical surface is parallel to and 40 cm from the picture plane. The nearest corner is 40 cm on the left of the spectator.
Distance of eye from the PP = 200 cm
Height of eye = 140 cm
Scale 1:20

7.85

Represent in perspective a right hexagonal prism, edges of ends 60 cm, axis 90 cm, when it stands on one of its ends on the ground plane with a rectangular surface parallel to and 20 cm from the picture plane. The nearest point is 30 cm on the left of the spectator.
Distance of eye from the PP = 200 cm
Height of eye = 140 cm
Scale 1:20

7.86

Represent in perspective a right pentagonal pyramid, side of base 60 cm, axis 90 cm, when it stands on its base on the ground plane with one edge parallel to and 20 cm from the picture plane. The axis of the pyramid is opposite the spectator.
Distance of eye from the PP = 160 cm
Height of eye = 120 cm
Scale 1:20

7.87

Represent in perspective a cube, side 120 cm, when it stands on the ground plane with its nearest corner 60 cm on the right of the eye and 20 cm from the picture plane. Two edges recede to the right at 30° from the picture plane.
Distance of eye from the PP = 180 cm
Height of eye = 140 cm
Scale 1:20

7.88

Represent in perspective a right equilateral triangular prism, edges of ends 90 cm, axis 120 cm, when it rests on a rectangular surface on the ground plane with its 120 cm edges receding to the left at 30° from the picture plane. The nearest point is 30 cm on the left of the eye and 30 cm behind the picture plane.
Distance of eye from the PP = 180 cm
Height of eye = 160 cm
Scale 1:20

7.89

Represent in perspective a cube, edges 100 cm, when it rests on the ground plane with its nearest corner 20 cm on the left of the eye and 20 cm behind the picture plane. The sides recede at 45° from the picture plane.
　　The cube is surmounted by a right square pyramid, axis 50 cm, whose base coincides with the top surface of the cube.
Distance of eye from the PP = 220 cm
Height of eye = 160 cm
Scale 1:20

7.90

Represent in perspective a right hexagonal prism, side 60 cm, axis 120 cm, when it rests on the ground plane on a rectangular surface with its nearest point 36 cm on the right of the eye and 40 cm behind the picture plane. The 120 cm edges of the prism recede to the right at 30° from the picture plane.
Distance of eye from the PP = 180 cm
Height of eye = 140 cm
Scale 1:20

7.91

Point A is 50 cm on the left of the eye and 20 cm behind the picture point. Point B is 50 cm on the right of the eye and 100 cm behind the picture plane. The line AB is the diameter of a circle contained in a vertical plane receding to the right. Complete the perspective of the circle.
Distance of eye from the PP = 220 cm
Height of eye = 160 cm
Scale 1:20

7.92

A disc 120 cm in diameter and 40 cm thick rests on a circular surface on the ground plane with its axis 30 cm on the right of the eye and 76 cm behind the picture plane. A hole 60 cm square is cut centrally in the disc. Show the disc in perspective when it is placed with one side of the hole parallel to the picture plane.
Distance of eye from the PP = 200 cm
Height of eye = 160 cm
Scale 1:20

7.93

A six-pointed star is made of two equilateral triangles, the points of the star being equally spaced around a circle. The length of side of each triangle is 120 cm. The star with its circumscribing circle is contained in a vertical plane receding to the right at 30° from the picture plane. The nearest point of the star is 80 cm to the right of the eye and 50 cm behind the picture plane. Draw the circle and the star in perspective.
Distance of eye from the PP = 220 cm
Height of eye = 180 cm
Scale 1:20

7.94

Two 80 cm diameter discs are cut from thin cardboard. One disc rests on its edge on the ground plane and is contained in a vertical plane that recedes to the right at 45° to the picture plane. The centre of the disc is 24 cm to the left of the eye and 76 cm behind the picture plane. The second disc rests on the first and is contained in a horizontal plane. Its centre is vertically above that of the first disc. Draw the perspective of both discs.
Distance of eye from the PP = 200 cm
Height of eye = 160 cm
Scale 1:20

Reflections using the ground plane as a mirror surface

The image of an object produced by a reflecting surface, such as a mirror, is called a *reflection*.

In Figure 7.32(a) a simple block is standing upright on a horizontal mirror surface. The reflection of the object appears to be below the mirror and directly under the object. The image is the same size as the object.

Figure 7.32(b) shows the perspective view of a cube and its reflection. The height of the cube above the ground line (GL) is equal to its reflected height below the ground line. The reflection can be enhanced by using vertical hatching, light rendering or shading.

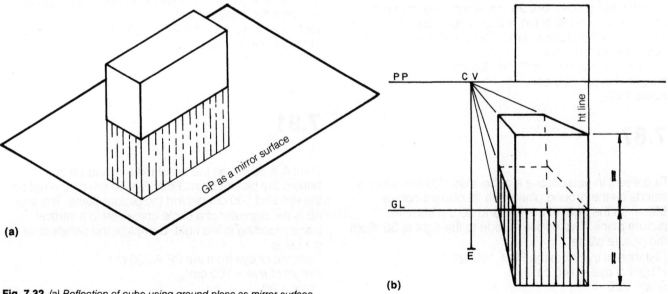

Fig. 7.32 *(a) Reflection of cube using ground plane as mirror surface*
(b) Perspective view of cube and its reflection

Exercises on reflections

The following constructions are given for reference. You should practise them and become familiar with the method and techniques.

7.95

1. Represent in perspective a square, side 25 mm, when it is contained in a vertical plane that recedes to the right at 30°. The nearest point is 30 mm to the right of the eye and 15 mm behind the picture plane.
Distance of eye from the PP = 70 mm
Height of eye = 40 mm
Scale 1:1

Solution
(a) Position the PP and GL and mark the CV and eye.
(b) Draw the top view of the square with the nearest point 30 mm on the right of the eye and 15 mm behind the PP.
(c) Find VP 60° and VP 30° by drawing vanishing parallels from the eye to the PP.
(d) Produce D, CA and B to the PP and project to the GL.
(e) Draw a line from a to VP 30°.

(f) Mark off on the height line points d, b and c and draw lines to VP 30°.
(g) Draw visual rays from D, CA and B to the PP and project down to meet the lines receding to VP 30°.
(h) Join points a, b, c and d to complete the perspective representation of the square.

2. Show the reflection of the square in the ground plane as a mirror surface.

Solution
(a) Reflect the height line a, d', b' and c'.
(b) Draw lines from d'b' and c' to VP 30°.
(c) Reflect verticals from d, c, a and b to intersect the lines to VP 30° at the points d', c' and b'.
(d) Join points a', d', b' and c' to complete the reflection of the square.

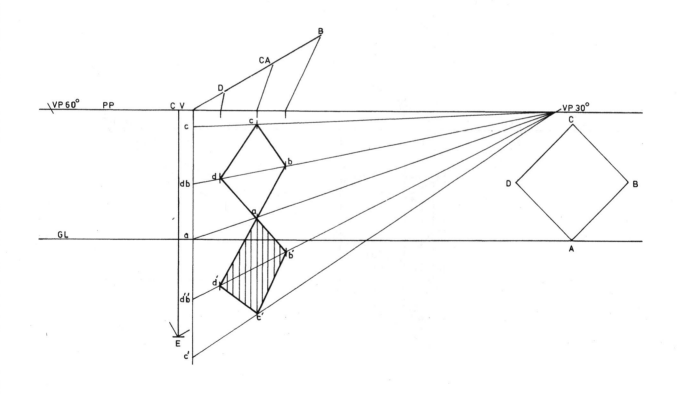

7.96

1. Represent in perspective a right square prism, edges of ends 60 mm, length 90 mm, when it rests on one of its rectangular faces on the ground plane with its longer edges inclined to the right at 30° to the picture plane. The nearest point is 15 mm to the right of the spectator and 20 mm behind the picture plane.
   Distance of eye from the PP = 150 mm
   Height of eye = 120 mm
   Scale 1:1

Solution
   (a) Position the PP and GL and mark the CV and eye.
   (b) Draw the top view of the prism with the nearest point 15 mm to the right of the eye and 20 mm behind the PP.
   (c) Find VP 60° and VP 30° by drawing vanishing parallels from the eye to the PP.
   (d) Produce AB to PP and project to the GL at point x.
   (e) Draw a line from x to VP 30°.

   (f) To determine a and b, visual rays are drawn from A and B to the PP and projected down to meet the line from X to VP 30°.
   (g) From a and b draw lines to VP 60° and a visual ray from D will determine d on the line from a.
   (h) Complete the rectangle by drawing a line from d to VP 30° to cut the line from b to VP 60°.
   (i) Mark off on the height line 60 mm and join to VP 30°.
   (j) Complete the rectangle defg in the same way in which the rectangle abcd on the ground was obtained.

2. Show the reflection of the prism in the ground plane as a mirror surface.

Solution
   (a) Reflect the height line xy'.
   (b) Draw a line from y' to VP 30° to intersect verticals from ae and bf at points e' and f'.
   (c) Join e' to VP 60° to intersect verticals from dh and gc at points h' and g'.
   (d) Join all points to complete the reflection of the prism.

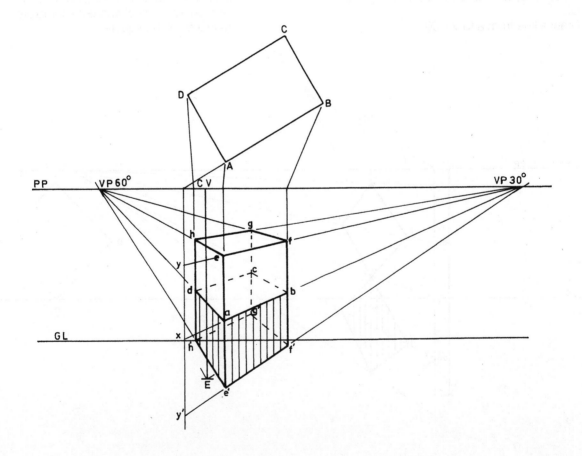

7.97

1. Represent in perspective a right square pyramid, side of base 30 mm, height 40 mm, when it stands on the ground plane with its base edges making equal angles with the PP. The nearest point is 13 mm behind the picture plane and 10 mm on the right of the eye.
   Distance of eye from the PP = 60 mm
   Height of eye = 45 mm
   Scale 1:1

Solution
(a) Position the PP and GL and mark the CV and eye.
(b) Draw the top view of the pyramid with the nearest point 10 mm on the right of the eye and 13 mm behind the PP.
(c) Obtain the vanishing points 45° by drawing vanishing parallels from the eye to touch the PP.
(d) Produce AB to the PP and project to the GL at point x.
(e) Draw a line from x to VP 45° right.
(f) To determine a and b, visual rays are drawn from A and B to the PP and projected down to meet the line from x to VP 45° right.
(g) Complete the square abcd by drawing a line from d to VP 45° right.

(h) The height line from the apex is obtained by drawing a line parallel to AB through the apex o to meet the GL. Project this point down to meet the GL.
(i) Mark off on this line 40 mm and draw to VP 45° right.
(j) From the intersection of the diagonals, draw a vertical line to point o. Join all points to complete the pyramid.

2. Show the reflection of the pyramid in the ground plane as a mirror surface.

Solution
(a) Reflect the height line ht'.
(b) Draw a line from T' to Vp 45° right.
(c) Reflect the axis of the pyramid to point o'.
(d) Join all points to complete the reflection of the pyramid.

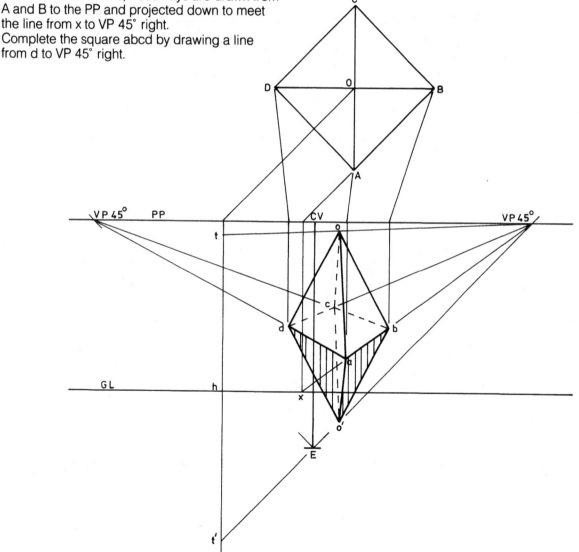

213

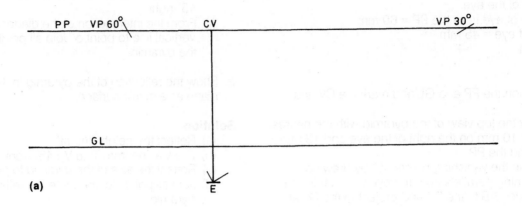

(a)

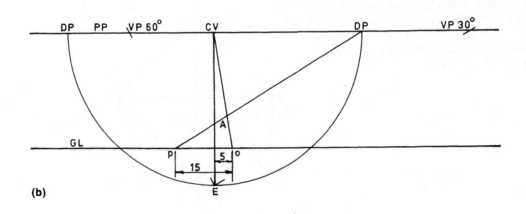

(b)

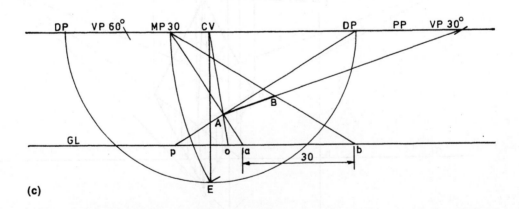

(c)

Fig. 7.33 *Measuring point perspective (a) initial layout (b) finding nearest point A (c) establishing measuring points*

214

Measuring point perspective

Measuring point perspective has several advantages over plan method perspective.

1. The top view of the object is not required and, therefore, there is no need for visual rays.
2. There are fewer projection lines.
3. Time and space is saved by not having to draw the top view.

The description of Figure 7.33 given below introduces the new terms and procedures as simply as possible to help the student gain a clear understanding of measuring point perspective.

Figure 7.33(a) shows the initial layout for measuring point perspective. All lines and vanishing points are the same as for plan method perspective. A line AB 30 mm long rests on the ground plane and recedes to the right at 30° from the picture plane. The nearest point is 5 mm on the right of the eye and 15 mm behind the picture plane. The eye is 40 mm from the picture plane at a height of 30 mm. The scale is 1:1.

Figure 7.33(b) illustrates the method used to find the nearest point A in measuring point perspective.

1. Locate the distance points (DP) by taking the centre of vision as centre and the eye as radius. Describe a semicircle to touch the picture plane at the two distance points.
2. Measure 5 mm along the ground line to point o on the right of the eye and join it to the centre of vision.
3. To find the distance behind the picture plane, measure 15 mm to the left of point o along the ground line to point p. Join p to the distance point.
4. The intersection of these two lines will locate point A in perspective.

In Figure 7.33(c) a line is drawn from A to VP 30°. To find the length of AB in perspective, a measuring point (MP 30) is required.

1. With the VP 30° as centre and the eye as radius, draw an arc to cut the picture plane at the measuring point, MP 30.
2. From MP 30 draw a line through A to touch the ground line at point A.
3. From a measure the length of the line AB, 30 mm along the ground line at point b.
4. Join b to MP 30 to intersect the line from A to VP 30° to locate point B in perspective.

Exercises finding the nearest point

7.98

Find point P on the ground plane 60 mm to the right of the eye and 80 mm behind the picture plane.
Distance of eye from the PP = 100 mm
Height of eye = 70 mm
Scale 1:1

7.99

Find point A on the ground plane 20 mm to the left of the eye and 30 mm behind the picture plane.
Distance of eye from the PP = 100 mm
Height of eye = 70 mm
Scale 1:1

7.100

Show point A when it is directly opposite the spectator and 50 mm behind the picture plane.
Distance of eye from the PP = 100 mm
Height of eye = 70 mm
Scale 1:1

7.101

Show point P on the ground plane 50 mm to the left of the eye and 50 mm behind the picture plane.
Distance of eye from the PP = 100 mm
Height of eye = 70 mm
Scale 1:1

Exercises on measuring point perspective

Exercises 7.102–7.108 have solutions provided for students to follow. No solutions are provided for 7.109–7.130 which can be used to test students' knowledge of the procedures.

7.102

Represent in perspective a rectangle 70 mm × 50 mm when it rests on the ground plane with its 70 mm edges receding to the left at 30° from the picture plane. The nearest point is 10 mm on the left of the eye and 30 mm behind the picture plane.
Distance of eye from the PP = 120 mm
Height of eye = 100 mm
Scale 1:1

Solution

1. Draw the perspective layout and find the nearest point A (Refer to Fig. 7.33(b)).

2. Join A to VP 30°.
3. Draw a line from MP 30 through A to touch the GL at point o.
4. From o measure 70 mm along the GL to point p.
5. Draw a line from p to MP 30 to locate point B.
6. Join A to VP 60°.
7. Draw a line from MP 60 through A to touch the GL at point r.
8. From r measure 50 mm along the GL to points.
9. Draw a line from s to MP 60 to locate point D.
10. Join B to VP 60° and D to VP 30° to complete the perspective view.

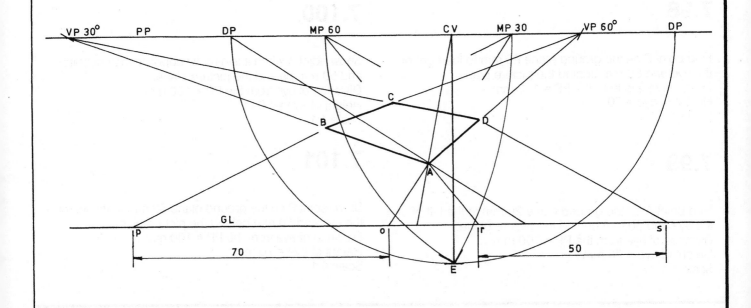

7.103

Show in perspective an equilateral triangle, side 100 cm, resting on the ground plane. The nearest corner is 20 cm to the right of the eye and 20 cm behind the picture plane. An edge from this point recedes to the left at 45° from the picture plane.

Distance of eye from the PP = 140 cm
Height of eye = 100 cm
Scale 1:20

Solution

1. Draw the perspective layout and find the nearest point A.
2. Join A to VP 45° left.
3. Draw a line from MP 45 left through A to touch the GL at a.
4. From a measure 100 cm along the GL to point b.
5. Draw a line from b to MP 45 left to locate point B.
6. Below the GL on ab construct an equilateral triangle abc. Draw the altitude co.
7. Draw a line from o to MP 45 left to meet the line AB at O.
8. Join O to VP 45° right.
9. Draw a line from MP 45 right through O to touch the GL at o'.
10. From o' set off the distance o'c' equal to oc.
11. Draw a line from c' to MP 45 right to locate point C.
12. Then ABC is the required triangle.

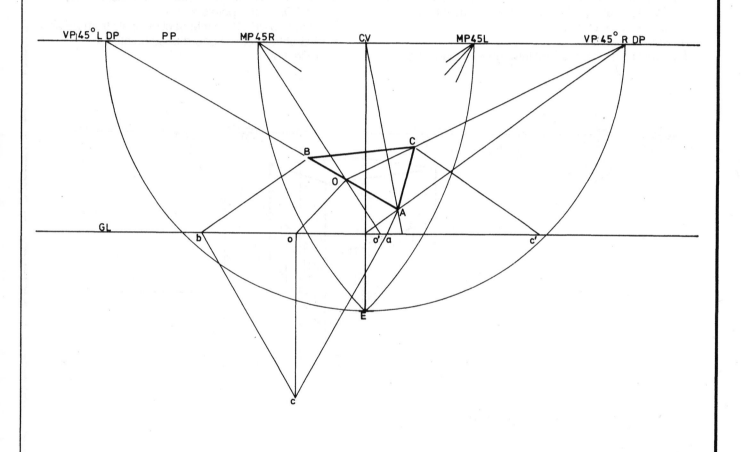

7.104

1. Represent in perspective a rectangular block 160 cm by 120 cm by 60 cm which rests on a 160 cm by 120 cm face on the ground plane with its 160 cm edges receding to the left at 30° from the picture plane. The nearest point is 20 cm to the left of the eye and 40 cm behind the picture plane.
Distance of eye from the PP = 140 cm
Height of eye = 100 cm
Scale 1:20

Solution
(a) Draw the perspective layout and find the nearest point A.
(b) Join A to VP 30°.
(c) Draw a line from MP 30 through A to touch the GL at o. Set off od equal to 160 cm. Join to MP 30 to locate point D.
(d) Join A to VP 60°.
(e) Draw a line from MP 60 through A to touch the GL at p. Set off pb equal to 120 cm. Join to MP 60 to locate point B.
(f) Join D to VP 60° and B to VP 30° to locate C.
(g) Produce DA to the GL at a and draw a vertical line. Set off ax equal to 60 cm.
(h) Draw a line from x to VP 30° to meet verticals from A and D to obtain E and H.
(i) Draw lines from E and H to VP 60° to meet verticals from B and C at F and G.

2. Show the reflection of the block in the ground plane as a mirror surface.

Solution
(a) Reflect the height line ax'.
(b) Join x' to VP 30° to intersect verticals from AE and DH at the points e and h.
(c) Join e to VP 60° to intersect verticals from BF and CG at the points f and g to complete the reflection.

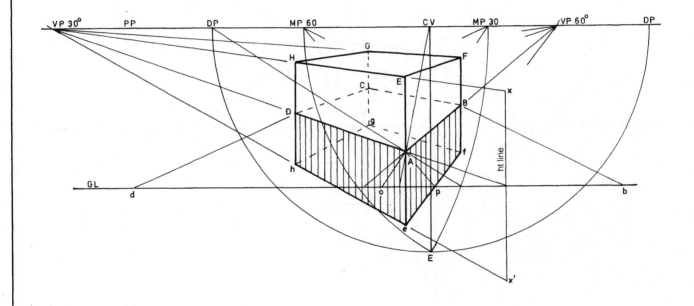

7.105

1. Represent in perspective an equilateral triangular prism, side 100 cm, axis 180 cm, when it rests on the ground plane on a rectangular face. The 180 cm edges recede to the right at 30° from the picture plane. Point P, the centre of the side AB, is 40 cm on the right of the eye and 60 cm behind the picture plane.
   Distance of eye from the PP = 140 cm
   Height of eye = 100 cm
   Scale 1:20

Solution
(a) Draw the perspective layout and find point P.
(b) Join P to VP 60° and extend to the GL at point o.
(c) Draw a line from MP 60 through P to touch the GL at p. Set off 50 cm either side of p to points r and s. Join to MP 60 to locate A and B.
(d) Draw a vertical line from o and set off ox equal to pc'. Join x to VP 60° to intersect a vertical from P at point C.

(e) Join points AB and C to complete one end of the prism.
(f) Draw lines from A, P, B and C to VP 30°.
(g) From MP 30 draw a line through P to touch the GL at t. Set off tv equal to 180 cm. Join t to MP 30 to locate R.
(h) Draw a line through R to VP 30° to locate D and E. A vertical line from R to intersect the line from C to VP 30° will locate point F.
(i) Join all points to complete the prism.

2. Show the reflection of the prism in the ground plane as a mirror surface.

Solution
(a) Reflect the height line ox'.
(b) Join x' to VP 60° to intersect a vertical from P at point c.
(c) Join c to VP 30° to intersect a vertical from R at point f.
(d) Join all points to complete the reflection of the prism.

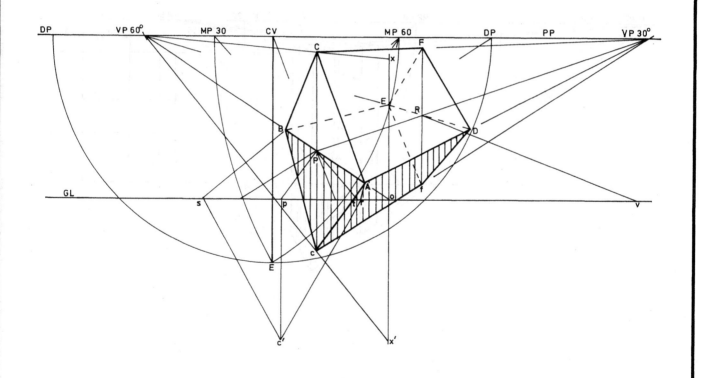

219

7.106

1. Represent in perspective a letter H cut from 30 mm material. The letter stands upright on the ground plane with its H-shaped faces contained in vertical planes that recede to the left at 30° from the picture plane. The nearest point is 10 mm on the left of the eye and 20 mm behind the picture plane.
Distance of eye from the PP = 130 mm
Height of eye = 100 mm
Scale 1:1

Solution

(a) Draw the perspective layout and find point A.
(b) Join A to VP 30°.
(c) Draw a line from MP 30 through A to touch the GL at r. Set off measurements representing the width at points s, t and v. Join to MP 30 to locate B, C and D.
(d) Produce the line from VP 30° through A to touch the GL at o.
(e) Draw a vertical line from o and set off the heights m, n and x.

(f) Draw lines from m, n and x to VP 30° to intersect verticals drawn from A, B, C and D to complete the front face of the H.
(g) Join A to VP 60°.
(h) Draw a line from MP 60 through A to touch the GL at p. Set off pq equal to 30 mm. Join q to MP 60 to locate l.
(i) Complete the back face of the H and line in, omitting dotted detail.

2. Show the reflection in the ground plane as a mirror surface.

Solution

(a) Reflect the height line om'n'x'.
(b) Join m', n' and x' to VP 30° to intersect verticals from AE, BF, CG and DH at points E, F, G and h.
(c) Join e to VP 60° to intersect a vertical from kl at point k.
(d) Join all points to complete the reflection.

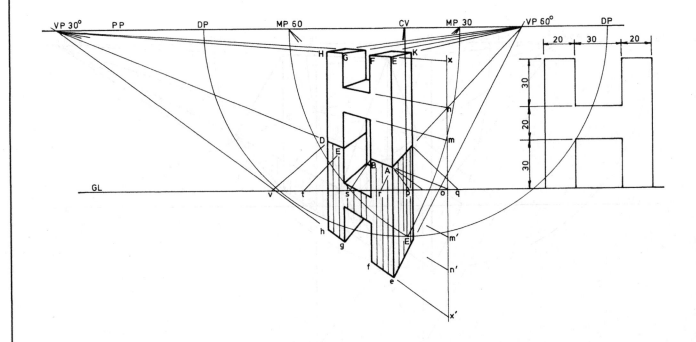

7.107

Represent in perspective a right square prism, side 80 cm, height 40 cm when it rests on the ground plane on a square face with a 80 cm by 40 cm surface parallel to and 20 cm behind the picture plane. The axis of the prism is directly opposite the eye.

Distance of eye from the PP = 120 cm
Height of eye = 80 cm
Scale 1:20

Solution

1. Draw the perspective layout. Measure 40 cm either side of the eye to locate points a and b and join to the CV.
2. From point a set off 20 cm to p. Join p to MP 90 to locate A.
3. Draw a horizontal line from A to intersect the line from b to the CV to locate B.
4. Draw a line from b to MP 90 to intersect the line from a to the CV to locate D.
5. Draw a horizontal line from D to intersect the line from b to the CV to locate C. ABCD is the square face on the GP.
6. Draw a height line from b and measure 40 cm to r. Join r to the CV.
7. Draw vertical and horizontal lines to complete the prism.

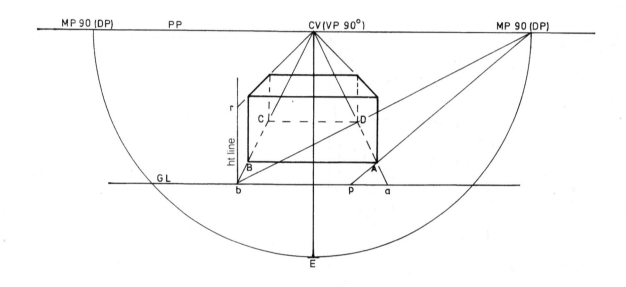

7.108

Represent in perspective a hexagonal prism, side 60 cm, axis 40 cm, when it rests on the ground plane on a hexagonal face with two edges parallel to the picture plane. The nearest point is 40 cm on the right of the eye and 20 cm behind the picture plane. The prism is on the left of this point.
Distance of eye from the PP = 160 cm
Height of eye = 100 cm
Scale 1:20

Solution
1. Draw the perspective layout and find the nearest point A.
2. Draw a line from the CV through A to touch the GL at a. Set off 60 cm to point b. Join b to the CV to intersect a horizontal line drawn from A to locate B.
3. Construct an aid on the side ab. Project s and t to the GL at c and f and join to the CV.
4. Draw a line from MP 90 through A to touch the GL at h. Set off hg and ge equal to ax.
5. Join g and e to MP 90 to intersect the line from A to CV at points g' and E.
6. Through g' draw a parallel to AB to intersect at the points C and F.
7. Through E draw a line parallel to AB to intersect the line from B to CV at D.
8. Join all points to complete the hexagonal face on the GP.
9. Erect a height line at point C and complete the perspective view of the hexagonal prism.

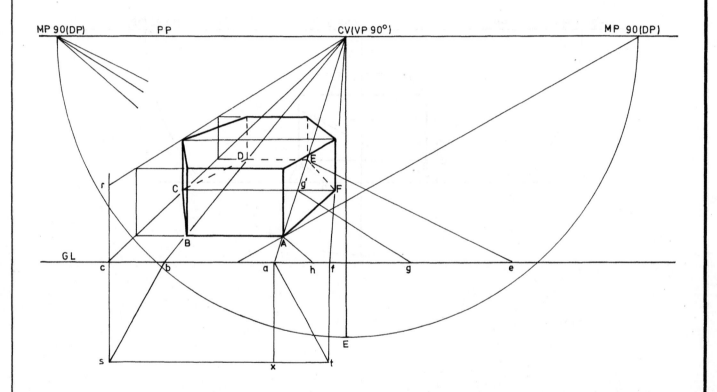

7.109

Show in perspective a line AB, 160 cm in length, which rests on the ground plane and recedes to the left at 30°. The nearest point is 40 cm to the right of the eye and 100 cm behind the picture plane.
Distance of eye from the PP = 160 cm
Height of eye = 100 cm
Scale 1:20

7.110

Show in perspective a line ST, 140 cm in length, which rests on the ground plane and recedes to the right at 40°. The nearest point is 60 cm to the left of the eye and 40 cm behind the picture plane.
Distance of eye from the PP = 140 cm
Height of eye = 100 cm
Scale 1:20

7.111

Show in perspective a line RS, 200 cm in length, which rests on the ground plane and recedes to the left at 45°. The nearest point is 100 cm to the right of the eye and 40 cm behind the picture plane.
Distance of eye from the PP = 180 cm
Height of eye = 120 cm
Scale 1:20

7.112

Show in perspective a line PQ, 140 cm in length, which rests on the ground plane and recedes to the right at 40°. The nearest point is 20 cm to the right of the eye and 80 cm behind the picture plane.
Distance of eye from the PP = 140 cm
Height of eye = 80 cm
Scale 1:20

7.113

Represent in perspective a rectangle 240 cm × 160 cm resting on the ground plane with its 240 cm edges receding to the left at 45° to the picture plane. The nearest point is 60 cm to the right of the eye and 60 cm behind the picture plane.
Distance of eye from the PP = 240 cm
Height of eye = 180 cm
Scale 1:20

7.114

Represent in perspective a rectangle 180 cm × 120 cm resting on the ground plane with its 180 cm edges receding to the left at 40° to the picture plane. The nearest point is directly opposite the spectator and 40 cm behind the picture plane.
Distance of eye from the PP = 240 cm
Height of eye = 180 cm
Scale 1:20

7.115

Represent in perspective a circle, diameter 80 mm, when it rests on the ground plane with its centre opposite the spectator and 90 mm behind the picture plane. A diameter from this point recedes to the right at 40° to the picture plane.
Distance of eye from the PP = 120 mm
Height of eye = 90 mm
Scale 1:1

7.116

Represent in perspective a right square pyramid, side 160 cm, axis 160 cm, when it rests on the ground plane with its nearest corner 100 cm to the left of the eye and 40 cm behind the picture plane. An edge from this point recedes to the left at 30° from the picture plane.
    Show the reflection in the ground plane as a mirror surface.
Distance of eye from the PP = 240 cm
Height of eye = 180 cm
Scale 1:20

7.117

Show in perspective a square slab, side 50 mm, height 20 mm, surmounted by a right square pyramid, side 50 mm, axis 30 mm. The edges of the solids make equal angles with the picture plane and the nearest point is opposite the spectator and 10 mm behind the picture plane.
Distance of eye from the PP = 120 mm
Height of eye = 90 mm
Scale 1:1

7.118

Represent in perspective a right hexagonal pyramid, side 60 cm, axis 120 cm, when it rests on the ground plane with two edges of the base receding to the right at 50° to the picture plane. The nearest corner is 30 cm on the left of the eye and 60 cm behind the picture plane.
    Show the reflection in the ground plane as a mirror surface.
Distance of eye from the PP = 240 cm
Height of eye = 180 cm
Scale 1:20

7.119

Show in perspective a hexagonal prism, side 60 cm, axis 200 cm, when it rests on the ground plane on a 200 cm edge that recedes to the right at 40°. The nearest point is 40 cm to the right of the eye and 50 cm behind the picture plane. A diagonal from this point is at right angles to the ground plane.
    Show the reflection in the ground plane as a mirror surface.
Distance of eye from the PP = 300 cm
Height of eye = 210 cm
Scale 1:20

7.120

Show in perspective a cylinder, diameter 120 cm, axis 200 cm, when it rests on the ground plane on its curved surface. The axis recedes to the left at 30° from the picture plane. The nearest point is 60 cm on the left of the eye and 100 cm behind the picture plane.
Distance of eye from the PP = 240 cm
Height of eye = 180 cm
Scale 1:20

7.121

The illustration shows the top and front views of an angle bracket. Draw the perspective view of the bracket when it rests on the ground plane with an edge AB parallel to and touching the picture plane. The nearest point A is 50 mm to the left of the spectator.
Distance of eye from the PP = 100 mm
Height of eye = 80 mm
Scale 1:1

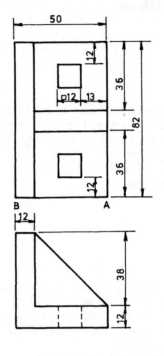

7.122

The illustration shows the top and front views of an angle bracket. Represent the bracket in perspective when it rests on the ground plane with an edge AB parallel to and touching the picture plane. The nearest point A is 40 mm on the right of the spectator.
Distance of eye from the PP = 100 mm
Height of eye = 80 mm
Scale 1:1

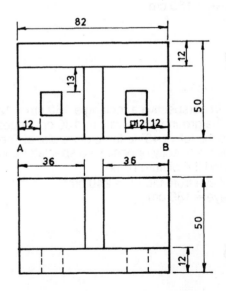

7.123

Represent the angle bracket shown in Exercise 7.122 in perspective when it rests on the ground plane with its longer edges receding to the left at 45° to the picture plane. The nearest point B touches the picture plane and is 20 mm on the right of the spectator.
Distance of eye from the PP = 110 mm
Height of eye = 80 mm
Scale 1:1

224

7.124

The illustration shows the top and front views of a shaped block. Represent the block in perspective when it rests on the ground plane with its longer edges receding to the right at 30° to the picture plane. Point A is 10 mm behind the picture plane and 15 mm on the right of the spectator.

Show the reflection in the ground plane as a mirror surface.

Distance of eye from the PP = 100 mm
Height of eye = 80 mm
Scale 1:1

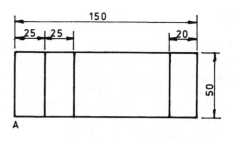

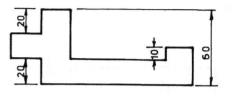

7.125

The illustration shows the top and front views of a shaped block. Draw the perspective view of the block when it rests on the ground plane with its longer edges receding to the left at 30° to the picture plane. The nearest point A is opposite the spectator and 10 mm behind the picture plane.

Show the reflection in the ground plane as a mirror surface.

Distance of eye from the PP = 130 mm
Height of eye = 110 mm
Scale 1:1

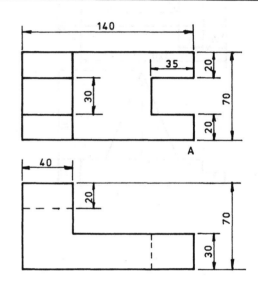

7.126

The illustration shows the top and front views of a machined block. Draw the perspective view of the block when it rests on the ground plane with its longer edges receding to the left at 45° to the picture plane. The nearest point A touches the picture plane and is 40 mm to the right of the spectator.

Distance of eye from the PP = 140 mm
Height of eye = 120 mm
Scale 1:1

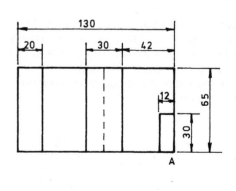

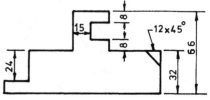

7.127

The illustration shows a rectangular slab surmounted by a right square pyramid. Represent these solids in perspective when the longer edges of the rectangular slab recede to the right at 30° to the picture plane. The nearest point A is 20 mm on the right of the spectator and 10 mm behind the picture plane.

Distance of eye from the PP = 130 mm
Height of eye = 100 mm
Scale 1:1

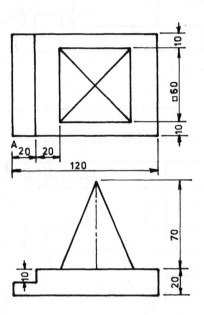

7.128

The illustration shows the top and front views of a modern building. Represent this building in perspective when the length of the building is parallel to the picture plane. The nearest point A is 25 mm on the right of the spectator and 10 mm behind the picture plane.

Distance of eye from the PP = 140 mm
Height of eye = 120 mm
Scale 1:1

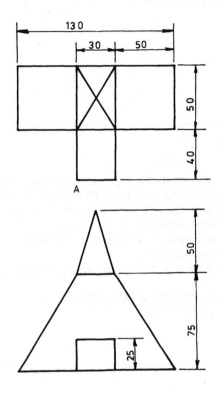

7.129

The illustration gives the left side view and front view of an 'A' frame display building. Draw the building in perspective when the longer edges recede to the right at 30° to the picture plane. The nearest point A is 20 mm on the left of the spectator and 30 mm behind the picture plane.

Distance of eye from the PP = 110 mm
Height of eye = 70 mm
Scale 1:1

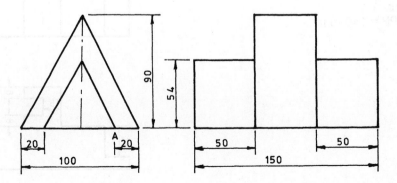

7.130

The illustration shows the top and front views of a letter T. Draw the letter T in perspective when the T-shaped faces are contained in vertical planes that recede to the left from the picture plane at 30°. The nearest point A is 20 mm to the left of the spectator and 40 mm behind the picture plane. Show all dotted detail.

Also show the reflection in the ground plane as a mirror surface.

Distance of eye from the PP = 130 mm
Height of eye = 100 mm
Scale 1:1

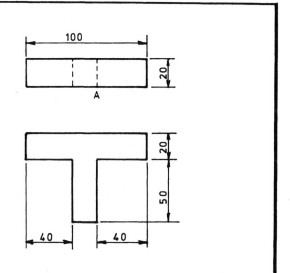

Working drawings 8

Detail and assembly drawing

During the design process an engineer records ideas by means of sketches and design drawings of prototypes and their development. Once engineer and client are satisfied with the degree of perfection, the sketches are handed over to the drafter who 'takes off' the detail and makes working drawings of the whole unit.

A set of working drawings for a machine would include detail drawings of the various parts and an assembly drawing showing how these parts are assembled to make up the complete machine.

Detail drawings

The detail drawing is used as the main reference in the manufacture of individual components. It should contain sufficient information to manufacture the part as well as suitable, fully dimensioned orthogonal views of each part, together with other information that may be required in the manufacturing process. A complete detail drawing should contain at least the following information (not necessarily in order of importance):

1. sufficient orthogonal views of the part concerned
2. dimensions and instructional notes
3. scale used
4. projection used, for example first or third angle
5. drafting standard reference, for example AS 1100 Part 101
6. name or title of drawing
7. relevant dimensional units
8. tolerances where necessary
9. surface finish requirements
10. special treatments needed (e.g. heat, metallic coatings, paint)
11. reference to a particular assembly, if applicable
12. type of material used
13. names of drafter, checker, approver and so on
14. relevant dates of action by those concerned
15. zone reference system when necessary
16. revisions or modifications
17. drawing sheet size
18. name of company or department, as applicable
19. drawing sheet reference, for example sheet 1 of 2

It is preferable to draw only one item on a single drawing sheet, the sheet size depending on the dimensions and number of views required. However, there are instances when multidetail drawings are used. Many of the exercises in this section are multidetail drawings as the individual parts are simple and it is more convenient to group them on one sheet.

It is common practice for firms to print their own drawing sheets with a drawing frame and title block in order to standardise the general information provided and to ensure that such information is included on all drawings. Figure 8.1 illustrates the layout of three separate detail drawings of parts of a machine screw jack.

While the title block is shown in the bottom right-hand corner (the preferable location) in Figure 8.1, AS 1100 Part 101 also provides that the title block may be located in the top right-hand corner and the revisions table in the top left-hand corner, when that is convenient for drawing layout.

In all cases the drawing number may be repeated in other corners or along the sides of the sheet to ensure that it is visible when the drawing is filed or folded.

A check of the drawings in Figure 8.1 against the list above will illustrate the points raised. Note that each of these drawings is referenced to the assembly drawing of the jack, shown in Figure 8.2.

Each drawing in Figure 8.1 was originally issued on 9.4.86, but a revision was carried out to the thread on the 'Spindle' and 'Jack body' drawings, changing from a Whitworth to a metric thread form. These revisions have been inserted on the drawing, and a record of them tabulated in the top right-hand corner on 1.6.87. Minor modifications to components are an everyday occurrence in a drawing office, and when such a modification does not affect the interchangeability of a part, the revision may be carried out on the old drawing. Where interchangeability is affected, a new drawing number should be raised.

Assembly drawings

Assembly drawings are primarily used to show how a number of components are fitted together to make a complete product unit. The term *subassembly* is commonly applied to a product unit that combines with other subassemblies to make an assembly. For example, an assembly drawing of a motor car engine would show a number of complete units, such as the distributor, generator, carburettor and so on. Each of these units is referred to as a subassembly of the engine assembly.

Assembly drawings may be divided into two categories depending on the proposed use:

1. *general assembly*—where the main purpose is to identify the individual components and show their working relationship
2. *working* or *detailed assembly*—a combined detail and general assembly drawing that fulfils the function of both types

Figure 8.2 is a general assembly of a machine screw jack. Detailed drawings of the individual parts of the jack are shown in Figure 8.1

Features of a *general assembly drawing* are:

1. Views that show how the parts fit together and how the unit may function are selected.
2. These views are often sectional views, which also eliminate the use of hidden detail lines where possible.
3. Dimensions that relate to the function of the unit as a whole are indicated. For example, Figure 8.2 indicates the maximum and minimum operating heights of the jack.
4. Individual components are identified by the use of numbers contained in circles. They are connected by leaders to the related parts.
5. A parts list relates to the numbers on the drawing and identifies the component.
6. A revisions table is provided to record modifications to individual components that may occur from time to time.
7. Some assemblies may be so large that it is necessary to draw different views of the assembly on separate sheets.

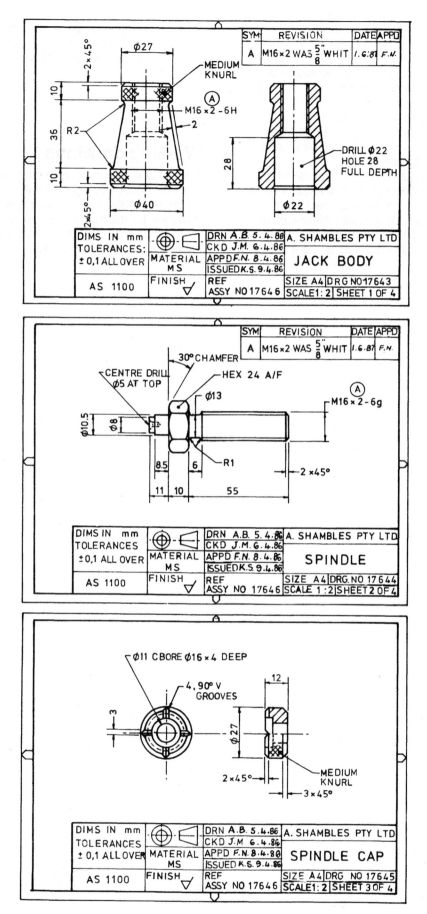

Fig. 8.1 *Detail drawings (sheet dimensions reduced to $\frac{A4}{3}$ approximately)*

Features of a *working assembly drawing* are:

1. Only simple assemblies are drawn in this manner, as views have to be chosen to show the assembly relationship as well as sufficient dimensional details of individual components to enable their manufacture.
2. These drawings are suitable for furniture construction drawings where the assembly views are not complex and details of joints may be enlarged and shown as partial views.

The information provided on a general assembly drawing is somewhat different from that required on a detail drawing. Information on the manufacture of individual parts such as surface finish, tolerances or treatments is not required. However, assembly instructions (see note in zone B2, Fig. 8.2) are required,

as are dimensions that may be used for installation purposes or that are relevant to the operation of the assembly as a working unit (see note in zone B4, Fig. 8.2).

Working drawings

A set of working drawings includes detail drawings of the individual parts together with an assembly drawing of the assembled unit. For example, a set of working drawings for the machine screw jack would include the three detail drawings shown in Figure 8.1 plus the assembly drawing, Figure 8.2.

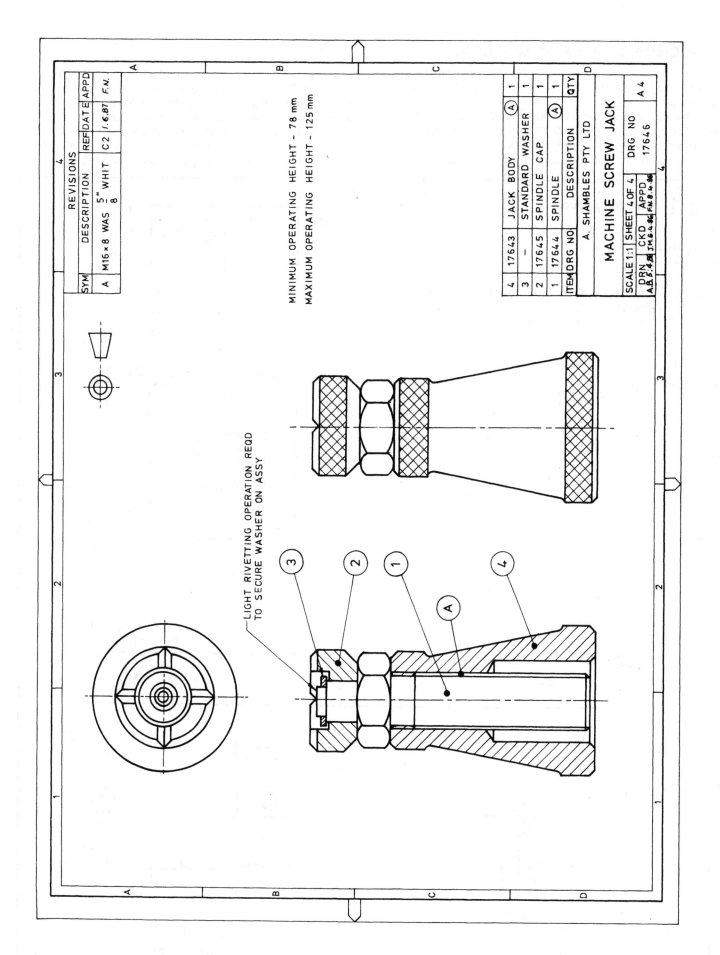

REVISIONS

SYM	DESCRIPTION	REF	DATE	APPD
A	M16 × 8 WAS $\frac{5}{8}$" WHIT	C2	1.6.87	F.N.

MINIMUM OPERATING HEIGHT – 78 mm
MAXIMUM OPERATING HEIGHT – 125 mm

4	17643	JACK BODY	Ⓐ	1
3	–	STANDARD WASHER		1
2	17645	SPINDLE CAP		1
1	17644	SPINDLE	Ⓐ	1
ITEM	DRG NO	DESCRIPTION		QTY

A. SHAMBLES PTY LTD

MACHINE SCREW JACK

SCALE 1:1	SHEET 4 OF 4	DRG NO
DRN	CKD	APPD	17646
A.B. 5.4.86	J.M. 6.4.86	F.N. 8.4.86	A 4

LIGHT RIVETTING OPERATION REQD
TO SECURE WASHER ON ASSY

Fig. 8.2 *Assembly drawing*

233

Exercises on working drawings

The exercises in this section are intended to provide the student with practice in producing detail and assembly drawings.

Standard size drawing frames should be used, along with standard title blocks and material lists. The layout of views within the frame area is an important consideration; it should be planned by the student and approved by the instructor before the drawing is commenced.

Dimensioning may be unidirectional or aligned as required. Surface finish requirements and tolerances have been omitted from the examples for convenience; where they are required they may be assessed in consultation with the instructor. The sheets of details show the quantities of parts required for one assembly. Such information is normally provided in a parts list.

8.1 Marking gauge

Details of the component parts of a marking gauge are given. Using them, draw the following general assembly views on a standard metric size sheet in third-angle projection:
1. a sectional front view
2. a side view

Scale 1:1

Provide a title block and material list.

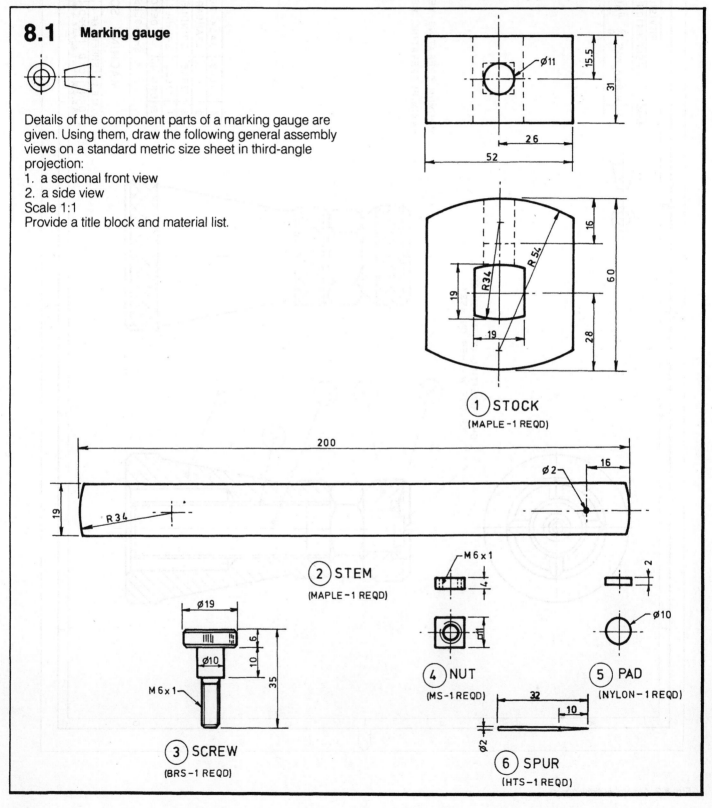

(1) STOCK
(MAPLE – 1 REQD)

(2) STEM
(MAPLE – 1 REQD)

(3) SCREW
(BRS – 1 REQD)

(4) NUT
(MS – 1 REQD)

(5) PAD
(NYLON – 1 REQD)

(6) SPUR
(HTS – 1 REQD)

8.2 Sliding bevel

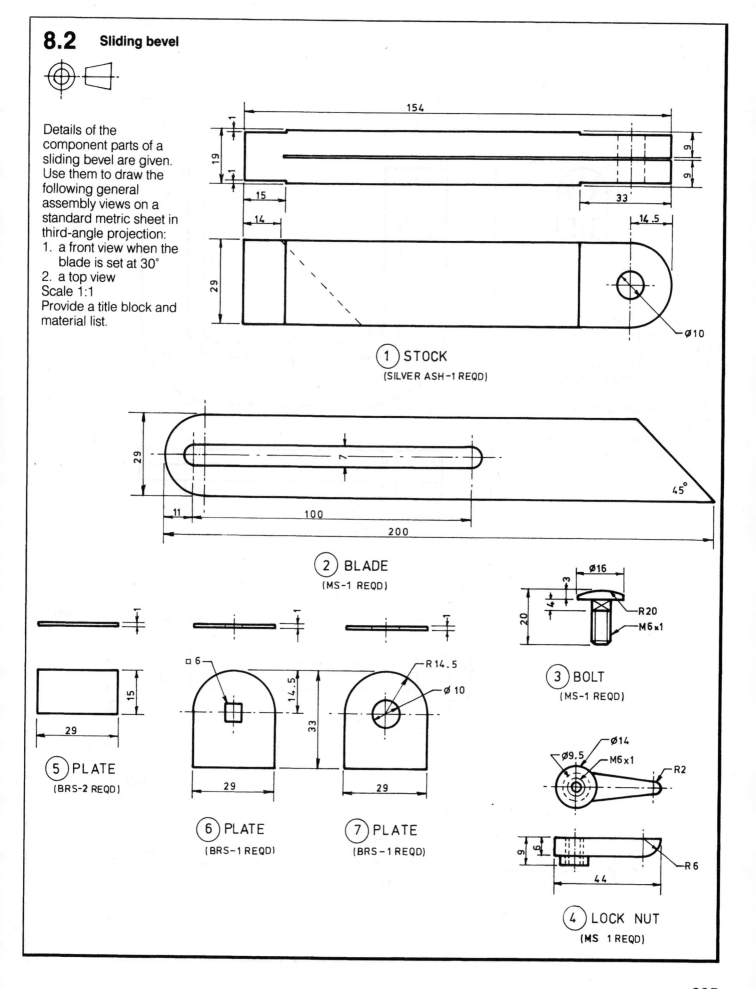

Details of the component parts of a sliding bevel are given. Use them to draw the following general assembly views on a standard metric sheet in third-angle projection:
1. a front view when the blade is set at 30°
2. a top view
Scale 1:1
Provide a title block and material list.

1 STOCK
(SILVER ASH–1 REQD)

2 BLADE
(MS–1 REQD)

3 BOLT
(MS–1 REQD)

5 PLATE
(BRS–2 REQD)

6 PLATE
(BRS–1 REQD)

7 PLATE
(BRS–1 REQD)

4 LOCK NUT
(MS 1 REQD)

235

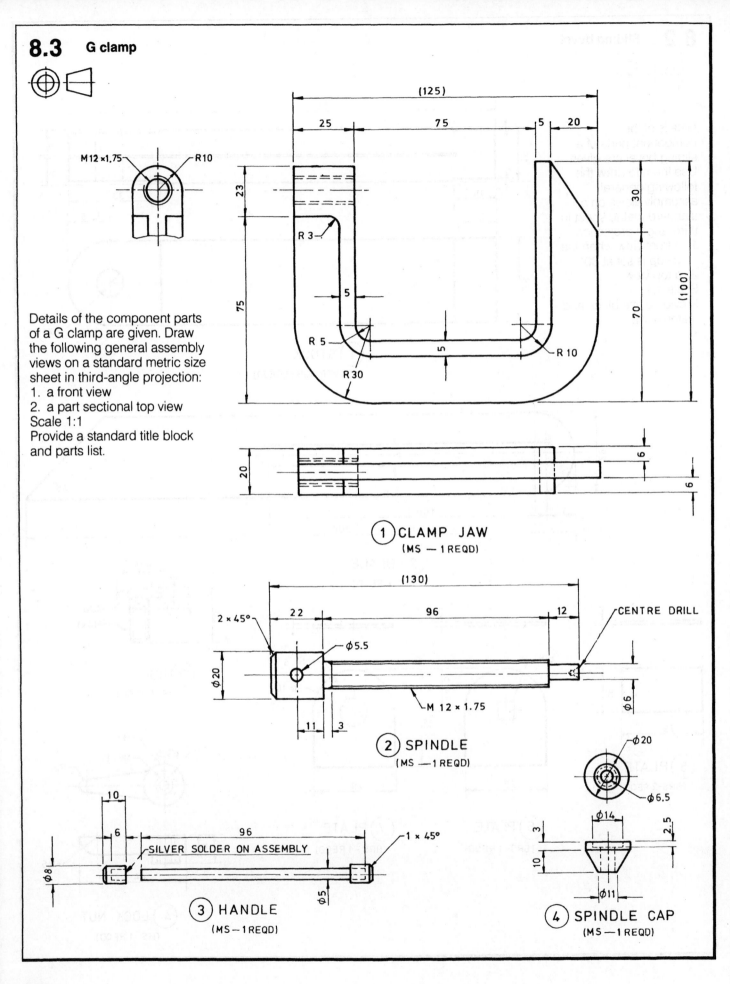

8.3 G clamp

M12×1.75 — R10

Details of the component parts of a G clamp are given. Draw the following general assembly views on a standard metric size sheet in third-angle projection:
1. a front view
2. a part sectional top view
Scale 1:1
Provide a standard title block and parts list.

(125)

25   75   5   20

23

R 3

75

5

R 5

5

R 30

R 10

30

(100)

70

20

6

6

① CLAMP JAW
(MS — 1 REQD)

(130)

2 × 45°

22   96   12

CENTRE DRILL

ø20

ø5.5

M 12 × 1.75

ø6

11   3

② SPINDLE
(MS — 1 REQD)

10

6

96

1 × 45°

SILVER SOLDER ON ASSEMBLY

ø8

ø5

③ HANDLE
(MS — 1 REQD)

ø20

ø6.5

3

ø14

2.5

10

ø11

④ SPINDLE CAP
(MS — 1 REQD)

236

8.4 Plumb-bob

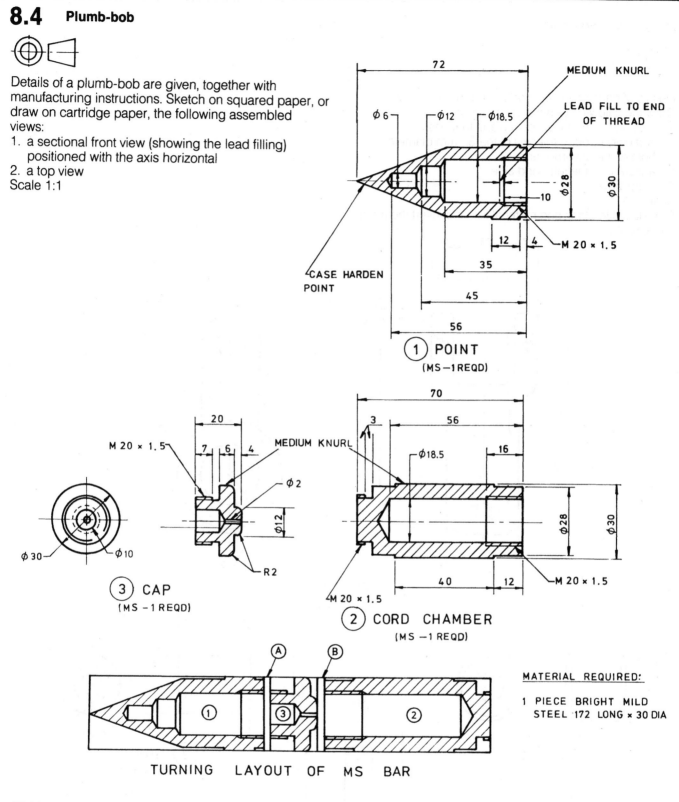

Details of a plumb-bob are given, together with manufacturing instructions. Sketch on squared paper, or draw on cartridge paper, the following assembled views:
1. a sectional front view (showing the lead filling) positioned with the axis horizontal
2. a top view

Scale 1:1

① POINT
(MS –1 REQD)

MEDIUM KNURL

LEAD FILL TO END OF THREAD

CASE HARDEN POINT

φ6   φ12   φ18.5   φ28   φ30   10   M 20 × 1.5

72   12   4   35   45   56

③ CAP
(MS –1 REQD)

M 20 × 1.5   MEDIUM KNURL   φ2   φ12   R2

20   7   6   4   φ30   φ10

② CORD CHAMBER
(MS –1 REQD)

φ18.5   M 20 × 1.5   M 20 × 1.5

70   3   56   16   φ28   φ30   40   12

TURNING LAYOUT OF MS BAR

Ⓐ   Ⓑ

①   ③   ②

MATERIAL REQUIRED:

1 PIECE BRIGHT MILD STEEL 172 LONG × 30 DIA

PROCEDURE:

1. TURN END AND BODY OF ②

2. PART OFF AT Ⓐ

3. DRILL AND TAP ① AND SCREW ② INTO.

4. TURN END OF ③ AND PART OFF AT Ⓑ

5. COMPLETE ②, SCREW ③ INTO AND FINISH

6. REVERSE JOB, TURN POINT AND KNURL

Details of parts of a governor arm are given below.
Draw the following general assembly views:
1. a sectional front view on a plane that passes vertically through the axis of the 32 mm diameter hole of the governor arm
2. a view from the left-hand side
3. a top view

Scale 1:1

Provide a title block and material list, and insert the main dimensions.

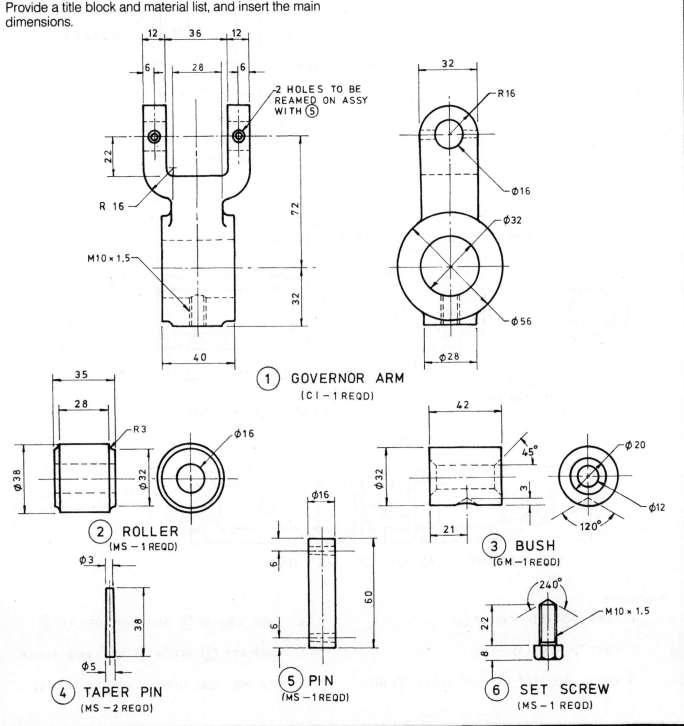

8.6 Machinist's jack

Details of components of a machinist's jack are given.
Draw the necessary views to show how these
components are assembled together. Two views should
be provided.
Scale 1:1
Use a standard size sheet and provide a title block and
material list.

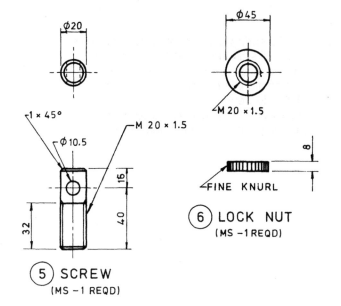

⌀20

1 × 45°
⌀10.5
M 20 × 1.5
16
40
32

⑤ SCREW
(MS – 1 REQD)

⌀45
M 20 × 1.5

8
FINE KNURL

⑥ LOCK NUT
(MS – 1 REQD)

TOLERANCES
±0.5 EXCEPT
AS STATED

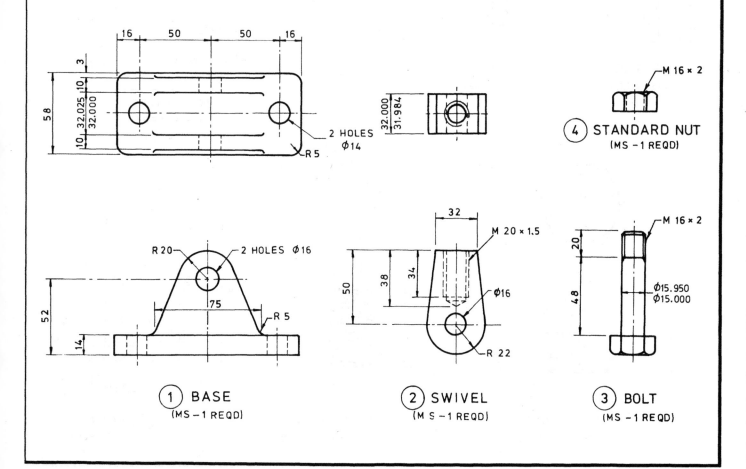

16   50   50   16
3
10
32.025
32.000
58
10

2 HOLES ⌀14
R 5

32.000
31.984

M 16 × 2

④ STANDARD NUT
(MS – 1 REQD)

R 20   2 HOLES ⌀16
75
R 5
52
14

① BASE
(MS – 1 REQD)

32
M 20 × 1.5
50
38
34
⌀16
R 22

② SWIVEL
(MS – 1 REQD)

M 16 × 2
20
48
⌀15.950
⌀15.000

③ BOLT
(MS – 1 REQD)

8.7 Spring-loaded safety valve

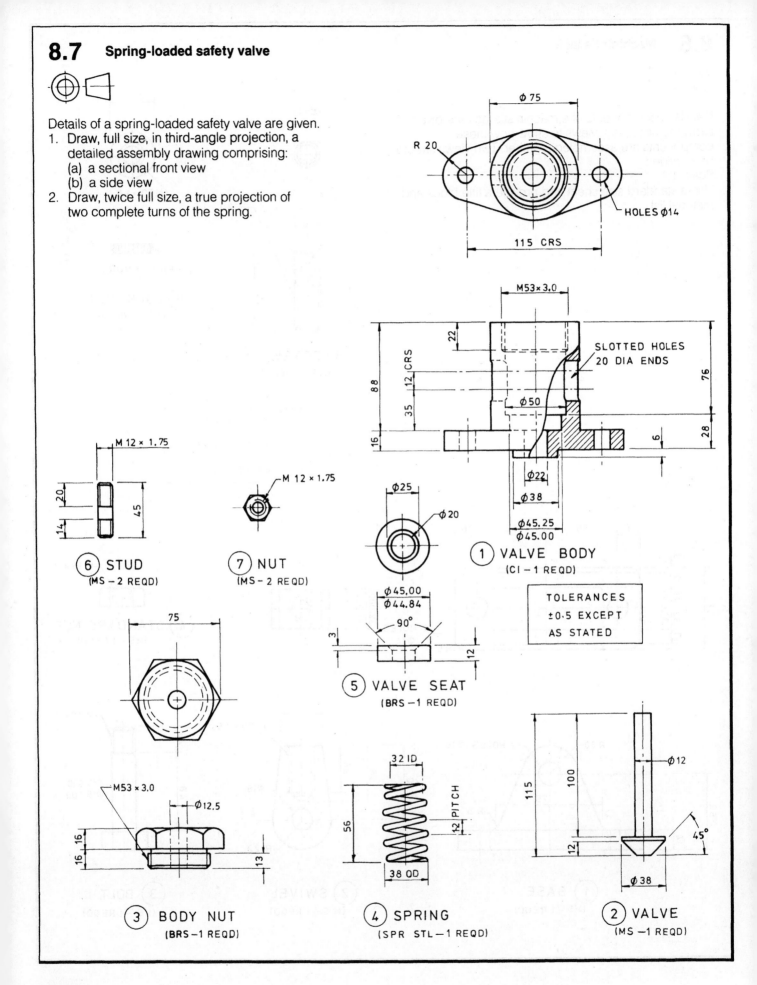

Details of a spring-loaded safety valve are given.
1. Draw, full size, in third-angle projection, a
   detailed assembly drawing comprising:
   (a) a sectional front view
   (b) a side view
2. Draw, twice full size, a true projection of
   two complete turns of the spring.

Ø 75

R 20

HOLES Ø14

115 CRS

M53 × 3.0

22

SLOTTED HOLES
20 DIA ENDS

88

12 CRS

35

76

Ø 50

16

28

6

Ø22

Ø 38

Ø45.25
Ø45.00

(1) VALVE BODY
(CI – 1 REQD)

M 12 × 1.75

20

45

14

(6) STUD
(MS – 2 REQD)

M 12 × 1.75

(7) NUT
(MS – 2 REQD)

Ø25

Ø 20

Ø 20

Ø45.00
Ø44.84

90°

3

12

(5) VALVE SEAT
(BRS – 1 REQD)

TOLERANCES
±0·5 EXCEPT
AS STATED

75

M53 × 3.0

Ø12.5

16 16

16

13

(3) BODY NUT
(BRS – 1 REQD)

32 ID

56

12 PITCH

38 OD

(4) SPRING
(SPR STL – 1 REQD)

115

100

Ø 12

12

45°

Ø 38

(2) VALVE
(MS – 1 REQD)

240

Details of a roller bracket are included below. Draw, full size, in third-angle projection, a general assembly drawing showing the following views:
1. a sectional front view on the centre line of the base
2. a top view

Provide also a standard title block and material list.

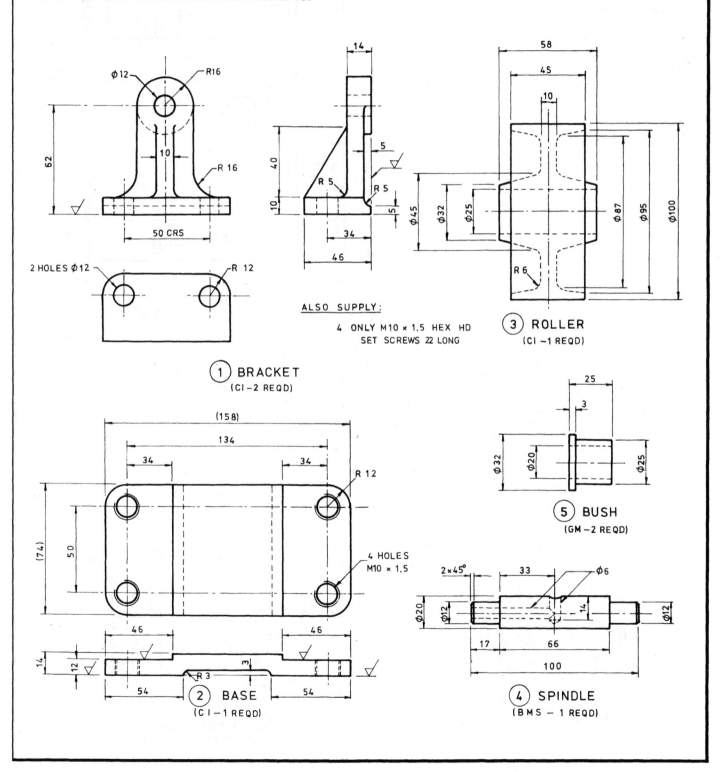

ALSO SUPPLY:

4 ONLY M10 × 1.5 HEX HD
SET SCREWS 22 LONG

① BRACKET
(CI–2 REQD)

③ ROLLER
(CI –1 REQD)

⑤ BUSH
(GM –2 REQD)

② BASE
(CI –1 REQD)

④ SPINDLE
(BMS – 1 REQD)

8.9 Machine vice

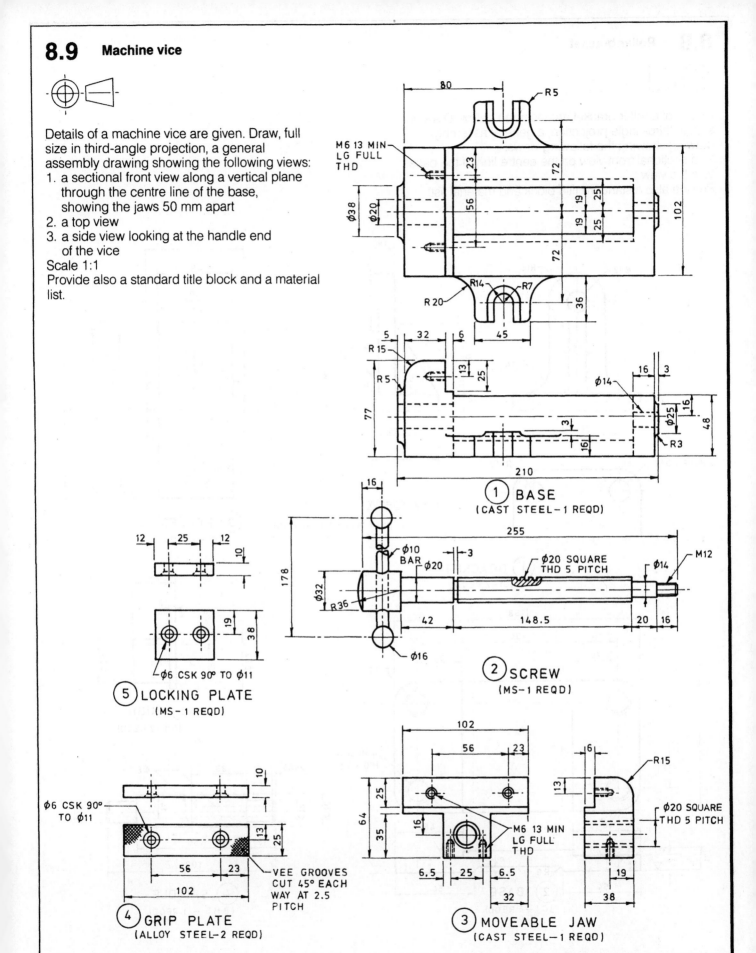

Details of a machine vice are given. Draw, full size in third-angle projection, a general assembly drawing showing the following views:
1. a sectional front view along a vertical plane through the centre line of the base, showing the jaws 50 mm apart
2. a top view
3. a side view looking at the handle end of the vice

Scale 1:1

Provide also a standard title block and a material list.

① BASE
(CAST STEEL – 1 REQD)

② SCREW
(MS – 1 REQD)

③ MOVEABLE JAW
(CAST STEEL – 1 REQD)

④ GRIP PLATE
(ALLOY STEEL – 2 REQD)

⑤ LOCKING PLATE
(MS – 1 REQD)

Surveying and setting out

Before civil engineering works such as highways, roads, dams and aerodromes can be designed, the design engineer must have drawings that accurately describe the natural state of the area. The data to produce these drawings is obtained by surveying.

This chapter describes the various methods surveyors can use to record data and the types of survey drawings produced from the data.

Introduction

Surveying may be defined as the technique of measuring a tract or area of land on the earth's surface and by means of a suitable scale transferring such measurements onto a drawing in order to produce a map or plan of the area. Surveying is concerned with the original investigation and collection of topographical and artificial data on a site in its natural or original state.

Setting out is the reverse process of surveying. Points on the ground are positioned according to predetermined data in order to lay out a city subdivision, a building site or other civil engineering project, such as a road, railway, dam or aerodrome. Setting out normally follows a survey when engineering designers have considered the survey data and then designed a course of action to achieve the final layout leading to the project completion. For example, in planning a new road, the centre line of the road and a strip each side is surveyed to obtain an accurate description of the rise and fall of the natural surface, the direction or line of traverse and other identifying natural features (creeks, rocky outcrops, timber and so on) necessary to enable the designers to set out the designed surface of the proposed roadway. Calculations of 'cut' and 'fill' are made so that earth is removed from high areas and deposited in low areas in order to make the final road surface as level as possible.

Survey classifications

Surveying may be classified according to the purpose of the survey or the techniques and equipment used.

Those classified according to their purpose include the following:

1. Cadastral surveys are concerned with the positioning of property boundaries and areas on plans for the purpose of determining land ownership and assigning official place names. This type of survey forms the basis of street directories and is commonly used by town planners, real estate agents and property buyers.
2. Geodetic surveys are highly accurate surveys which take into account the curvature of the earth's surface, and major topographical and artificial features, such as rivers, mountain ranges, roads and dams. They also provide fixed reference points for lower accuracy surveys of smaller areas.
3. Topographical surveys record the location and description of natural and existing man-made features in relation to the line of traverse of the survey.
4. Engineering surveys record preliminary data on a site before the design of an engineering project such as a road, dam, aerodrome or railway can take place.

Survey classifications based on techniques and the equipment used include the following:

1. Linear or chain surveying involves the preparation of survey plans by the use of linear measurements only. No horizontal angular measurements are taken.

With modern lightweight and inexpensive equipment readily available this technique is fast being superseded.

2. Theodolite surveying involves the accurate measurement of lengths and angles using a high precision instrument called a theodolite.
3. Aerial surveying is conducted from an aircraft equipped with special cameras capable of taking continuous photographs of the terrain below from a known altitude. Each photograph overlaps the last one so that a continuous map of the terrain can be formed. A technique known as photogrammetry can also be used. In photogrammetry pairs of photographs of the same area are taken from different positions and then analysed by an instrument called a stereoplotter. This enables detail such as houses, roads, rivers and contours to be plotted. (Contours are a series of imaginary continuous lines plotted on the earth's surface by connecting points of the same height above a reference level, for example sea level.) Aerial surveying is appropriate when surveying large, remote and inaccessible areas.
4. Satellite surveying utilises images of the earth's surface taken from outer space by sensors mounted on an orbiting satellite. One such satellite, known as Landsat, has been used to map state boundaries and to locate large natural resources, such as forests and rivers. The mining industry has developed special techniques for identifying areas on the earth's surface that contain particular ore-bearing deposits, such as iron and coal.

Surveying methods

When recording and plotting information from a survey, surveyors use field books and level books.

The *field book* is used by the surveyor to record major physical and topographical information along the line of traverse of a survey. The positions of houses, trees, fences, creeks, roads, tracks and other identifying landmarks are measured and recorded in relation to the line of traverse. Later a survey plan can be drawn from the information recorded in the field book.

The *level book* is used by the surveyor to record levels taken during the survey. These levels are referred to as 'reduced levels', that is, the heights of selected points on the ground relative to a given datum, usually taken as sea level. The level book enables drawings of the longitudinal and cross-sections of a survey to be drawn up.

Using the field book

The field book usually takes the form shown in Figure 9.1, which illustrates a single page of the book. The column in the centre of the page represents the traverse line within which the chainages (distances along the survey from the start, in metres) and bearings (direction of the survey relative to north measured in degrees, minutes and seconds, clockwise from north) are noted. Data is entered on the page beginning at the bottom and proceeding to the top.

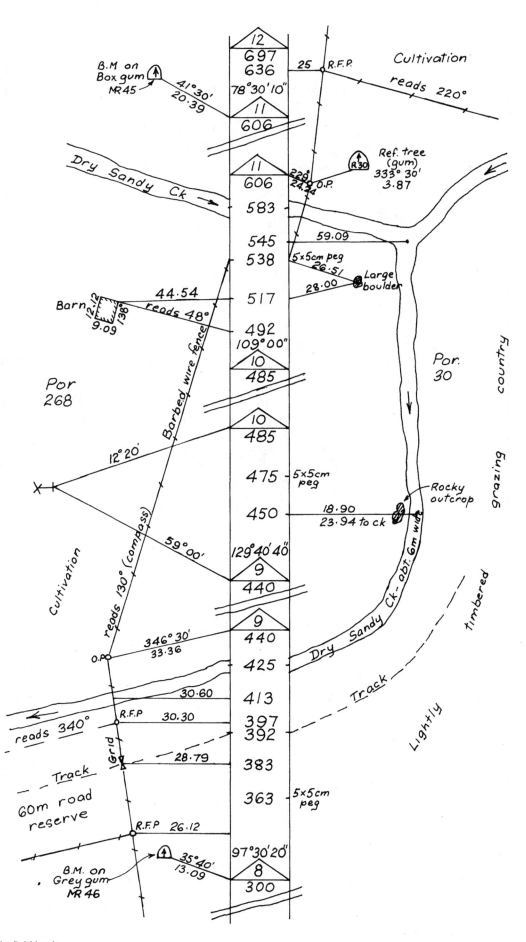

Fig. 9.1 *Page of a field book*

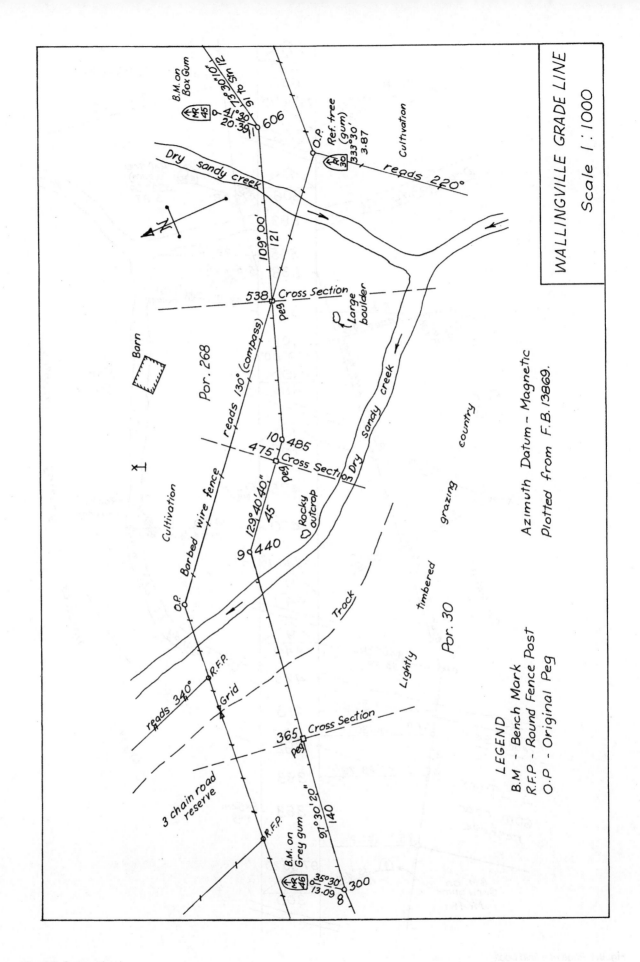

Fig. 9.2 *Survey plan*

All chainages in the centre column are reduced, that is, if two points are at different levels the recorded distance between them is the horizontal distance, which will have to be calculated. Data appearing on the right or left of the centre column corresponds to the features found on the right or left of the line of traverse of the survey. Features are located relative to the centre line in one of three ways.

1. Offsets are made at right angles to the centre line; usually they are shown in the field book by a light line drawn at right angles from the centre line chainage point out to the feature. The distance is written on the line, for example in Figure 9.1 the rocky outcrop and the centre of the creek are respectively 18.90 and 23.94 from chainage 450.

2. Where a feature crosses the line of traverse chainages are noted on the centre line. For example in Figure 9.1 Dry Sandy Creek crosses at chainage 425 and again at 583, and the fence crosses at chainage 538.

3. Distances or bearings are located from chainage points on the centre line, for example:
   (a) The windmill is located by two bearings, 12°20′ and 59° from chainage 485 (station 10) and 444 (station 9), respectively.
   (b) The barn is located by a distance to the corner (44.54) from chainage 517 and a bearing of 48° from chainage 492 in line with the end of the barn. The dimensions of the barn are also given.
   (c) The large boulder is located by two distances, 26.51 and 28.00 measured from chainages 538 and 517, respectively.

Where the line of traverse changes its bearing the column is intersected by two parallel lines and the chainage at the point of change is indicated on both sides of the parallel lines. To further clarify this point it may be given a progressive number located in a triangle immediately above the chainage. In Figure 9.1 there are four such bearing changes to the line of traverse at 8, 9, 10 and 11.

The survey plan

Using the data plotted by the surveyor in the field book, a civil engineering drafter is able to draw a survey plan. Figure 9.2 is the survey plan drawn from the data shown in Figure 9.1.

First, a suitable scale is selected, usually 1:1000 or 1:2000 for large areas and 1:500 for small areas. The drafter then positions the direction of north (N) such that the line of traverse is roughly left to right across the drawing sheet. Using the chainages and the bearings between station points, the line of traverse is plotted. The accompanying data to the right and left of the line of traverse are added according to the location prescribed in the field book. All data is entered and the drawing is finally given a title block and other peripheral information.

As previously stated, bearings are always plotted such that the angle is measured clockwise from north. Figure 9.3 shows five station points (17 to 21) with the method of measuring the bearing angle clearly indicated in each case.

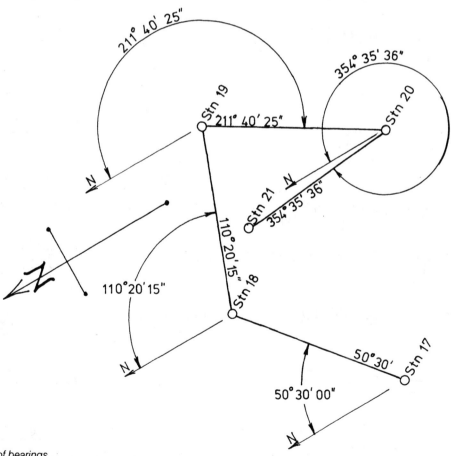

Fig. 9.3 *Determination of bearings*

Table 9.1 *Representation of common topographical features*

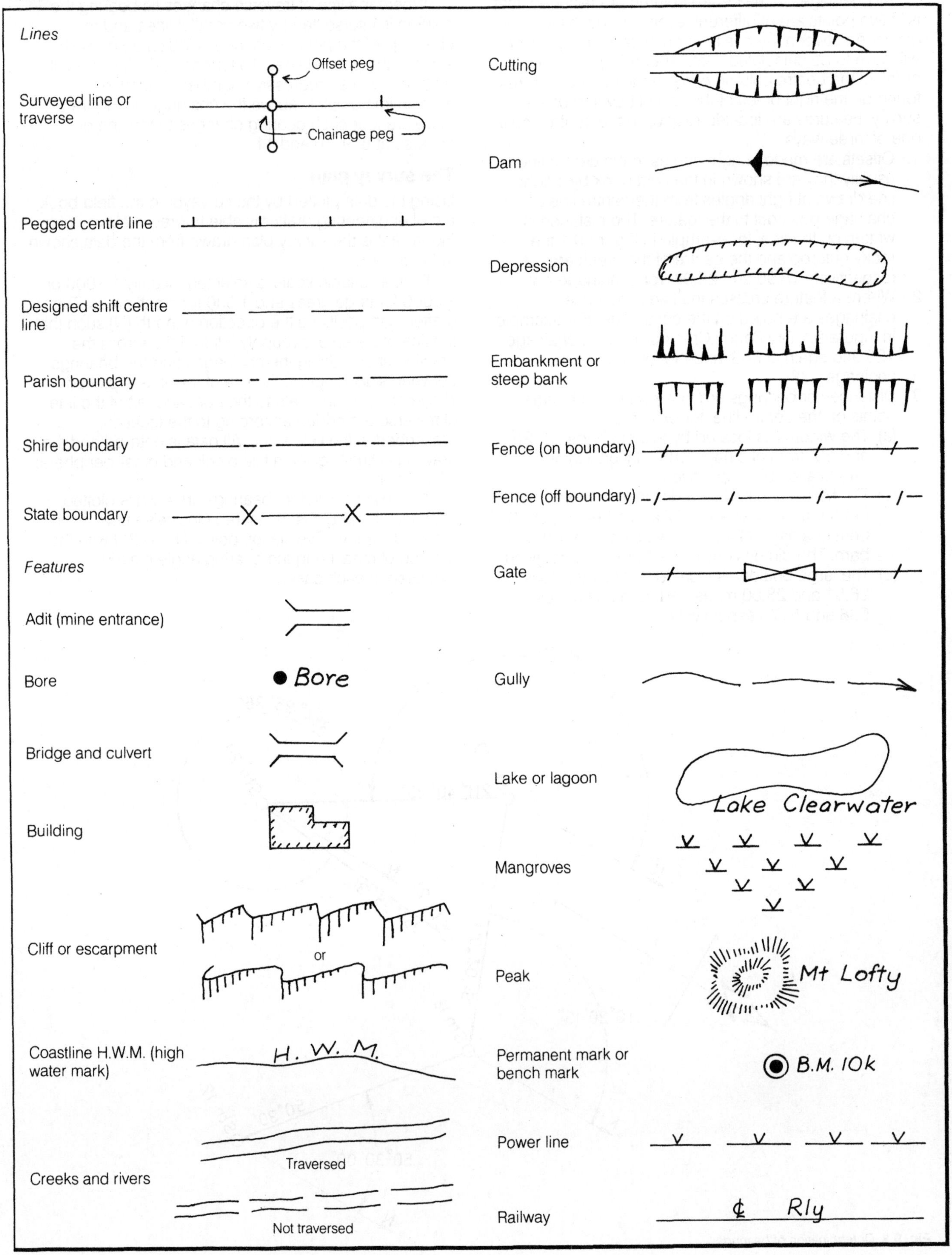

Lines

Surveyed line or traverse — Offset peg, Chainage peg

Pegged centre line

Designed shift centre line

Parish boundary

Shire boundary

State boundary —X—X—

Features

Adit (mine entrance)

Bore ● *Bore*

Bridge and culvert

Building

Cliff or escarpment   or

Coastline H.W.M. (high water mark) *H. W. M.*

Creeks and rivers Traversed Not traversed

Cutting

Dam

Depression

Embankment or steep bank

Fence (on boundary)

Fence (off boundary)

Gate

Gully

Lake or lagoon *Lake Clearwater*

Mangroves

Peak *Mt Lofty*

Permanent mark or bench mark ◉ *B.M. 10k*

Power line

Railway ₵ *Rly*

Table 9.1 *(continued)*

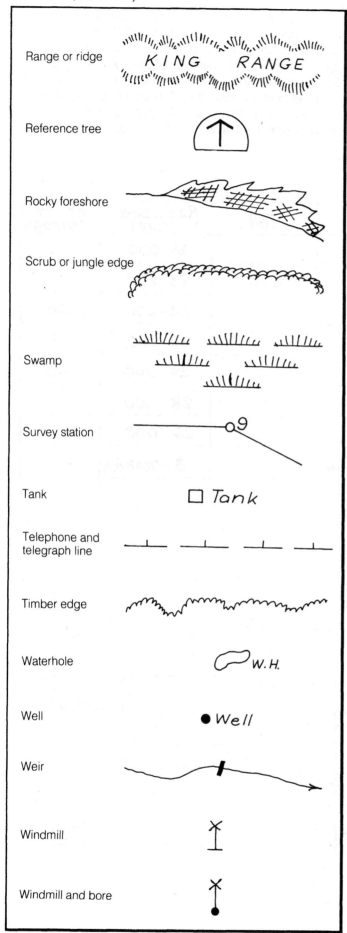

Range or ridge	*KING RANGE*
Reference tree	
Rocky foreshore	
Scrub or jungle edge	
Swamp	
Survey station	9
Tank	□ *Tank*
Telephone and telegraph line	
Timber edge	
Waterhole	*W.H.*
Well	● *Well*
Weir	
Windmill	
Windmill and bore	

Standard lines and features

When drawing survey plans it is essential that drafters use standard lines and features to ensure consistent interpretation of drawings. Table 9.1 shows the most commonly used symbols, and Table 9.2 gives a list of commonly used abbreviations.

Table 9.2 *Standard abbreviations*

Bench mark	B.M.
Permanent mark	P.M.
Recovery mark	R.M.
Tangent point	T.P.
Intersection point	I.P.
Edge of bitumen	E. of B.
Table drain	T.D.
Top of bank	T. of B.
Fence line	F.L.
Chainage	ch.
Centre line	¢
Stop valve	S.V.
Hydrant	Hyd.
Ground	Grd.
Reduced level	R.L.
Offset peg	O.P.
Original	Orig.
Station	Stn.
Round fence post	R.F.P.
Centre round fence post	C.R.F.P.
Field book	F.B.
Level book	L.B.
Natural surface (grd)	N.S.

Using the level book

The level book provides information that enables the surveyor to give an accurate picture of the slope and the rises and falls of the surface of the terrain along the line of traverse. It contains:

1. a series of levels taken along the traverse line from which a civil engineering drafter can plot a longitudinal section;
2. a series of levels taken at right angles on both sides of the traverse line to enable cross-sections to be taken at these points. In Figure 9.2 three cross-sections are to be taken, at chainages 365, 475 and 538.

Figure 9.4 shows a small part of a page from a level book. Figure 9.5 illustrates how the surveyor obtains these levels. The line of traverse proceeds from left to right. The surveyor may be using a dumpy, tilting or automatic level together with a 4 m staff graduated in centimetres. The survey begins at point O, at the base of the first staff position, and a reduced level of 25 m is assumed. Levels are to be recorded at 25 m intervals along the line of traverse. The level is positioned approximately midway between the staff positions O and A. A reading called the backsight is taken with the level focused on the staff at position O. A reading of 2.200 is noted and recorded in the backsight column of the level book (see Fig. 9.4). The level is rotated 180° to the staff's second position at A, which is 25 m (measured horizontally) further on. A reading of 4.000 is

Backsight	Foresight	Rise	Fall	Reduced level	Staff Chainage
2.200				25.000	0
3.500	4.000		1.800	23.200	25
3.750	3.250	0.250		23.450	50
3.500	1.500	2.250		25.700	75
	1.200	2.300		28.000	100
12.950	9.950	4.800		28.000	
9.950		1.800		25.000	
3.000(rise)		3.000(rise)		3.000(rise)	

Level checks { (for the last three rows)

Fig. 9.4 *Part of a page from a level book*

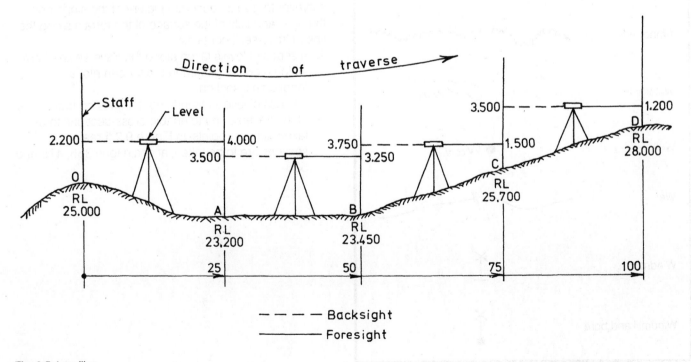

Fig. 9.5 *Levelling*

250

noted and recorded in the foresight column. The difference between these two readings indicates, in this case, a fall of 1.800 and this is recorded in the fall column. The fall is subtracted from the reduced level at chainage O (25.000) to give the reduced level at A of 23.200, which is entered in the reduced level column. The chainage at A (25.000) is noted in the distance column.

The staff is maintained at A while the level is taken further along the traverse to approximately midway between A and B. The procedure is repeated to obtain chainage row 50 and then chainages 75 and 100. Note that when the foresight is less than the backsight the ground 'rises', as is the case for chainage positions 50, 75 and 100, and the values are inserted in the rise column. These values are each added to the previous reduced level to give the new reduced level for the row of data concerned.

The level book may be checked for accuracy. For example, the difference between the sum of the backsight and foresight columns should be equal to the difference between the rise and fall columns, which in turn should be equal to the difference between the first staff position O and the last staff position D at chainage 100. These checks are shown at the foot of the level book page (see Fig. 9.4).

Plotting longitudinal sections

The longitudinal section is a graphical plot of all the reduced levels taken along the traverse line. When plotted, it gives the line of intersection of a vertical plane and the surface of the ground traversed. Figure 9.6(a) shows the survey plan of five points, O, A, B, C and D, which are 25 m apart. The longitudinal section (Fig. 9.6(b)) is plotted by erecting an ordinate diagram on a base line, marked in 5 m intervals at the same scale as the survey plan. It is normal to exaggerate the vertical scale of the ordinate diagram by ten times the horizontal scale. A suitable datum level is selected so that the longitudinal profile lies above the datum, in this case 23.000. The reduced level of point O is positioned on the left-hand side of the diagram and the remaining levels are plotted on their respective ordinates.

The example used determines levels at equal intervals along the traverse line. If the surveyor observes a significant rise in the terrain between two of these points, say B and C, an intermediate reading is taken with the staff located at the highest point of the rise. A row of level data for this reading is inserted in the level book. The point can then be plotted and the curve on the longitudinal section adjusted (see the dashed line in Fig. 9.6(b)).

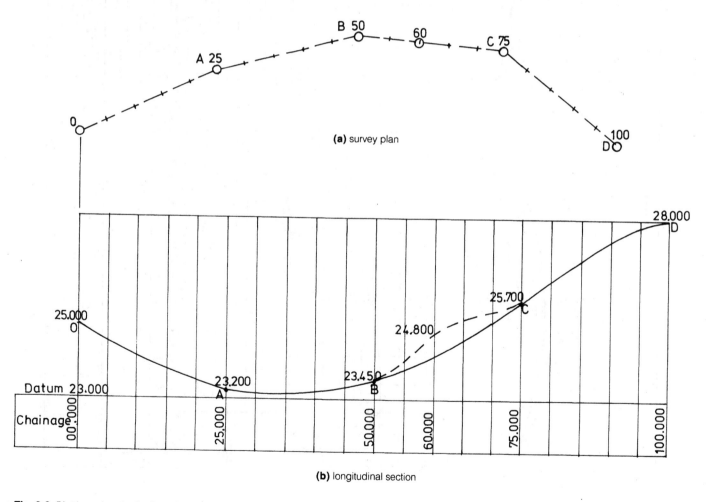

(a) survey plan

(b) longitudinal section

Fig. 9.6 *Plotting a longitudinal section*

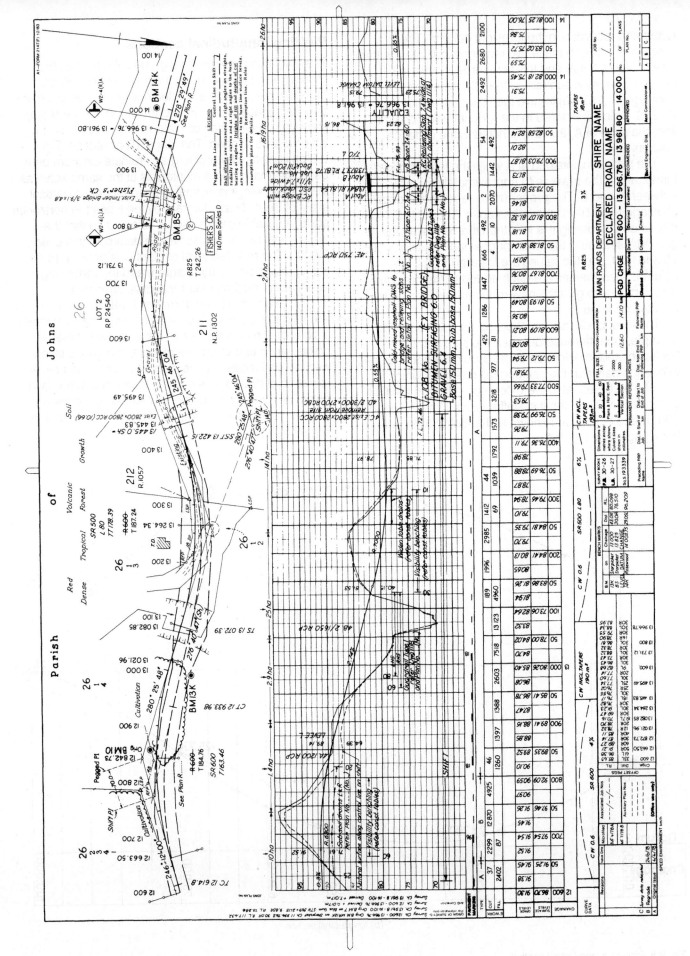

Fig. 9.7 Survey plan with longitudinal section   DEPT OF MAIN ROADS, QUEENSLAND

Study of an actual survey drawing

Figure 9.7 shows an actual survey plan and its accompanying longitudinal section, courtesy of the Department of Main Roads, Queensland. At the top is the survey plan, which shows the bearing lines (a light line pegged at bearing changes), the base line (a heavy line shown divided into 50 m segments) and a shift line (a heavy dash line), which commences at chainage 12 607 and merges with the base line at chainage 13 445. Other features to be identified on the survey plan are:

1. an existing gravel road
2. four bench marks
3. bridge over a creek
4. three steep banks or escarpments
5. cultivation
6. dense forest growth
7. total chainage covered by this drawing
8. name of the area through which the survey traverses
9. boundary fence
10. off boundary fence
11. telephone or telegraph line
12. direction of north

The longitudinal section is drawn below the survey plan with the left-hand end vertically in line with that of the plan. The longitudinal section is longer than the plan because it is equal to the stretched out length of the plan. Three surfaces are indicated on the section:
1. the natural surface along the base line (full line)
2. the natural surface along the shift line (dashed line)

3. the designed surface of the road with the gradients indicated (2.74% means that the road is designed to slope 2.74 m over a distance of 100 m measured horizontally in the direction of the road)

The longitudinal section also provides information on the removal or addition (cut or fill) of earth necessary to transform the natural surface along the shift line to that of the designed surface of the road. The amounts of cut and fill given in the row of data at the bottom of the longitudinal section should be compared with the two surfaces in question to appreciate this concept. Other data provided include surface levels (along the base line), grade levels (along the designed road surface), curve data for the survey plan showing curve radii and straight sections of the proposed road. Other information in the title block includes revision information, microfilm identification, scales, chainage units, offset peg data, bench mark locations and other typical title block entries.

Plotting cross-sections

Cross-sections provide a short profile of the terrain taken at right angles to the centre line. They are viewed along the centre line in the direction of the survey traverse. Figure 9.8(a) shows a cross-section of the natural profile only. Figure 9.8(b) shows a cross-section of the natural profile with the designed roadway profile imposed on it. Vertical and horizontal scales are normally the same on cross-sections and the scale normally used is 1:100.

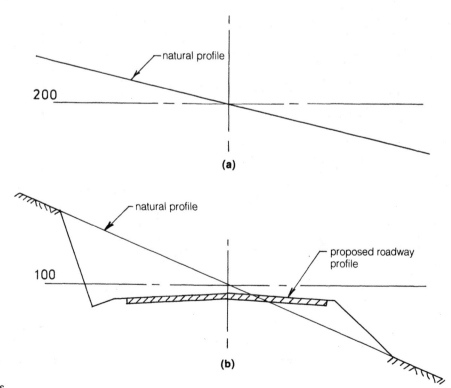

Fig. 9.8 *Cross-sections*

253

Plotting contours

A contour line is an imaginary line on the earth's surface connecting all points of equal height above some reference plane or datum surface, usually sea level. A contour interval is the vertical distance between horizontal planes passing through successive contours.

Figure 9.9 illustrates contour lines on the top view of a mountain. These contours represent lines of intersection of a series of horizontal planes cutting through the mountain at contour intervals of 5 m. The figure clearly illustrates that where the contour lines are more widely spaced (on the right-hand side) the slope of the terrain is much flatter. Contour lines that are evenly spaced indicate a uniform slope. A contour line must always ultimately close on itself, although not necessarily within the limits of the map.

The most common method of drawing contour lines on a given area of terrain is to plot a grid of levels on the actual area concerned. Using surveying instruments, the surveyor locates the grid and measures the reduced levels of the grid intersections. These levels are then noted on the grid map. Figure 9.10(a) shows a grid prepared by the surveyor. A drafter can then draw contour lines over the grid at whatever contour interval is chosen. In Figure 9.10(b) a contour interval of 2 m is used, commencing at the lowest contour line of 100. The lines then rise to 118 in steps of 2. Use the following method when assessing the intersection of a contour line with a grid line. Consider the grid points 98 and 104 at the top of the map. The difference between these two values, 6, means that the slope rises 6 m between these two points. The contour line 100 can,

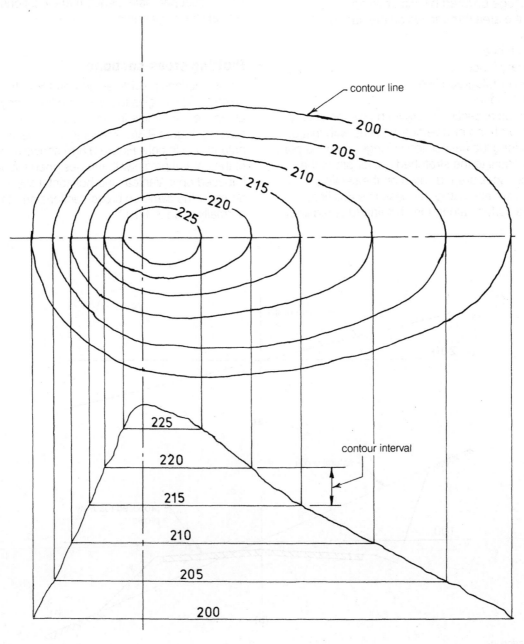

Fig. 9.9 Contours

therefore, be considered to intersect the grid line at $\frac{2}{6}$ the distance between 98 and 104 from grid point 98. Similarly contour line 102 is $\frac{2}{6}$ of the same distance from 104.

Between grid points 98 and 115 there is a rise of 17 m. Each 2 m contour line is, therefore, spaced $\frac{2}{17}$ of the grid distance between grid points 98 and 115,

starting from point 98. All intersections of contour lines with grid lines between specific points have to be individually assessed by this method. The use of a calculator greatly speeds up the task. Contour lines can be drawn in freehand, or with a French curve if a better quality line is required.

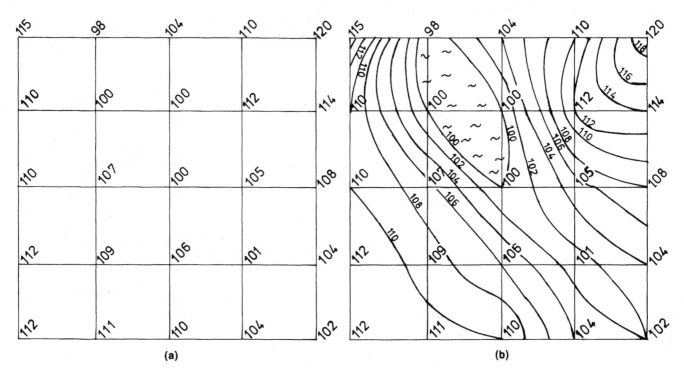

Fig. 9.10 *Contour plotting—grid method*

Exercises on surveying

9.1

A traverse commences at point A and proceeds to B, C and D with the following chainages and bearings.

	Chainage (m)	Bearing
AB	105	30°
BC	100	120°
CD	47.5	210°

Draw the traverse to D to a scale of 1:1000. Join D to A and determine the length and bearing of DA.

9.2

The following traverse is to be drawn to a scale of 1:1000.

	Chainage (m)	Bearing
AB	83	64° 45′
BC	97	345° 15′
CD	121	275° 30′
DE	98	170° 30′
EA	69.5	129° 30′

If drawn accurately, the traverse should close.

9.3

The illustration shows a page from a surveyor's field book commencing at station 4 and finishing at station 7. Draw a survey plan of this page to a scale of 1:2000. In order to position the plan horizontally on the drawing sheet, north should be towards the top of the drawing sheet. Allow 150 mm clear space below the plan to draw the longitudinal section (see Exercise 9.4).

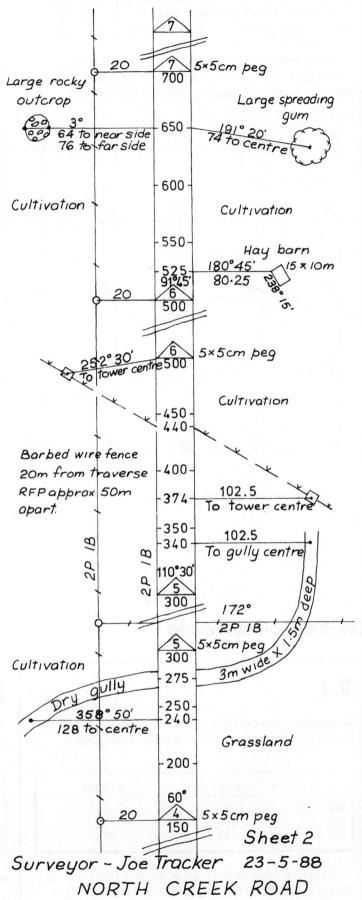

9.4

The illustration shows the level book information for the survey plan mentioned in Exercise 9.3. Using this information, plot a longitudinal section of the traverse directly underneath the survey plan. Use a horizontal scale of 1:2000 and a vertical scale of 1:100.

Backsight	Foresight	Rise	Fall	Reduced level	Chainage
2·300				102·100	150
3·350	1·500	0·800		102·900	200
1·750	2·050	1·300		104·200	250
4·000	2·450		0·700	103·500	275*
3·100	1·900	2·100		105·600	300
2·500	2·500	0·600		106·200	350
2·800	3·400		0·900	105·300	400
2·750	3·200		0·400	104·900	450
2·550	3·450		0·700	104·200	500
2·650	3·350		0·800	103·400	550
2·100	3·350		0·700	102·700	600
3·750	2·000	0·100		102·800	650
	2·250	1·500		104·300	700

*Intermediate value for dry gully

9.5

Draw a grid 150 mm square divided into 25 mm intervals. Using the levels shown at the grid intersections on the diagram provided (they are in metres) and commencing at contour 130, draw contour lines at 2 m intervals up to contour 144.

	144.7	139.2	143.1	C 144.6	144.3	143.5	142.2
	142.5	138	139	141.3	142.7	139.3	139.1
	140.7	137.5	136.1	138.6	138	136.1	137.2
A	138.8	136.5	135	136.2	135.7	135.9	136.1 B
	139.1	136.4	134.6	133.5	133.7	134.1	135.8
	135.3	134.5	133	132.7	132	131.9	132.3
	135.9	134	132.7	131.3 D	130.8	129.6	131.5

9.6

Using the contour plot prepared in Exercise 9.5, project vertical sections along grid lines AB and CD using a vertical scale of 1:200.

Note: The grid intersection levels along AB and CD, as well as the contour intersections, should be plotted separately for each section.

Cabinet drawing 10

Cabinet drawing has become an effective means of communication between the designer and the cabinet maker. For the student, it provides the opportunity to make decisions regarding the choice of materials, method of construction, types of joints, shape, fitness of purpose and finish. It is important that students be able to express their decisions in visual form so that they achieve a satisfactory design and, ultimately, a practical item of furniture. The degree of ability with which the student can do this is, therefore, an important factor in cabinet drawing.

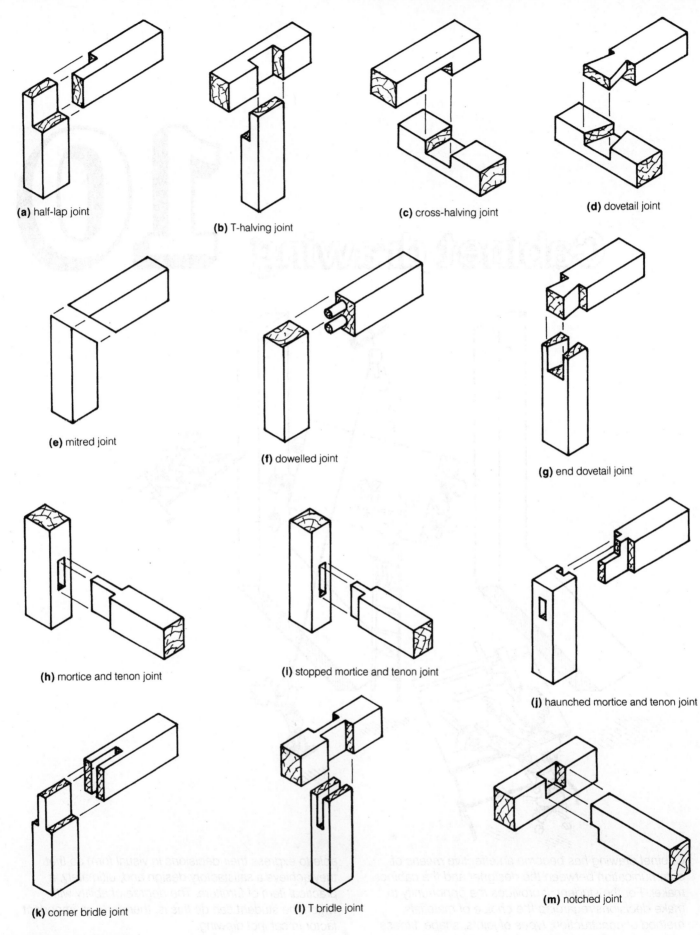

(a) half-lap joint

(b) T-halving joint

(c) cross-halving joint

(d) dovetail joint

(e) mitred joint

(f) dowelled joint

(g) end dovetail joint

(h) mortice and tenon joint

(i) stopped mortice and tenon joint

(j) haunched mortice and tenon joint

(k) corner bridle joint

(l) T bridle joint

(m) notched joint

Fig. 10.1 *Framing joints*

Joining timber

A knowledge of the various types of woodwork joints, and where to use them, is essential to the successful design of any woodwork project.

Framing joints

The five main types of framing joints are half-lapped, dowelled, mortice and tenon, bridle and notched. The most common and strongest form are mortice and tenon joints although bridle joints are sometimes used where a stronger joint than a halving joint is required.

1. **Half-lapped joint** The pieces of timber, which meet or cross each other, are halved in their thickness so that when assembled the faces are flush. Figure 10.1(a)–(d) shows the four halving joints. They are used for such framing jobs as furniture bases and carcases.

2. **Mitred joint** In Figure 10.1(e) the pieces of timber are mitred across their width, butted together and held together by nails or glue. This joint is mainly used in picture frames, architraves and for edging around solid or particle board table tops.

3. **Dowelled joint** This form of butt joint is made by boring holes in the pieces to be joined and glueing in round pieces of timber called dowels (Figure 10.1(f)). It is used where a strong carcase frame is required, for joining rails to the legs and tables and for widening joints.

4. **End dovetail joint** This framing or angle joint (Fig. 10.1(g)) is a very strong joint for joining narrow pieces with their faces or edges at right angles. This joint is used for brackets, rails of trays and frames.

5. **Common mortice and tenon** This (Fig. 10.1(h)) is used when joining pieces of the same thickness, and where a rail meets a stile some distance from the end. It is chiefly used for panelled frames for doors, furniture framework and for joining rails to legs of tables and chairs.

6. **Stopped mortice and tenon** This (Fig. 10.1(i)) is used when the tenon does not pass through the material. It is used in quality cabinet work so that the end grain of the tenon is not seen on the edge of the stile.

7. **Haunched mortice and tenon** This (Fig. 10.1(j)) is essentially a corner joint. The tenon is made narrower than the width of the rail, leaving the outside end of the mortice closed. It is used where the rail meets the stile at its end, for example for door frames and where the rail meets the leg on a table.

8. **Corner bridle joint** This (Fig. 10.1(k)) is used as a substitute for a haunched mortice and tenon at corners of frames.

9. **Tee bridle joint** This (Fig. 10.1(l)) is used as a substitute for the common mortice and tenon joint where one member meets the other some distance from the end. It is sometimes used on small tables and stools.

10. **Notched joint** This (Fig. 10.1(m)) is used mostly in carcase construction. Notched joints are similar to halved joints with the timber positioned on edge.

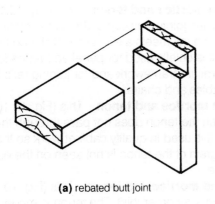

(a) rebated butt joint

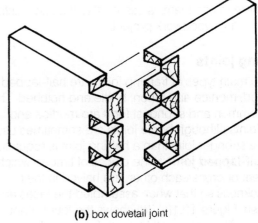

(b) box dovetail joint

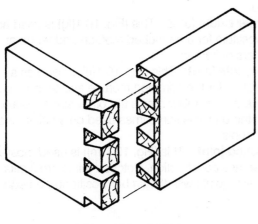

(c) lapped dovetail joint

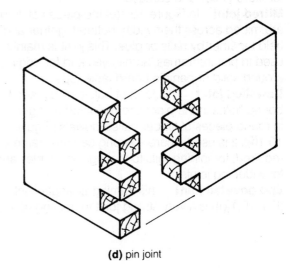

(d) pin joint

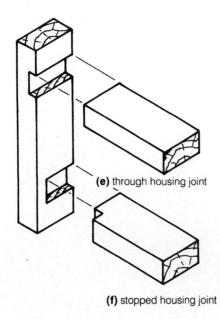

(e) through housing joint

(f) stopped housing joint

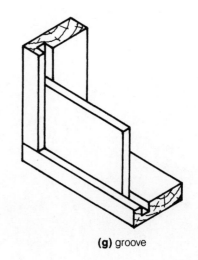

(g) groove

Fig. 10.2 *Corner angle and housed joints*

Corner angle joints

These joints are generally used for fixing together pieces that have their faces at right angles and they are are chiefly used in box-like constructions, such as cabinets and drawers.

1. **Rebated butt joint**   This (Fig. 10.2(a)) has the end of one piece of timber fitted into a rebate on the second piece. The joint is glued, nailed or screwed. It is used mainly in boxes, less expensive cabinets, bookshelves and drawers.
2. **Box dovetail joint**   This (Fig. 10.2(b)) has two or more dovetails and is the strongest form of angle joint. This joint is mainly used in boxes, carcase construction and drawers.
3. **Lapped dovetail joint**   This (Fig. 10.2(c)) is designed so that the lap, usually one-third the thickness of the timber, extends between the pins and covers the end grain of the dovetail. It is used where both strength and appearance are important, especially for carcase construction and drawer fronts.
4. **Pin joint**   This (Fig. 10.2(d)) is sometimes referred to as a box pin-joint. It consists of a series of alternate square pins of the same width interlocking in the assembled joint. It is chiefly used in box construction.

Housed joints

These joints are constructed by cutting a trench in the supporting member to take the full thickness of the other piece. Their main use is fixing shelves or divisions in bookcases and partitions in cabinets.

1. **Through housing joint**   This (Fig. 10.2(e)) has a trench cut in the supporting member to take the full thickness and full width of the shelf or partition.
2. **Stopped housing joint**   This (Fig. 10.2(f)) has the end of the trench stopped back from the front edge and the end of the shelf notched to suit, so that in the assembled joint the trench is not seen.
3. **Grooves**   These (Fig. 10.2(g)) are used mainly in drawer fronts and sides to allow the drawer base to be supported.

Producing a cabinet drawing

Cabinet drawings can be multiview orthogonal drawings showing external details, section views or a combination of both. They may also include pictorial views, open and in line for assembly, of one or more joints to show additional detail.

An integral part of the drawing is the cutting list, which gives the finished sizes of all members so as to simplify material selection. Figure 10.3 shows a completed cabinet drawing.

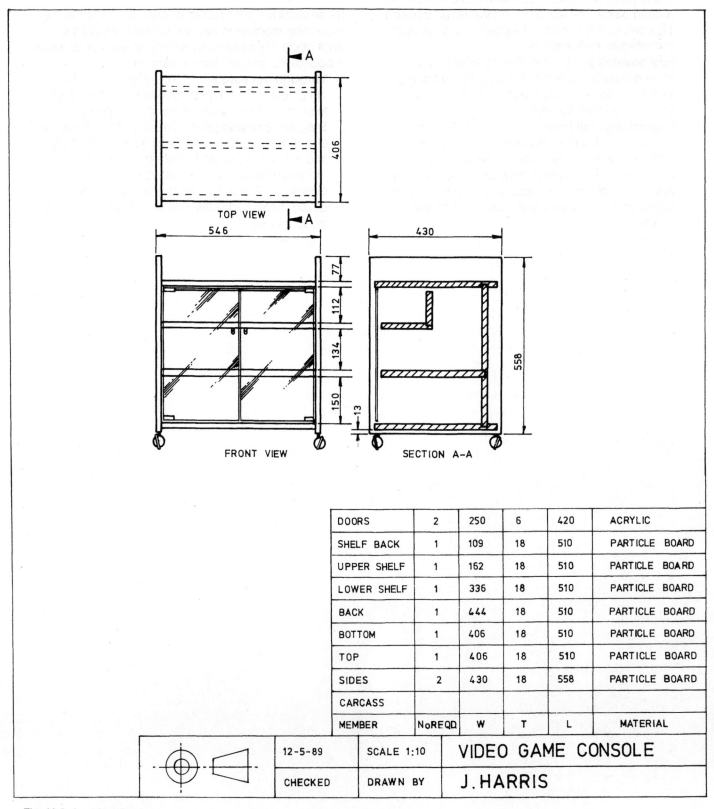

MEMBER	NoREQD	W	T	L	MATERIAL
CARCASS					
SIDES	2	430	18	558	PARTICLE BOARD
TOP	1	406	18	510	PARTICLE BOARD
BOTTOM	1	406	18	510	PARTICLE BOARD
BACK	1	444	18	510	PARTICLE BOARD
LOWER SHELF	1	336	18	510	PARTICLE BOARD
UPPER SHELF	1	162	18	510	PARTICLE BOARD
SHELF BACK	1	109	18	510	PARTICLE BOARD
DOORS	2	250	6	420	ACRYLIC

	12-5-89	SCALE 1:10	VIDEO GAME CONSOLE
	CHECKED	DRAWN BY	J. HARRIS

Fig. 10.3 *A cabinet drawing*

Exercises on cabinet drawing

10.1

The illustration shows the pictorial view of a dressing stool. Prepare a cabinet drawing of the stool.
Scale 1:5
Include a title block and cutting list.

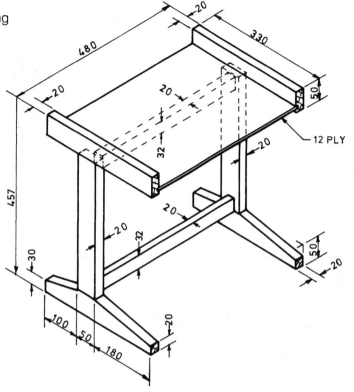

10.2

The illustration shows the pictorial view of a bar stool with the fabric seat removed for clarity. Prepare a cabinet drawing of the bar stool.
Scale 1:5
Include a title block and cutting list.

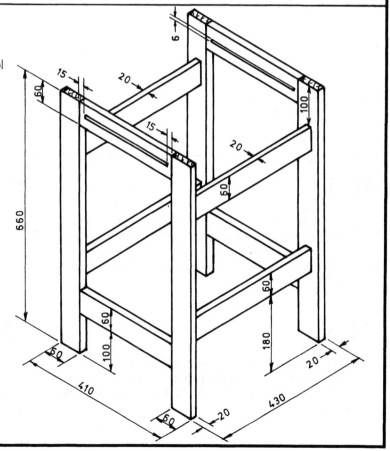

10.3

The illustration shows the pictorial view of a television stand. Prepare a cabinet drawing of the stand.
Scale 1:5
Include a title block and cutting list.

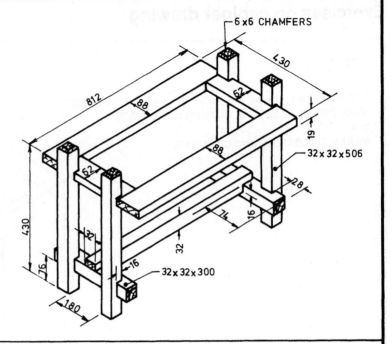

10.4

The illustration shows the pictorial view of a wall unit. Prepare a cabinet drawing of the unit.
Scale 1:5
Include a title block and cutting list.

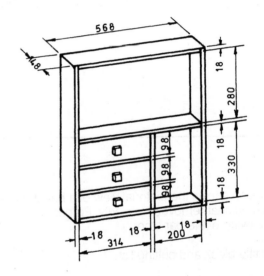

10.5

A set of bookshelves is to be designed for a study. The bookshelves are to be freestanding, 1050 mm wide and contain no more than four shelves. Prepare a cabinet drawing of the bookshelves. Include a title block and cutting list.

10.6

Design a wine rack to hold twelve bottles of wine. The rack is to be portable and made from hoop pine. Prepare a cabinet drawing of the rack and include a title block and cutting list.

10.7

Design an occasional table complete with magazine rack for a lounge room. The table top can be made from veneered particle board. Prepare a cabinet drawing of the table and include a title block and cutting list.

10.8

Design a pair of bedside cabinets for a guest room. Each cabinet is to have shelf space and a single drawer. Prepare a cabinet drawing of one of the cabinets and include a title block and cutting list.

Architectural drawing and light construction

The architect plans, designs and supervises the construction of buildings, from houses to skyscrapers, and the spaces around them that create urban and suburban communities. The aim is to encourage and heighten all human activity, that is, how people live, work and play. Architecture has to serve three functions. First, it has to satisfy a social purpose, to reflect the functional patterns of human activity. Second, it must be well engineered—materials must be well selected and structural members appropriate. Third, it must be appealing in design and quality.

Residential planning

Residential planning has to satisfy the unlimited variety of tastes and needs of individuals. Realistic planning, therefore, becomes a matter of selection—choosing features that will contribute to the client's comfort and that the client feels he or she would like to have.

Factors to be considered when planning include the provision of:
1. sleeping areas—according to the number, ages and sex of family members and the necessity of providing guest facilities
2. service areas—kitchen, laundry and bathroom(s)
3. formal areas—lounge, dining room
4. informal areas—family room, games room
5. special interest areas—study, hobby areas

The designer may submit several designs to the client, each in the form of a floor plan (Fig. 11.1(a)) and a perspective view (Fig. 11.1(b)). The perspective view or presentation drawing is the designer's graphic concept of a building in its natural setting, and it should represent the structure realistically and artistically and in a manner easily understood by the prospective client.

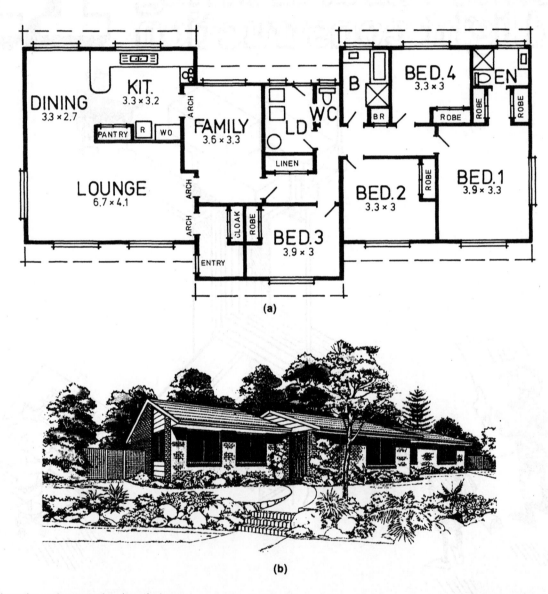

Fig. 11.1 *Floor plan and perspective view of a house*

Classification of documents

For clarity in communication and to enable information to be found quickly, documents are classified according to the type of information to be presented. Each should contain only that information appropriate to its category. Drawings should not contain information that can be better conveyed by schedules or specifications, and vice versa.

Drawings

Drawings at design stage fall into two categories:
1. schematic drawing—a preliminary design drawing, sketch or diagram showing in outline form the designer's general intention
2. development drawing—a design drawing developed to show the building and site as envisaged by the designer and from which production drawings can be produced

Drawings at production stage fall into three categories:
1. location drawing—a drawing produced in order that drawing users may:
   (a) gain an overall picture of the layout and shape of the building
   (b) determine setting-out dimensions for the building as a whole
   (c) locate and identify the spaces and parts of the building, for example rooms, doors, cladding panels, drainage
   (d) pick up references that lead to more specific information, particularly about junctions between the parts of the building

   Each group of location drawings will almost always include site plans, floor plans, elevations, sections and, very often, drainage plans, but there will be occasions when further categories, such as joist layouts, reflected ceiling plans or the enlargement of complex areas may be necessary.
2. component drawing—a drawing to show the information necessary for the manufacture and application of components. Information on basic sizes, and system or performance data are often also shown
3. assembly drawing—a drawing to show in detail the construction of buildings and junctions in and between elements and/or components

Other documents

Other documents identified at production stage are:
1. specification—a precise description of materials and workmanship needed for a project or parts thereof but not shown on drawings or in schedules
2. schedule—tabulated information on a range of similar items differing in detail, such as doors, windows and so on

3. bill of quantities—a complete measure of the quantities of material, labour and any other items required to carry out a project based on the specification, drawing and schedules

Architectural drafting

Some of the information presented in this chapter has been used with permission of Standards Australia (AS 1100 (Technical Drawing) Part 301 (Architectural drawing); AS 1233 (Glossary of Terms for Dimensional Coordination) and AS 1234 (Recommendations for Coordinated Preferred Dimensions in Building); AS 1684 (Code of Practice for Construction in Timber Framing); AS 2870 (Residential Slabs and Footings); AS 2870.1 (Residential Slabs and Footings Construction); SAA MH2 (Metric Data for Building Designers)).

Drafting techniques

Line techniques are the most important skills for architectural design and detail. Different thicknesses should be used for clarity and the range of line thicknesses on each drawing should be kept to a minimum.

On any particular drawing it is recommended that lines from only one of the groups in Table 11.1 be used. The thickness of border and other lines should be as shown in Table 11.2.

Table 11.1 *Groups of line thicknesses*

Size	L 1	L 2	L 3	L 4
Thick	1.0	0.7	0.5	0.35
Medium	0.7	0.5	0.35	0.25
Thin	0.5	0.35	0.25	0.18

Table 11.2 *Line thickness by use*

Features	Thickness of lines		
	Sheet size A0	Sheet size A1	Sheet sizes A2 A3, A4
Border lines	1.4	1.0	0.7
Camera alignment	0.7	0.5	0.35
Fold lines	0.7	0.5	0.35
Principal lines in title block	1.0	0.7	0.5
Minor lines in title block	0.5	0.35	0.35
Grid lines	0.7	0.5	0.35
Projection symbol	1.0	0.7	0.5

Table 11.3 *Application of lines*

Drawing and application	Size
Schematic drawings:	
outline of new buildings and site boundaries	thick
outline of existing building	medium
reference lines, dimension lines and hatching	thin
Development drawings:	
outline of site and new building	thick
general building works and landscaping	medium
reference grids, dimension lines and hatching	thin
Location drawings:	
primary elements in horizontal or vertical section, outlines requiring emphasis	thick
components and assemblies in plan, section and elevation	medium
reference grids, dimension lines and hatching	thin
Component and assembly drawings:	
profiles in horizontal or vertical section	thick
profiles in plan or elevation	medium
reference grids, dimension lines and hatching	thin

Table 11.4 *Representation of materials*

Material	General location drawings—section (scale 1:50 or less)	Large scale drawings—section
Brickwork		
Concrete		
Concrete block		
Earth		
Structural steel		
Stud walls		

Table 11.3 gives recommended continuous lines for specific purposes. Hatching should normally be in the finest line thickness of the chosen group.

Lettering

Architects use traditional and contemporary styles of lettering, both of which are common and acceptable. The lettering style should be uniformly maintained throughout any one series of drawings. The style of lettering should be simple and rounded, and type faces without serifs are preferred.

Hatching

Hatching or symbolic representation of material is shown in Table 11.4. It is time consuming and should only be used to avoid confusion. The conventions used should be in accordance with *AS 1100* Part 301 (see Table 11.4).

Levels

Levels should be expressed to the nearest multiple of 5 mm and the numerals for the required level should be enclosed in a drawn rectanglar box. Where the level of an existing feature is to be varied, the existing level should be placed directly above the box containing the required level.

Where levels might be confused with other numerals on a drawing, such as room numbers, linear dimensions or grid references, the use of the prefix RL (reduced level) or FFL (finished floor level), as applicable, is recommended.

The three general methods for indicating levels on plan views are:

1. **Job datum level**   The job datum level is indicated by the symbol followed by the numerals for that level.

The position of the job datum should be clearly marked on site plans together with a short description and its assumed level (chosen so that all the reduced levels specified for a given job will be positive numbers).

2. **Spot levels**   Spot levels are used to indicate the required level for a specific point or limited area and consist of the symbol + placed at the exact spot to which the level applies, followed by the numerals for the proposed and/or existing level.

3. **Contour lines** Contour lines are used to indicate the slope and shape of the ground surface. They shall be drawn to pass through all points on the site having the same RLs. The RL represented is placed at the end of each line at the site boundary, or in the case of large sites may be placed at intermediate points in or on the line (see Fig. 11.2).

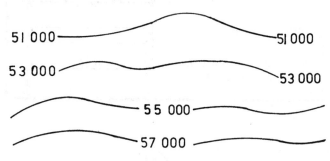

Fig. 11.2 *Placing levels on contour lines*

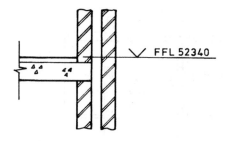

Fig. 11.3 *Placing levels on a section*

On sections or elevations, levels, existing or proposed, should be projected clear of the detail by means of an extension line, as shown in Figure 11.3. A box is not required when a level is written in this way.

Orientation of plans

A north point should appear on every plan. Wherever practicable, all plans, including the site plan, should be drawn with the same orientation. The site location plan should be drawn with the north at the top of the sheet irrespective of the orientation of the other plans.

Scales

Residential design and details are developed with the aid of reduction scales. For example, a site plan may be drawn at 1:200, a floor plan at 1:100 and a footing detail at 1:20.

The scale selected for a particular drawing should be determined by considering:
1. the type of information to be communicated
2. the need for the drawing to adequately and accurately communicate the information necessary for the particular work to be carried out
3. the need for economy of time and effort in drawing production

Dimensioning

Dimensions should be used on drawings to indicate the distance between planes or surfaces. Only those dimensions essential for carrying out the work should be shown; duplication of dimensions should be avoided.

Dimension figures should be written immediately above and parallel with the dimension lines to which they relate. Their position on the dimension line should be towards the centre of the space, as shown in Table 11.5. They should be written to be viewed from the bottom or right-hand edges of the sheet.

Figure 11.4 shows a traditional dimensioning system for a floor plan of a residential building.

Table 11.5 *Conventions for showing dimensions*

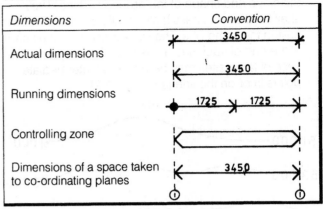

Dimensions	Convention
Actual dimensions	3450
Running dimensions	3450
	1725 · 1725
Controlling zone	
Dimensions of a space taken to co-ordinating planes	3450

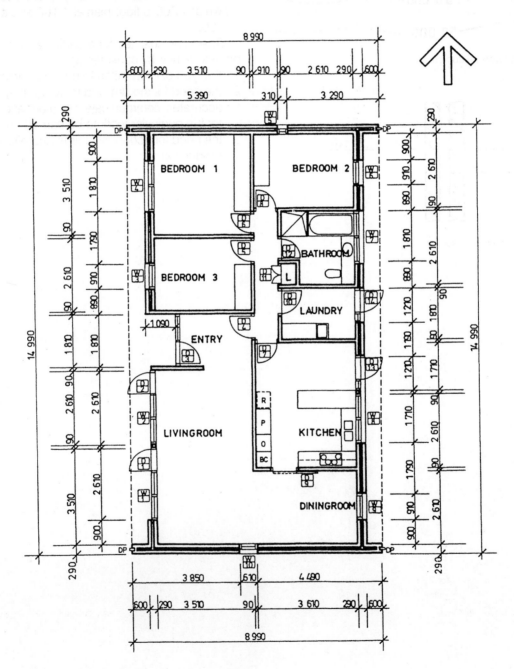

Fig. 11.4 *General location drawing (scale 1:100)*

Architectural conventions

Windows

On elevations of windows hinged along one edge (whether a side, top or bottom view), the opening/closing edge is shown by the point of an indication triangle, as shown in Figure 11.5(a)–(c). The same convention is used for pivoting windows (Fig. 11.5(d)–(e)). A sliding sash is indicated by an arrow drawn on it pointing in the direction of opening, as shown in Figure 11.5(f)–(g).

A window in plan view is indicated as shown in Figure 11.5(h).

Doors

To indicate the hinging and opening of doors in elevation, the opening/closing edge of the door is shown by the point of an indicating triangle, as shown in Figure 11.6(a)–(b).

In plan, single-swing doors are indicated in accordance with the conventions illustrated in Figure 11.6(c)–(d) and double-acting doors in accordance with the conventions in Figure 11.6(e) and (f). The conventions for sliding doors are shown in Figure 11.6(g) and (h).

Vertically opening doors are shown as in Figure 11.6(i). The dashed line indicates the door.

Folding doors and partitions are indicated in accordance with the conventions in Figure 11.6(j)–(m).

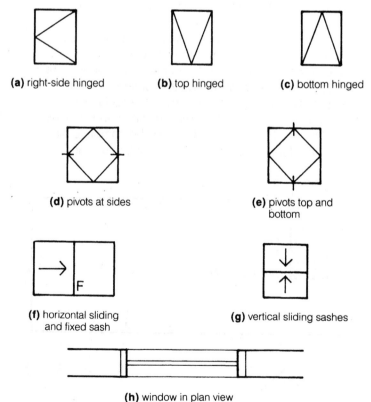

(a) right-side hinged   **(b)** top hinged   **(c)** bottom hinged

(d) pivots at sides   **(e)** pivots top and bottom

(f) horizontal sliding and fixed sash   **(g)** vertical sliding sashes

(h) window in plan view

Fig. 11.5 *Windows*

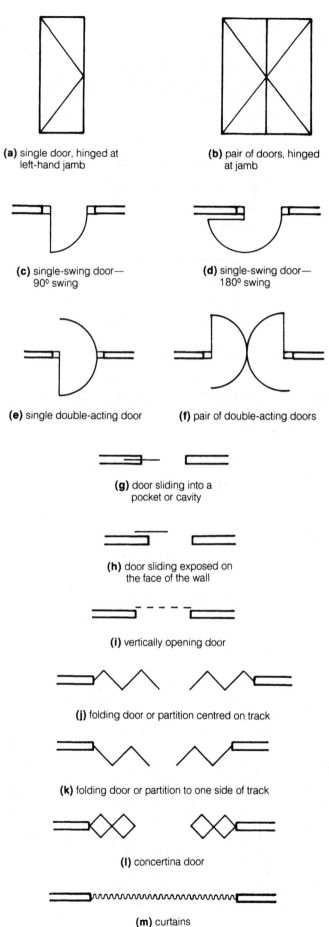

(a) single door, hinged at left-hand jamb   **(b)** pair of doors, hinged at jamb

(c) single-swing door— 90° swing   **(d)** single-swing door— 180° swing

(e) single double-acting door   **(f)** pair of double-acting doors

(g) door sliding into a pocket or cavity

(h) door sliding exposed on the face of the wall

(i) vertically opening door

(j) folding door or partition centred on track

(k) folding door or partition to one side of track

(l) concertina door

(m) curtains

Fig. 11.6 *Doors*

273

Stairs and escalators

On stairs, ramps and escalators, an arrow is used to indicate the direction of rise and the top of the flight. No further labelling of the arrow should be necessary. The actual number of risers from floor to floor should be shown, except where a break-line crosses the flight (see Fig. 11.7).

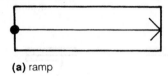

(a) ramp

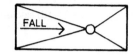

(b) floor slope shown with floor drain

Fig. 11.8 *Ramps and sloping surfaces*

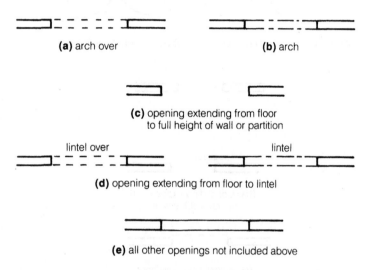

(a) arch over   **(b)** arch

(c) opening extending from floor to full height of wall or partition

lintel over   lintel

(d) opening extending from floor to lintel

(e) all other openings not included above

Fig. 11.9 *Archways and openings*

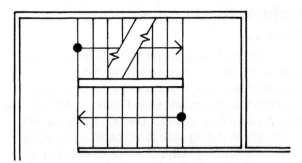

Fig. 11.7 *Stairs*

Ramps

Ramps are indicated in accordance with the convention illustrated in Figure 11.8(a), that is, an arrow indicates the direction of rise.

Where an arrow is used to indicate direction of fall, the arrow shall be clearly labelled with the word 'FALL' to avoid confusion.

Arrows are sometimes required to indicate the direction of slope, as illustrated in Figure 11.8(b). This is usually necessary for floor conditions but can also be used for roofs or other sloping surfaces.

Archways and openings

Archways can be indicated in either of the ways illustrated in Figure 11.9(a) and (b). Other openings through walls or partitions can be indicated in accordance with the conventions illustrated in Figure 11.9(c)–(e), whichever is most appropriate.

Plans

A block plan (Fig. 11.10) is used to locate the building site within the general district.

The site plan (Fig. 11.11) illustrates an ordinary building block and locates the house on the block. It should give complete and accurate dimensions, contour lines, set out point, driveways and other pertinent information required by local council regulations.

The north symbol is always shown on these plans.

Figures 11.12 and 11.13 show the complete plans required for a building in one area of Queensland.

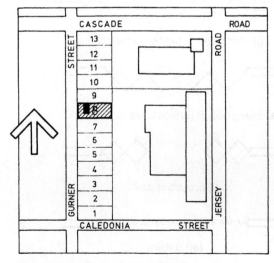

Fig. 11.10 *Block plan (scale 1:5000)*

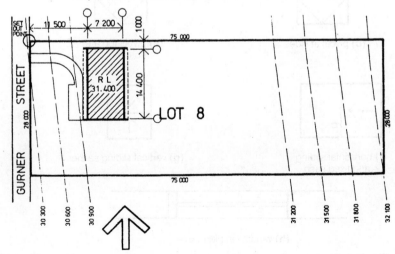

Fig. 11.11 *Site plan (scale 1:500)*

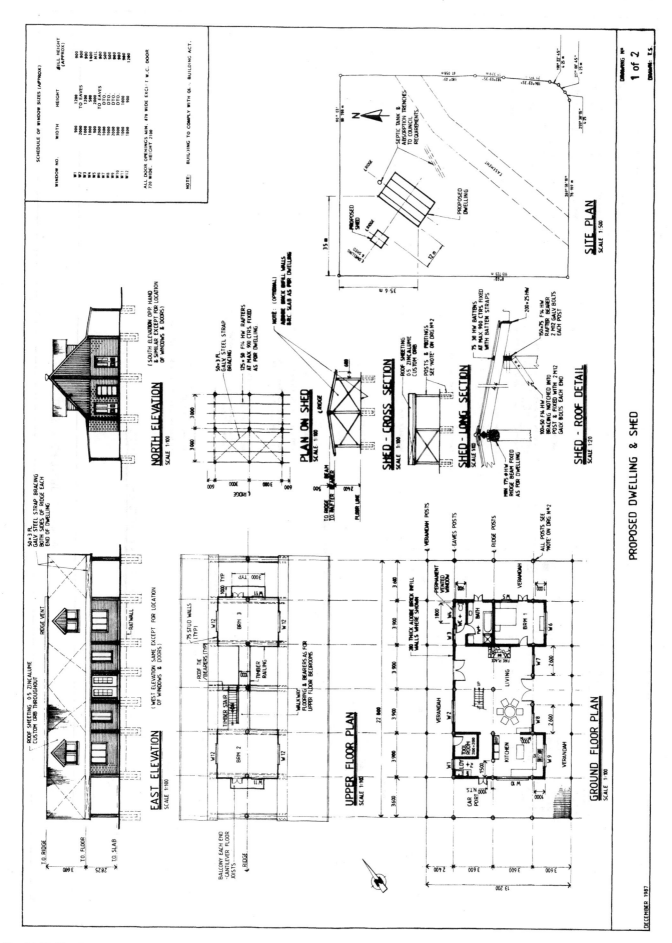

Fig. 11.12 Plan for dwelling and shed

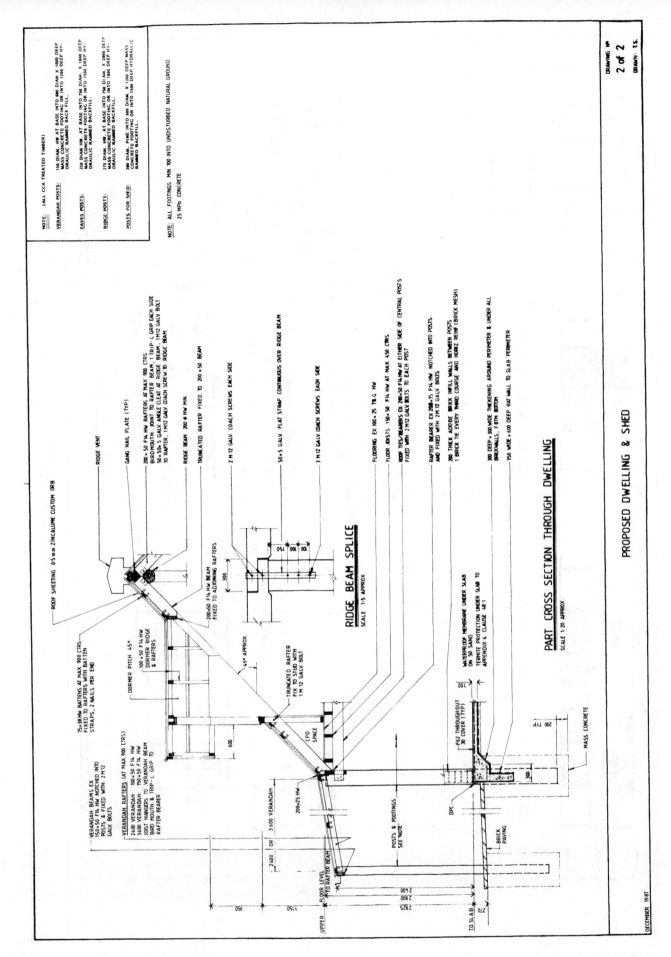

Fig. 11.13 *Drawing of detail for dwelling*

Construction details

Slab-on-ground construction

If the ground floor is to be a slab-on-ground in stable soils the design and construction of the footing system for brick veneer will be similar to that shown in Figure 11.14(a) and the system for cavity brick constructions will be similar to that shown in Figure 11.14(b). In both of these diagrams, edge beam depths may be increased to comply with local health regulations.

Figure 11.14(c) illustrates a footing designed to comply with building regulations in the Caloundra area of Queensland.

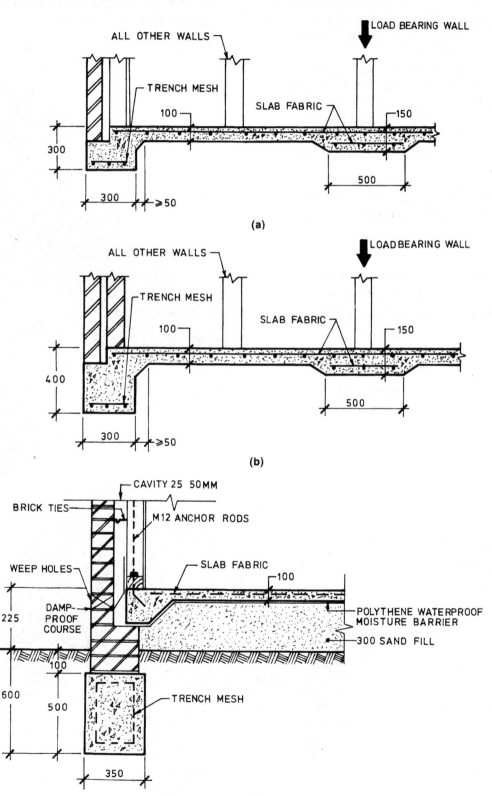

Fig. 11.14 *Residential slabs and footings*

Timber ground floor construction

Brick veneer and cavity brick have become popular methods of construction for residential homes. These types of constructions greatly reduce the amount of external maintenance of buildings and improve their appearance.

When a timber floor is used in brick veneer or cavity brick constructions, openings must be provided to ventilate the air space under the floor. The distance between the ground and the underside of the bearers should not be less than 200 mm.

In brick veneer buildings brick piers are located under all loadbearing walls. Attached piers on external walls are built with the veneer wall (Fig. 11.15(a)).

Cavity brick construction has reinforced concrete foundations to support both the external and internal walls. Attached piers are built on internal walls to support the bearers (Fig. 11.15(b)).

To help prevent termite attack, continuous antcapping should be provided under all bearers. The damp-proof membrane is usually located directly under the antcapping.

1	CONCRETE FOOTING	8	INTERNAL WALL
2	DETACHED PIERS	9	BOTTOM PLATE
3	DAMP-PROOF COURSE	10	SKIRTING BLOCKS
4	ANTCAPPING	11	FLOORING
5	BEARER	12	INTERNAL WALL LINING
6	JOIST	13	SKIRTING BOARD
7	RAKING	14	GROUND LEVEL

1	CONCRETE FOOTING FOR CAVITY BRICK WALL
2	CONCRETE FOOTING FOR SINGLE BRICK WALL
3	DAMP-PROOF COURSE
4	AIR VENT
5	ATTACHED PIER
6	ANTCAPPING
7	BEARER
8	JOIST
9	ACCESS OPENING
10	FLOORING
11	RACKING
12	GROUND LEVEL

Fig. 11.15 *Timber ground floor construction*

Wall framing

Figure 11.16 illustrates the main framing members used in a timber dwelling. They include:

1. Bearers—the heavier timbers that rest on the stumps or piers to take the floor joists
2. Bottom wall plates—load-carrying members laid flat on top of the floor joists. Studs, the vertical wall members, are positioned on top of the bottom plates and are nailed to them

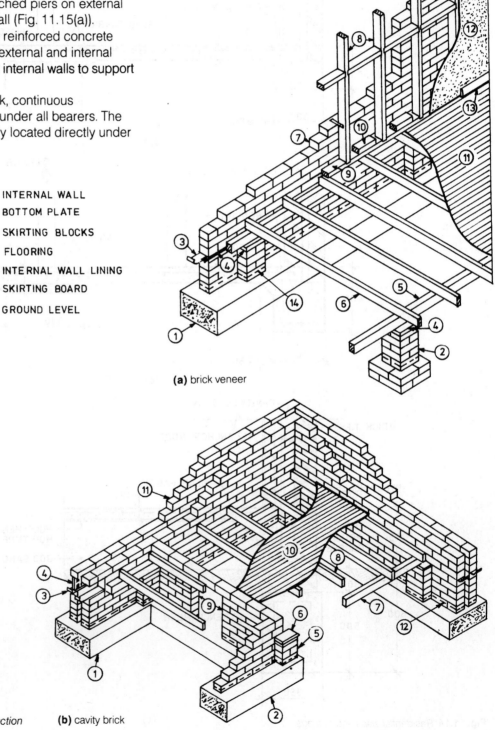

(a) brick veneer

(b) cavity brick

3. Top wall plates—load-carrying members at the top of a wall to provide seating and fixing for ceiling and roofing timbers. Studs are fixed to the top plates as for bottom plates
4. Studs—the vertical members in a wall. They fit between the top and the bottom plates in one continuous length and are nailed securely
5. Jamb studs—studs positioned beside door and window openings and trenched to receive a lintel-carrying load
6. Jack studs—short, common studs fixed above and below openings. They are positioned at the same spacings as common studs
7. Lintels—members over an opening. They are designed to carry any load from above that would normally be carried by common studs
8. Sill trimmers—members that fit between jamb studs or secondary jamb studs across the opening below the window
9. Noggings—members positioned between studs, approximately in the centre of the height of the wall. They also give support to the studs and provide nailing positions for wall sheeting
10. Braces—diagonal pieces of timber checked into the studs and plates of an open wall frame. Bracing is required in a wall to prevent lateral forces such as winds or internal construction pressures from collapsing the wall.

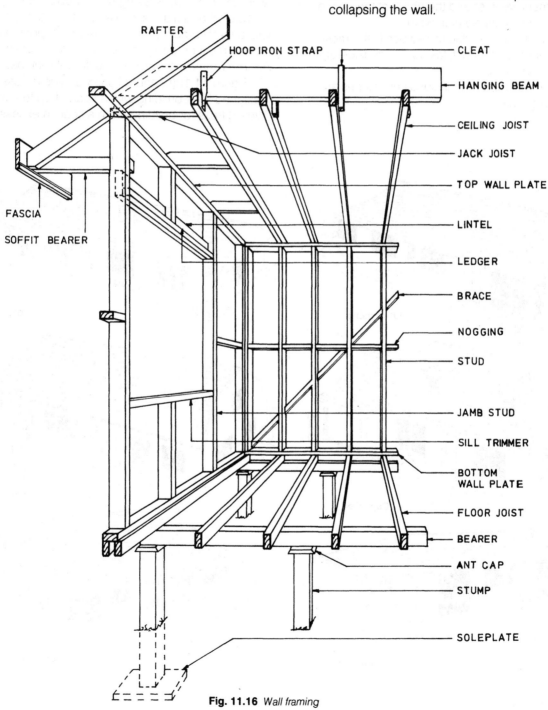

Fig. 11.16 Wall framing

279

Roofs

Roof pitch

The pitch of a roof is its slope in proportion to its span. It can vary from a very slight to a very steep incline. As an example, a roof having a 9 metre span and rising to a height of 3 metres would be described as having a pitch of 1:3, or a slope of approximately 18° in relation to the horizontal.

Pitched roofs offer a number of advantages. They can be used with either flat or cathedral ceilings and with or without exposed beams. Selection of the pitch does not, however, depend only on appearance. Priority must be given to the provision of adequate protection against inclement weather and the ability of the roof to resist the effects of possible high winds.

Most roof material manufacturers specify the most suitable pitch for their product, taking into consideration local climatic conditions.

Examples of different roof designs and pitches are shown in Figure 11.17.

Construction

Roofs constructed on site (Fig. 11.18(a)) are framed by using ceiling joists as part of each set of framing. Other members are rafters and a ridge. Underpurlins are used to reinforce long spans of rafters supported by struts from plates or ceiling joists on internal walls. Collar ties are sometimes used as additional strengthening members. They are placed midway between the top wall plate and the ridge, positioned on top of the underpurlins.

Prefabricated roof trusses offer many advantages in roof framing of small or average size residences. They are generally designed to be supported on the outer walls, the inner walls being non-loadbearing. Figure 11.18(d) shows a truss in position, fixed to the top plate with a metal bracket. The wall frames supporting the trusses must be designed for the correct roof load. *AS 1684* makes provision for different roof loads.

Figure 11.18(b) and (c) show a Type A truss designed for a building with a span of up to 8 m and a Type B truss for buildings with a span in excess of 8 m.

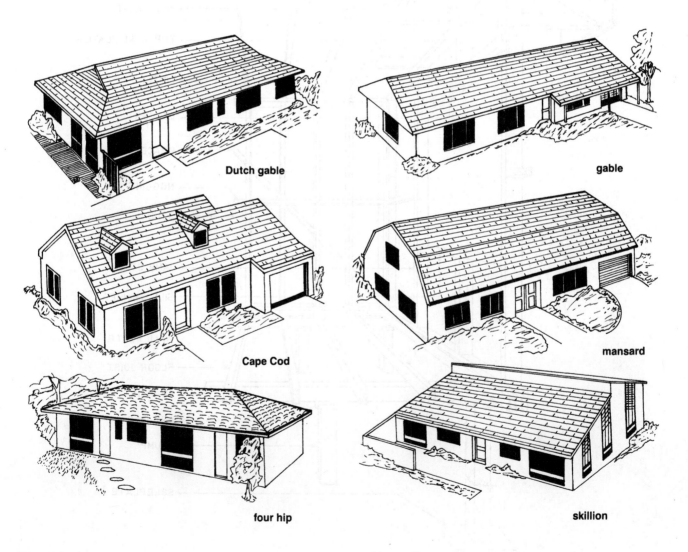

Dutch gable

gable

Cape Cod

mansard

four hip

skillion

Fig. 11.17 *Pitched roofs*

280

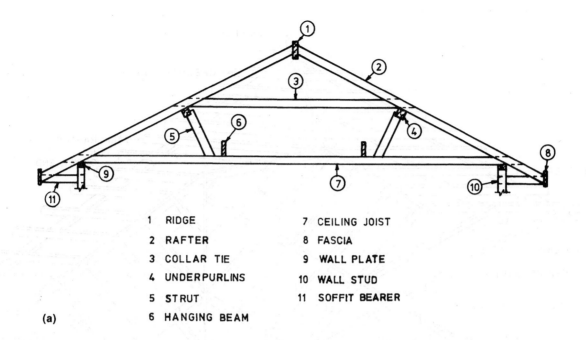

1 RIDGE
2 RAFTER
3 COLLAR TIE
4 UNDERPURLINS
5 STRUT
6 HANGING BEAM
7 CEILING JOIST
8 FASCIA
9 WALL PLATE
10 WALL STUD
11 SOFFIT BEARER

(a)

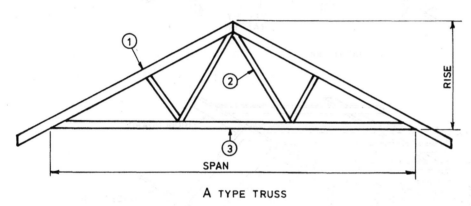

A TYPE TRUSS

1 RAFTER
2 TRUSS WEBB
3 CEILING JOIST

(b)

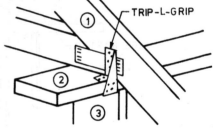

1 TRUSS
2 WALL PLATE
(d)   3 STUD

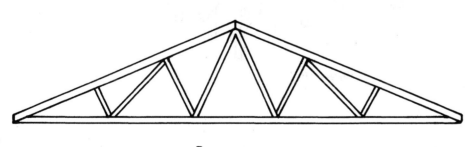

B TYPE TRUSS

(c)

Fig. 11.18 *Roof construction*

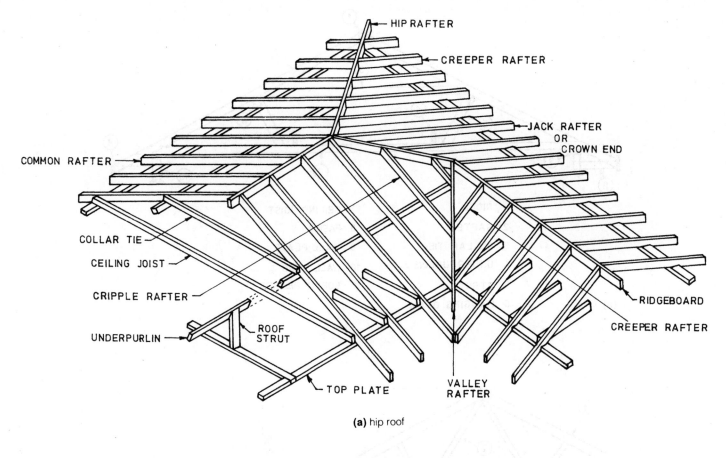

(a) hip roof

HIP RAFTER

CREEPER RAFTER

JACK RAFTER
OR
CROWN END

COMMON RAFTER

COLLAR TIE

CEILING JOIST

CRIPPLE RAFTER

UNDERPURLIN

ROOF
STRUT

TOP PLATE

VALLEY
RAFTER

RIDGEBOARD

CREEPER RAFTER

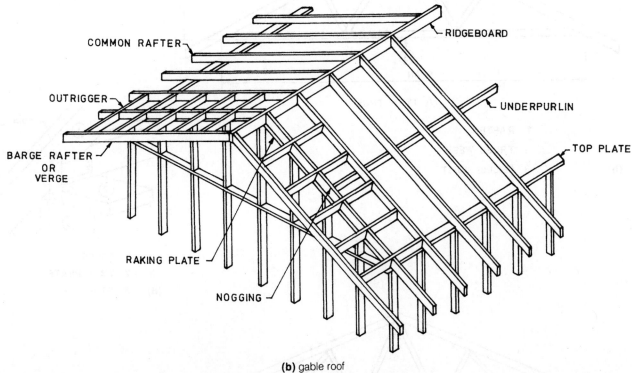

(b) gable roof

COMMON RAFTER

RIDGEBOARD

OUTRIGGER

UNDERPURLIN

BARGE RAFTER
OR
VERGE

TOP PLATE

RAKING PLATE

NOGGING

Fig. 11.19 *Roofing terms*

Roofing terms

Framing members for hip and gable roof constructions are shown in Figure 11.19. They include:

1. Top wall plate—the top member of a wall, on which a roof is fixed
2. Ceiling joist—the member that rests on the top wall plate and to which the ceiling is fixed
3. Common rafter—the main inclined framing timber in the roof
4. Ridge board—the uppermost member in the roof, against which the tops of rafters are fixed
5. Underpurlin—the member running parallel to the top wall plate
6. Hanging beam—a horizontal beam that provides intermediate support for ceiling joists
7. Strut—a vertical or inclined member used to give support to the centre of the rafter
8. Collar tie—a timber member tying a pair of rafters, usually placed midway between the top wall plate and the ridge, and positioned on top of the purlin
9. Roof batten—a timber member fixed to the top of the rafters and to which the roof covering is fixed
10. Trimmer—a cross member inserted between ceiling joists to provide fixing for ceiling sheeting
11. Roofing span—the horizontal width of a building measured from the outside of one top wall plate to the outside of the other wall plate
12. Pitch—the angle of inclination that the surface of the roof makes with the horizontal
13. Eaves—the projecting edge of a roof that overhangs the walls
14. Fascia—a flat, on-edge member finishing the edge of a roof and to which the guttering is fixed
15. Sarking—a secondary protective sheeting beneath the roof
16. Soffit—the underside of the eave or gable overhang, usually lined
17. Tilting batten—a special batten fixed at the eaves of a pitched roof to maintain the slope of tilting at the eaves course at the same plane as the remainder of the roof (the fascia board is generally used for this purpose)
18. Jack rafter or creeper—a member joining the hip or valley in roof construction
19. Barge rafter—the board fixed to the outer side of the gable overhang
20. Anti-ponding board—a metal or timber member fixed from the top of the tilting batten to the rafter under the sarking to prevent ponding of water behind the fascia board
21. Hip rafter—a rafter following the line of the external intersection of two roof surfaces
22. Valley rafter—a rafter following the line of the internal intersections of two roof surfaces

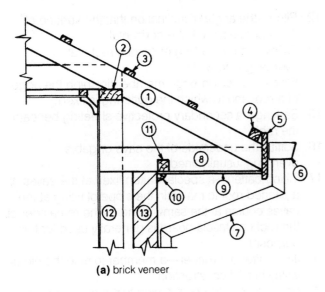

(a) brick veneer

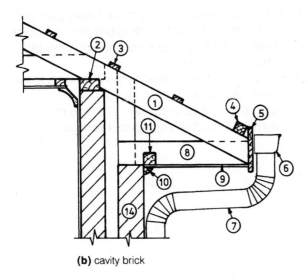

(b) cavity brick

1	RAFTER	8	SOFFIT BEARER
2	TOP PLATE	9	EAVES LINING
3	TILE BATTEN	10	COVER MOULD
4	TILTING BATTEN	11	BATTEN
5	FASCIA	12	STUD
6	GUTTERING	13	BRICK VENEER
7	DOWN PIPE	14	CAVITY BRICK WALL

Fig. 11.20 *Eaves construction*

Eaves construction

Roof rafters are projected beyond the supporting walls to form overhanging eaves.

If eaves are to be boxed, soffit bearers are used in the construction and are spaced to suit the eaves lining. Brick veneer buildings (Fig. 11.20(a)) have the inner ends of the soffit bearers fixed to the frame and cavity brick ones (Fig. 11.20(b)) have soffit bearers located by means of hangers from rafters or wall plates.

The building of eaves is simplified if slotted fascia boards are used. This allows an edge of the eaves lining to be supported by the fascia board.

Tiled roofs

A section of a tiled roof is shown in Figure 11.21. Sarking material is laid so that the overlaps occur beneath a tile batten. The overlap of material must be adjusted to suit the batten spacing, and the overlap should never be less that 150 mm. The sarking material should run over the fascia board far enough to provide drip into the gutter.

Anti-ponding boards must be of sufficient width and strength to ensure a fall towards the gutter and support the sarking without sagging. The fascia board and/or tilting batten must project above the top of the rafter 25 mm further than the thickness of the tile batten. This will maintain the slope of tiling at the eaves course at the same plane as the remainder of the roof.

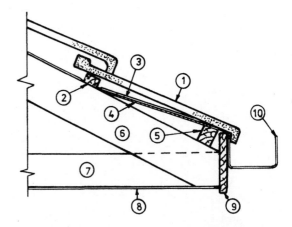

1	TILES	6	RAFTER
2	TILE BATTEN	7	SOFFIT BEARER
3	SARKING	8	EAVES LINING
4	ANTI PONDING BOARD	9	FASCIA
5	TILTING BATTEN	10	GUTTERING

Fig. 11.21 *Tiled roofs*

Exercises on architectural drawing

11.1

Using a scale of 1:100 draw the floor plan of your own home.

11.2

The illustration shows the outline of a floor plan for a brick veneer home. Use it to prepare an architectural drawing of the floor plan, given the following information:

Scale 1:100
external walls: 250 mm thick
internal walls: 70 mm thick
eaves: 600 mm wide

windows: lounge
bedrooms
dining
family } 1200 × 1800 mm

kitchen
bathroom
ensuite
entrance } 1200 × 1200 mm

laundry
WC } 900 × 600 mm

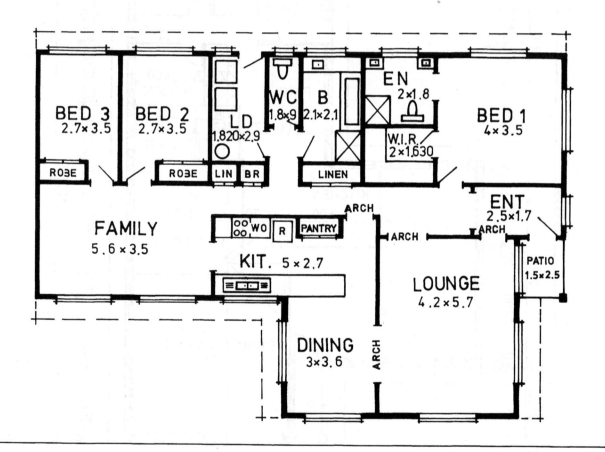

11.3

Draw a site plan showing the position of your home on the allotment. Take into consideration local council regulations.

11.4

The illustration shows the floor plan of a brick veneer residential home. Re-draw it and project south-west and north-west elevations of the home. The pitch of the roof is 21°.

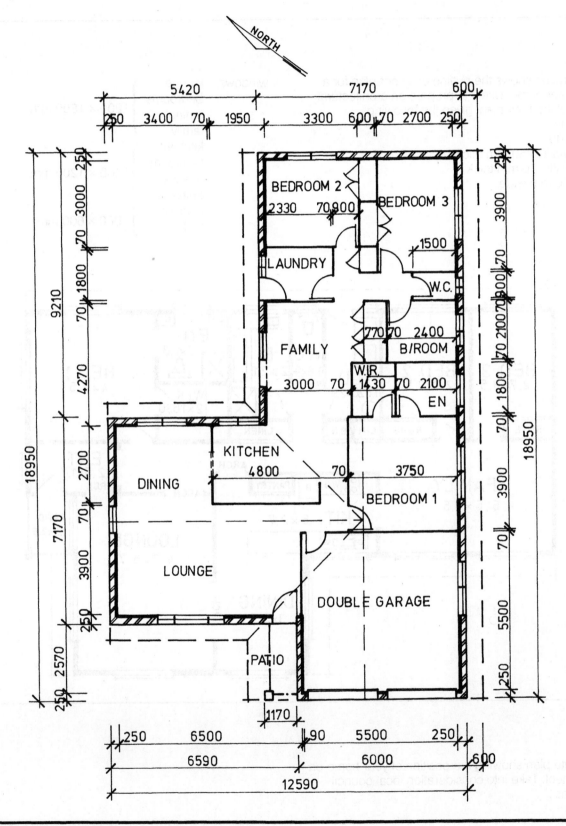

11.5

Design a footing system and slab-on-ground flooring for a cavity brick home. The diagram needs to show only structural proportions.

11.6

Draw a footing system suitable for a cavity brick home with a slab-on-ground floor. The drawing must comply with local building regulations.

11.7

Draw the elevation of the outer wall of a timber dwelling showing framing details. The wall is to include window and door openings.

11.8

Draw the isometric view of a section of a brick veneer home with a timber ground floor. Show the following details:
(a) attached pier
(b) air vent
(c) antcapping
(d) damp-proof course
(e) bearer
(f) floor joist
(g) detached pier
(h) studs

11.9

The brick veneer residential home in Exercise 11.4 has a cement tiled roof at a 21° pitch. Draw a prefabricated Type B truss of the type that would be used in the roof construction. Also show the eaves construction and details for the tiled roof.

Developments and intersections of solids 12

In the engineering trade a surface often has to be formed into a pipe, duct, chute or some other geometrically shaped form. These articles have their surfaces cut from flat sheet, and are then rolled, bent or formed into the desired shape. The material cut from the flat sheet is called the development or the pattern of the final object.

Very often an object is composed of two geometrically shaped forms and their intersection must be established before the development of either form can be obtained. This section deals with the development of both single and intersecting surfaces.

Development of prisms

Rectangular right prism

Figure 12.1(a) shows a pictorial view of a rectangular prism with open ends. This prism consists of four rectangular sides which, when folded out onto a flat surface, form the area necessary to make the prism. This area is called the *development of the prism* or the *pattern for the prism*. Figure 12.1(b) is a view showing the prism unfolding onto a flat surface, and Figure 12.1(c) is the complete layout of the surface of the prism when it is unfolded. It can be seen that the development of the rectangular prism is a rectangle the dimensions of which are the perimeter of the end and the length of the prism.

Truncated right prism

Figure 12.2 illustrates the development of a truncated right prism shown on the left of the figure. To obtain the development, follow these steps:

1. Draw the orthogonal views of the truncated prism as an aid, showing the line of truncation and the joint XX, which is usually positioned midway along the shortest side.

   Note: Only one orthogonal view (e.g. the side view) is normally required. In this case the others are included for clarity.

2. Number the corners 1, 2, 3 and 4 on the orthogonal views.
3. Project horizontally to the right (or left) from the side view. These projectors define the heights of the development.
4. Commencing at joint XX, mark off the sides of the prism along the bottom projector, making sure to finish with joint X. These distances are best taken from the top view.

   Note: X1 and 4X are together half of the side 1–4.

5. Draw vertical lines to intersect the other projectors at X, 1, 2, 3, 4 and X as shown.
6. Join the points X, 1, 2, 3, 4 and X to complete the development.

   Lines 1–1, 2–2, 3–3 and 4–4 are called *fold lines*, that is, the flat development is 'folded' or 'bent' along these lines to form the required prism.

Rectangular prism pipe elbow

A practical application of a truncated prism is shown in Figure 12.3, which illustrates an elbow in rectangular pipe. The development of one half of the elbow is shown on the right. In an elbow of this nature, the junction of the two branches of the elbow is on a line that bisects the total angle of the elbow, in this case 120° as shown on the side view. This is the only angle that will ensure the cross-sectional shape of each piece of the elbow is the same so that they match.

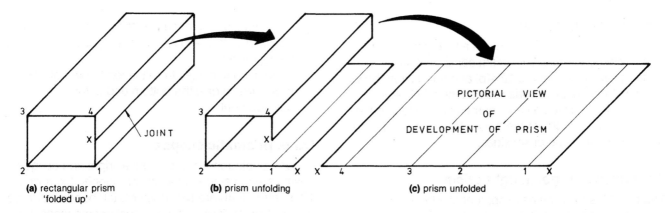

(a) rectangular prism 'folded up'

(b) prism unfolding

(c) prism unfolded

Fig. 12.1 *Rectangular right prism*

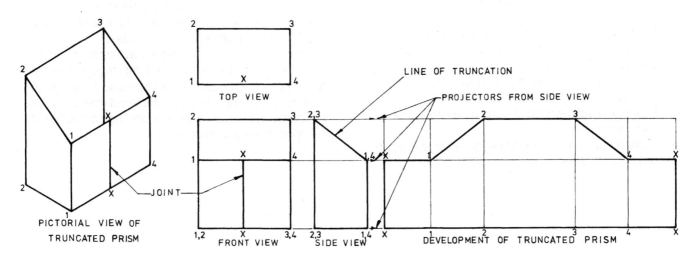

Fig. 12.2 *Truncated right prism*

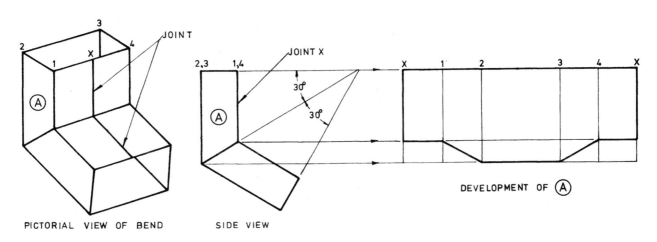

Fig. 12.3 *Rectangular prism pipe elbow*

Hexagonal right prism

Figure 12.4(a) is a pictorial view of a hexagonal right prism with open ends. This prism consists of six rectangular sides. Figures 12.4(b) and (c) illustrate how the development of this prism is obtained. The area required for its development consists of a rectangle the dimensions of which are the perimeter of the prism end and the length of the prism sides.

Truncated hexagonal right prism

Figure 12.5 shows the development of a truncated hexagonal right prism. It is constructed in a similar manner to the development of the truncated right prism in Figure 12.2.

Truncated oblique hexagonal prism

Figure 12.6 shows the development of a truncated oblique hexagonal prism. To obtain it, follow these steps:
1. Draw the front view as an aid, showing the true lengths of the prism sides. The top view is also shown.
2. Number the side corners at the top 1, 2, 3 and the joint line 0, as shown.
3. Project the side lengths at right angles to the sides of the front view.
4. Commencing at 0, mark off the base edge lengths 0–1, 1–2, 2–3 and so on to the appropriate

projectors, and join the points to give the top of the development. The base edge lengths are taken from the top view.
5. Project the fold lines from the top to the bottom projectors to give the bottom end of the development.

Other prismatic shapes

Square, pentagonal and octagonal, right and oblique prisms are developed in a similar manner. Exercises 12.39 and 12.40 are two lobster-back bends made up of truncated square and hexagonal prisms respectively, called segments. In Exercise 12.39 there is a half segment at each end of the bend, and in Exercise 12.40 the bend consists of three whole segments.

At the centre of each full segment in Exercise 12.39, the cross-sectional area of the bend is the same as at the inlet and outlet. In Exercise 12.40, the inlet and outlet cross-sectional areas are the same as at the junction of the segments, and the cross-sectional area halfway along each segment is smaller.

Hence if it is not desirable to have a reduction in cross-sectional area of the bend, the segments must be designed and fitted as those in Exercise 12.39. More segments may be inserted in the bend than are shown, in order to make the change of direction smoother and to approximate a radial bend.

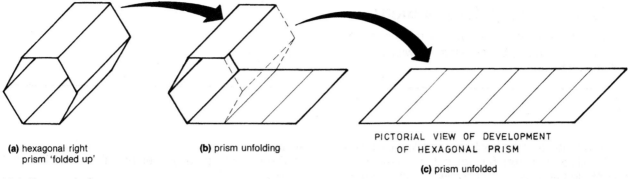

(a) hexagonal right prism 'folded up'

(b) prism unfolding

PICTORIAL VIEW OF DEVELOPMENT OF HEXAGONAL PRISM

(c) prism unfolded

Fig. 12.4 *Hexagonal prism*

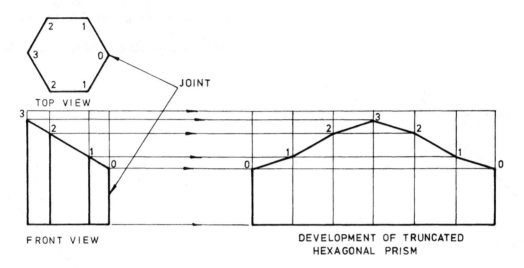

TOP VIEW

JOINT

FRONT VIEW

DEVELOPMENT OF TRUNCATED HEXAGONAL PRISM

Fig. 12.5 *Truncated hexagonal right prism*

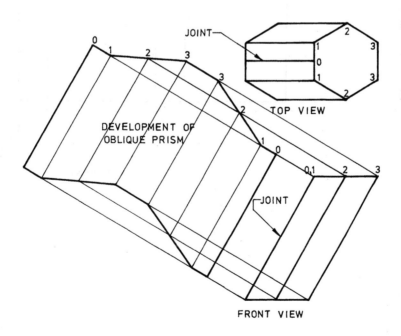

JOINT

TOP VIEW

DEVELOPMENT OF OBLIQUE PRISM

JOINT

FRONT VIEW

Fig. 12.6 *Oblique hexagonal prism*

Exercises on prism development

The following constructions are given for reference. You should practise them and become familiar with the method and techniques.

12.1

Given the top view and the front view of a square prism, edges of ends 55 mm, axis 65 mm, cut by a vertical plane at 45° to the vertical plane, draw the development of the prism.

Solution

1. Produce two parallel stretchout lines from the front view.
2. Draw aa′ parallel to the axis of the prism at a convenient distance from the front view.
3. Set off the distances ab, bc, cd, de and ea from the top view along the stretchout line.
4. Join a′a, b′b, c′c, d′d, e′e and a′a.
5. Add the ends abcde and a′b′c′d′e′ equal in shape to the top view.
6. Line in as shown to complete the development.

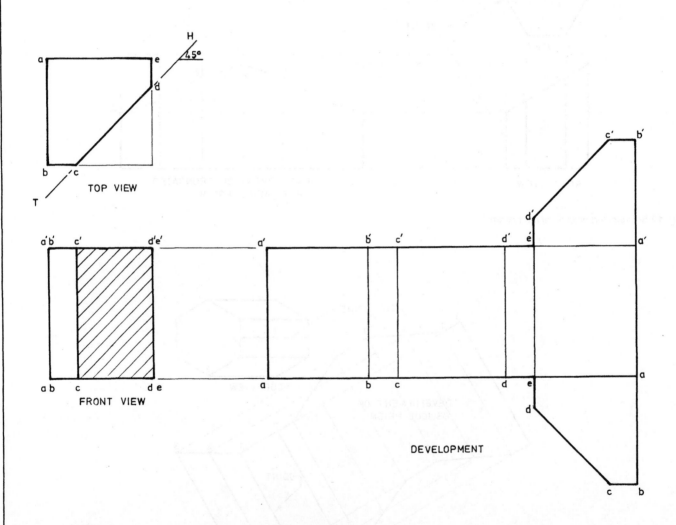

TOP VIEW

FRONT VIEW

DEVELOPMENT

12.2

Given the top view and the front view of a square prism, edges of ends 50 mm, axis 60 mm, cut by an inclined plane at 45° to the horizontal plane, draw the true shape made by the cut and the development of the prism.

Solution

1. To obtain the true shape, project lines at right angles to the cutting plane from a′b′ and c′d′.
2. Draw a′d′ parallel to a′b′, c′d′.
3. Set off from a′ and d′ the length of the side of base to obtain b′ and c′.
4. Join a′, b′, c′ and d′ and cross-hatch to complete the true shape.
5. To obtain the development, produce the stretchout lines from the front view.
6. Step off one side of the square base four times along the stretchout lines and letter each point.
7. Draw vertical lines from b, c, d, a and b to intersect the stretchout line at b′, c′, d′, a′ and b′.
8. Add the true shape and square end.
9. Line in as shown to complete the development.

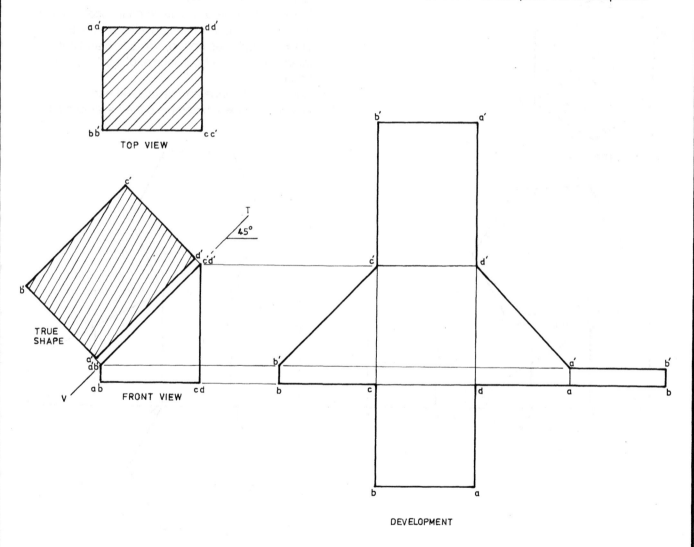

TOP VIEW

TRUE SHAPE

FRONT VIEW

DEVELOPMENT

12.3

Given the top view and the front view of a hexagonal prism, side of end 40 mm, axis 60 mm, cut by an inclined plane at 30° to the horizontal plane, draw the true shape made by the cut and the development of the prism.

Solution

1. To obtain the true shape, project lines at right angles to the cutting plane from a'b', c'f' and d'e'.
2. At any convenient distance draw a centre line parallel to the section line.
3. Transfer the distance of a, a' to the centre line in the top view, either side of the centre line on the true shape, to obtain points a' and b'.
4. The distances from f, f' and e, e' to the centre line in the top view are transferred either side of the centre line on the true shape to give points c'f' and e'd'.
5. Join a', b', c', d', e' and f' and cross-hatch to complete the true shape.
6. To obtain the development, produce stretchout lines from the top view.
7. Step off one side of the hexagonal base six times along the stretchout line and letter all points.
8. Draw vertical lines from d, e, f, a, b, c and d to intersect the stretchout lines at d', e', f', a', b', c' and d'.
9. Add the true shape and hexagonal end.
10. Line in as shown to complete the development.

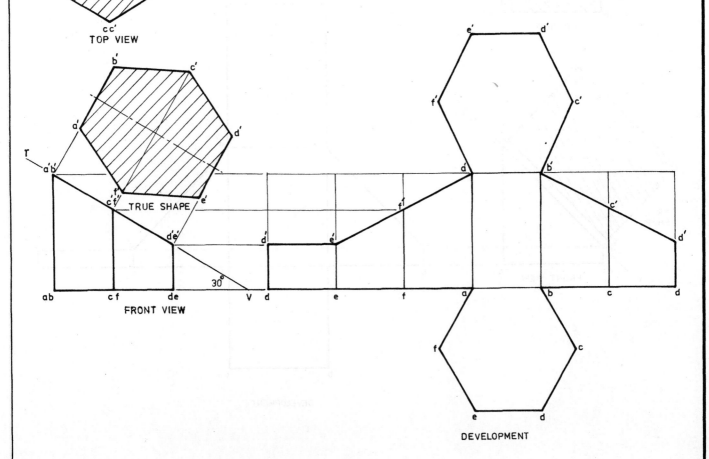

TOP VIEW

TRUE SHAPE

FRONT VIEW

DEVELOPMENT

12.4

Given the top and front views of an oblique octagonal prism, edges of ends 25 mm, axis 75 mm, draw the development of the prism.

Solution
1. Draw the given views and letter all points.
2. From each end of the front view, project lines at right angles to the sides.
3. At a convenient distance from the front view, draw the line aa' parallel to the side of the prism.
4. With a as centre and radius equal to the side of the prism, mark off points b, c, d, e, f, g and a on their appropriate projectors.
5. Join all points to give the top of the development.
6. Project the fold lines from the top to the bottom projectors to give the bottom end of the development.
7. Add the top and bottom of the prism to the development and letter all points.

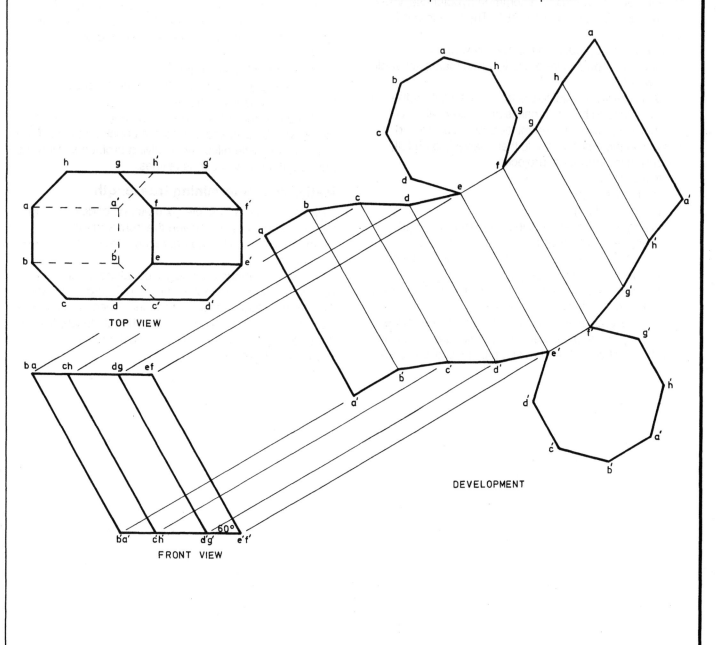

TOP VIEW

FRONT VIEW

DEVELOPMENT

True length and inclination of lines

It is necessary at this stage to introduce a topic that has a very important bearing on the subject of development, namely the relationship the front and top views of a line have to its true length.

Consider the pictorial view of a line AB situated in space in the third dihedral angle, as illustrated in Figure 12.7. This figure shows the projection of the top view, ab, and the front view, a'b'. It can also be seen that the line, projected on, penetrates the vertical and horizontal planes at two points called the vertical trace (VT) and the horizontal trace (HT) of the line. The line AB is inclined at an angle α to the horizontal plane and at an angle β to the vertical plane, as shown. The figure also shows the formation of two right-angled triangles which have the actual line AB as a common hypotenuse; these are triangle ABD and triangle ABC. They are formed as follows:

1. BD is drawn parallel to the front view and is, therefore, equal in length to it; it also makes an angle β with AB.
2. AC is drawn parallel to the top view and, therefore, is equal in length to it; it makes an angle α with AB.

There are seven important facts about a line and its position in the dihedral angle that enable it to be fully described in orthogonal projection:

1. its true length
2. its front view length
3. its top view length
4. its angle of inclination to the horizontal plane (α)
5. its angle of inclination to the vertical plane (β)
6. the vertical difference in height of the ends of the line below the horizontal plane
7. the horizontal difference in the distances of the ends of the line from the vertical plane

In the exercises in this chapter, some of the above facts are given, and it is necessary to find the others. In development work, the front and top views of a line are generally given, and it is necessary to find its true length in order to use it on the development.

A knowledge of the composition of the two right-angled triangles ABC and ABD will enable all of the above seven facts about the line to be solved. These two triangles are now described in detail. Figure 12.8 represents triangle ABC and four of the above seven facts about the line are represented on it. They are:

1. AB, the true length
3. AC, the top view length
4. α, the angle of inclination of the line to the horizontal plane
6. BC, the vertical difference

An important property about this right-angled triangle is that it can be solved geometrically by knowing any two of the four facts represented on it.

Similarly, the right-angled triangle ABD (Fig. 12.9) can be solved geometrically by knowing any two of the following four facts, which are represented on it:

1. AB, the true length
2. AD, the front view length
5. β, the angle of inclination to the vertical plane
7. DB, the horizontal difference

If you can remember and understand the origin of these two triangles and are able to construct them, you will have very little difficulty in solving problems involving true length and inclinations of lines.

Methods of determining true length

Figures 12.10–15 illustrate six methods of determining the true length of a line, given the front and top orthogonal views a'b' and ab, respectively, in third-angle projection.

Various methods are shown, but all determine one or other of the two triangles shown in Figures 12.8 and 12.9. In development work, it is usually necessary to find the true length of the line only, but the full description of the true length triangle is given in each case for recognition purposes.

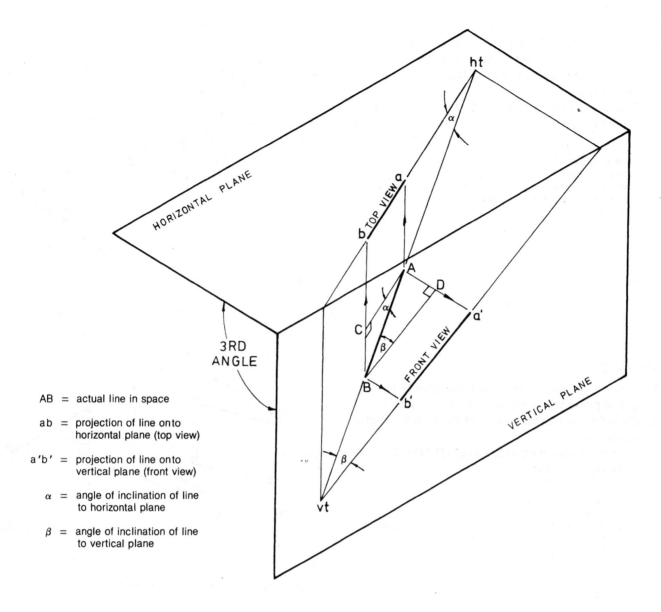

Fig. 12.7 *Relationship between a line, its projections and inclinations in the third dihedral angle*

AB = actual line in space

ab = projection of line onto horizontal plane (top view)

a′b′ = projection of line onto vertical plane (front view)

α = angle of inclination of line to horizontal plane

β = angle of inclination of line to vertical plane

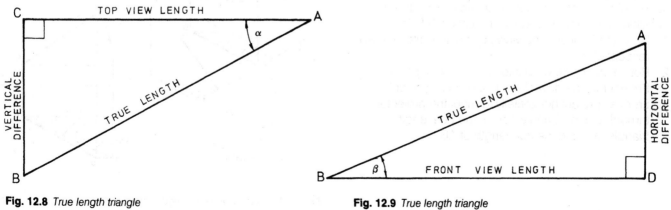

Fig. 12.8 *True length triangle*

Fig. 12.9 *True length triangle*

Method 1

This method is used when given a line parallel to the vertical plane and inclined to the horizontal plane.

1. Draw the front and top views of the lines a'b' and ab respectively.
2. Then a'b' is also the true length of AB. Note the true length triangle a'b'c.

TL = true length of AB
TVL = top view length
VD = vertical difference in ends of line from horizontal plane
α = angle of inclination to horizontal plane
AB = actual line

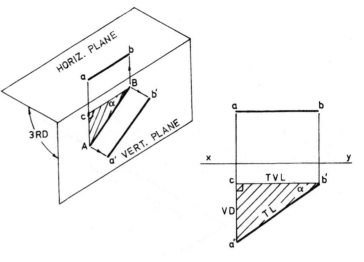

Fig. 12.10 *Determining true length—method 1*

Method 2

This method is used when given a line parallel to the horizontal plane and inclined to the vertical plane.

1. Draw the front and top views of the lines a'b' and ab respectively.
2. Then ab is also the true length of AB. Note the true length triangle abc.

TL = true length of AB
FVL = front view length
HD = horizontal difference in ends of line from vertical plane
β = angle of inclination to vertical plane
AB = actual line

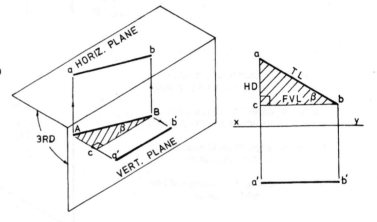

Fig. 12.11 *Determining true length—method 2*

Method 3

This method may be used when given a line inclined to both the horizontal and vertical planes. The aim is to construct one of the true length triangles, and this is possible in a number of ways, one of which follows.

1. Draw the front and top views of the lines a'b' and ab respectively.
2. Rotate the top view ab into the horizontal position.
3. Project the line down level to a'b', that is, to cb'.
4. From a', project horizontally to meet the projector through c at d. Triangle cdb' is the true length triangle, and b'd the true length of AB.

AB = actual line

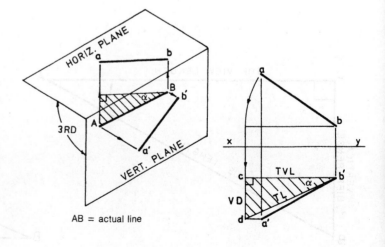

Fig. 12.12 *Determining true length—method 3*

Method 4

This method may be used when given a line inclined to both the horizontal and vertical planes. It is similar to method 3 except that the other true length triangle is constructed.

1. Draw the front and top views of the lines a'b' and ab respectively.
2. Rotate the front view a'b' into the horizontal position.
3. Project the line up level to b, that is, to cb.
4. From a, project horizontally to meet the projector to c at d. Triangle bcd is the true length triangle, and bd is the true length of AB.

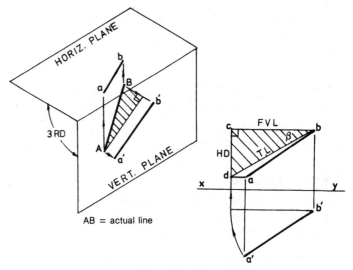

Fig. 12.13 *Determining true length—method 4*

Method 5

This method may be used when given a line inclined to both the horizontal and vertical planes. This method uses an auxiliary view which determines the true shape of the triangle cross-hatched in the pictorial view.

1. Draw the front and top views of the lines a'b' and ab respectively.
2. Project at right angles from a' and b' the horizontal distances of a and b respectively from the vertical plane (these distances are shown bracketed).
3. Join the ends of these projectors c and d to give the true length of AB. Note the true length triangle (hatched).

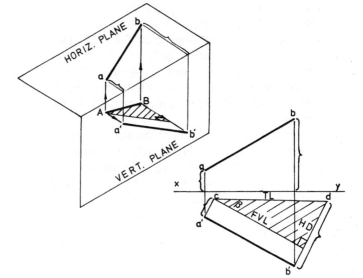

Fig. 12.14 *Determining true length—method 5*

Method 6

This method may be used when given a line inclined to both the horizontal and vertical planes. This method uses a different auxiliary view from method 5, as shown in the pictorial view.

1. Draw the front and top view of the lines a'b' and ab respectively.
2. Project at right angles from a and b the vertical distances of a' and b' respectively below the horizontal plane (these distances shown bracketed).
3. Join the ends of these projectors c and d to give the true length of AB. Note the true length triangle (hatched).

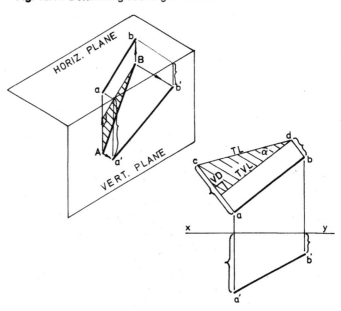

Fig. 12.15 *Determining true length—method 6*

Line of intersection—cylinders and cones

The line of intersection of two or more intersecting surfaces has to be determined in order to develop any of the surfaces. Methods used in drawing lines of intersection are as follows.

Element method

This involves the use of line elements drawn on the surfaces of the intersecting shapes and passing through the area where the line of intersection occurs.

Cone and cylinder intersection (Fig. 12.16)

1. Draw the front, side and top views of the cone, showing the intersection of the cylinder on the side view, the intersection being a circle.
2. Draw two identical sets of three elemental lines on the side view from the apex to cut the base at a, b and c and cut the cylinder at 0, 1, 2, 3, 4, 5 and 6. (The lines to c are tangential to the circle at 3.)
3. Project these elemental lines onto the top view and then down onto the front view as indicated by the arrows.
4. Project points 0, 1, . . . 6 from the side view to intersect the elemental lines on the front view and then up to the top view to give corresponding points 0, 1, . . . 6 on the line of intersection of the cylinder and cone on each view.
5. Draw a smooth curve through the points to give the line of intersection on each view.

Cutting plane method

This involves drawing a series of horizontal cutting planes, each of which cuts through both the intersecting surfaces, for example a cone (to give a circle) and a cylinder (to give a rectangle).

Cone and cylinder intersection (Fig. 12.17)

1. Draw the front, side and top views of the cone, showing the intersection of the cylinder on the side view, the intersection being a circle.
2. Divide the end view of the cylinder into twelve equal parts numbered 0, 1, 2, 3, 4, 5 and 6.
3. Project these points across to the front view to represent a series of horizontal cutting planes through the cylinder and cone.
4. Project the cutting planes from the front view onto the top view, where they are represented by circles.
5. Project the points 0, 1, . . . 6 from the side view up to the top view and along the cylinder. The distances between lines of similar numbers, for example 1–1 on the top view, is the width of the cylinder section at that level.
6. The intersections of lines 0, 1, . . . 6 drawn along the cylinder on the top view with the circles drawn by projecting the cross-sections of the cone at these levels represent points on the line of intersection. Join these points with a smooth curve to give the line of intersection.
7. Project the points 0, 1, . . . 6 from the line of intersection on the top view down to the front view to intersect the corresponding line on that view to give points on the line of intersection. Join them with a smooth curve.

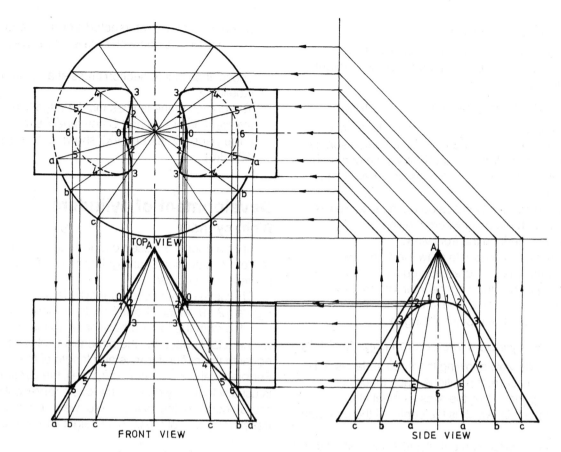

Fig. 12.16 *Line of intersection of cylinder and cone—element method*

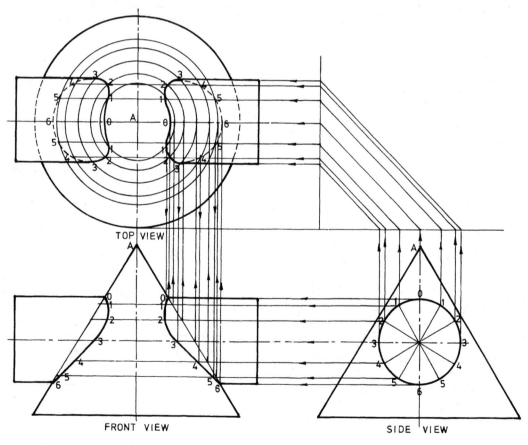

Fig. 12.17 *Line of intersection of cylinder and cone—cutting plane method*

Common sphere method

When intersecting cylinders and cones envelop a common sphere, the line(s) of intersection are straight when viewed from the side.

Cone and cylinder intersection (Fig. 12.18(a))

1. Draw the front and side views of the cylinder and cone, showing the two views of the common sphere touching both surfaces in each view. (The common sphere in the side view is also the end view of the cylinder.)
2. The point of tangency indicated on the side view is horizontal with the point where the lines of intersection meet on the front view.

   It is not really necessary to draw the side view in order to find the line of intersection on the front view. The two straight lines forming the intersection may simply be drawn from one side to the other as shown, and they will cross at the point of tangency.

Cone and two cylinders intersection (Fig. 12.18(b))

1. Draw the side view showing the three surfaces A, B and C in their correct positions, each touching the common sphere. Note the axes all meet at a common point O, which is the centre of the common sphere.
2. Project each surface on until it intersects both of the other two surfaces. These intersections are shown as points a, b, c, d, e and f.
3. The two cylinders B and C alone would have intersected along ad, the cylinder B and cone A

alone would have intersected along be, and cylinder C and cone A alone would have intersected along fc.
4. These lines of intersection cross at a common point labelled X.
5. The portion of these lines that form the line of intersection of A, B and C combined is made up of three parts (aX, cX and eX), shown outlined in Figure 12.18(b).

Development of cylinders
Right cylinder

A right cylinder is a closed circular surface. The shape of the cross-section of all right cylinders at right angles to the axis is circular.

The development of a right cylinder is illustrated in Figure 12.19, which shows how the cylinder can be unrolled from the formed position (a) to the flat rectangular surface (c), the dimensions of which are equal to the circumference and length of the cylinder. For practical development purposes, the circumference is usually found by dividing the circle representing the cross-section of the cylinder into twelve equal parts and transferring one of these to a straight line twelve times.

Truncated right cylinder

A truncated right cylinder is one that is cut at an angle to the axis. A practical example of its use is in the construction of the cylindrical elbow shown in Figure

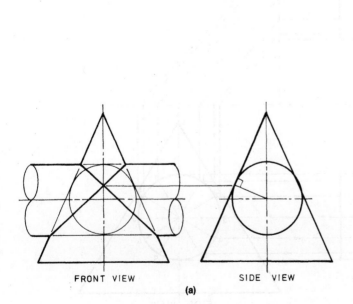

FRONT VIEW   SIDE VIEW

(a)

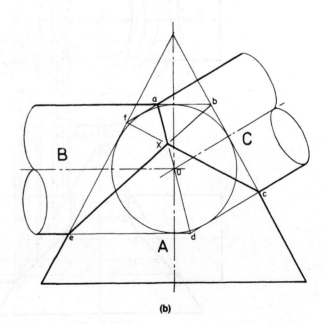

(b)

Fig. 12.18 *Line of intersection of cylinder and cone—common sphere method*

12.20(a). This is made up of two truncated right cylinders joined together along the axis of truncation to form the included angle of the bend, in this case 120°. The development is more complicated than for the plain right cylinder because the line of truncation on the development is curved and has to be plotted. It is obtained by following these steps:

1. Draw the front view of the cylindrical elbow as an aid (Fig. 12.20(a)).
2. Divide the surface of one arm of the elbow into twelve equal sections.
3. Draw the development of the overall cylinder, showing on it the twelve equal sections (Fig. 12.20(b)).
4. Project the points of intersection of the line of truncation XX and the section lines across to intersect the corresponding section lines on the development.
5. Join these points with a smooth curve to complete the development.

The joint line is normally located along the shortest section of the surface.

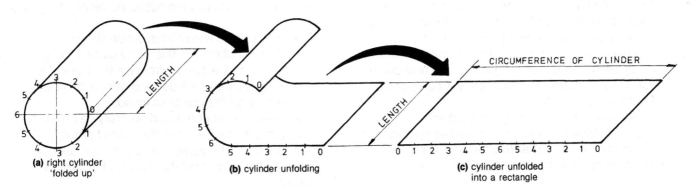

(a) right cylinder 'folded up'   **(b)** cylinder unfolding   **(c)** cylinder unfolded into a rectangle

Fig. 12.19 *Right cylinder*

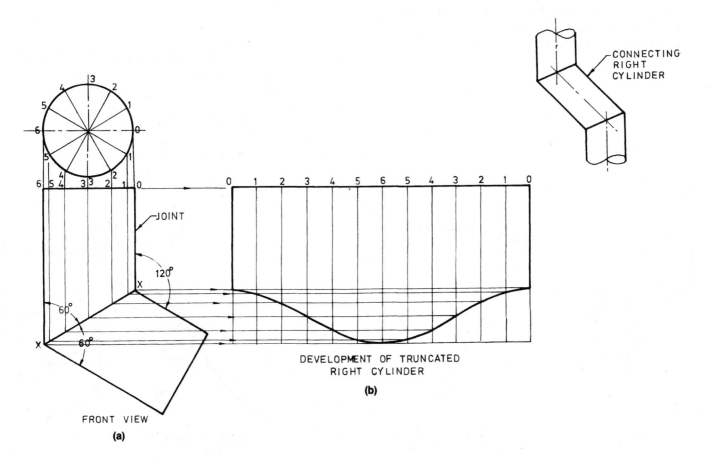

FRONT VIEW
(a)

DEVELOPMENT OF TRUNCATED RIGHT CYLINDER
(b)

Fig. 12.20 *Truncated right cylinder*

305

Oblique cylinder

An oblique cylinder can be defined as a closed curved surface in which the shape of the cross-section at right angles to the axis is elliptical. One particular cross-section at an angle to the axis is circular, and it is at this cross-section that the joining of right cylindrical pipes takes place. See Figure 12.21(a).

To obtain the development, follow these steps (refer to Fig. 12.21(b)):
1. Draw the front view of the oblique cylinder as an aid.
2. Divide the surface into twelve equal sections.
3. Project the joint line 6–6 to the side of the front view parallel to the axis of the oblique cylinder.
4. From one end of 6–6, say the top, strike an arc equal to one-twelfth the circumference of the base, and project point 5 across from the front view to intersect the arc to give point 5 on the development.
5. Continue striking arcs equal to one-twelfth the base circumference and projecting across the next point around the circumference until the top curve of the development is plotted.
6. Draw a smooth curve through these points.
7. To plot the bottom curve, project the base points across from the front view to intersect lines drawn from corresponding points on the top curve.

Development of T pieces

The development of cylindrical T pieces involves finding the line of intersection of the two cylinders, and then drawing the development as described on page 305.

Oblique T piece—equal diameter cylinders

To develop both branches of the oblique T piece, follow these steps and refer to Figure 12.22:
1. Draw the front view of the T piece as an aid to drawing the development. (A side view is also shown.)
2. Draw the line of intersection of the two branches, in this case two straight lines meeting on the centre line at d (common sphere method).
3. Divide both branches A and B into twelve equal parts, and draw surface element lines which will meet the line of intersection at b, c, d, e and f.
4. The development of B (a rectangle) is projected below the front view, and the surface element lines that meet the line of intersection, namely 3, 4, 5 and 6, are drawn.
5. The shape of the line of intersection on the development of branch B is found by projecting points a, b, c, d, e, f and g from the front view onto the relevant lines of the development; a curve is drawn through the points so found.
6. The development of A would normally be projected above the front view in the direction of arrow X at right angles to the centre line, but here it has been placed at the bottom of the page for the sake of layout.

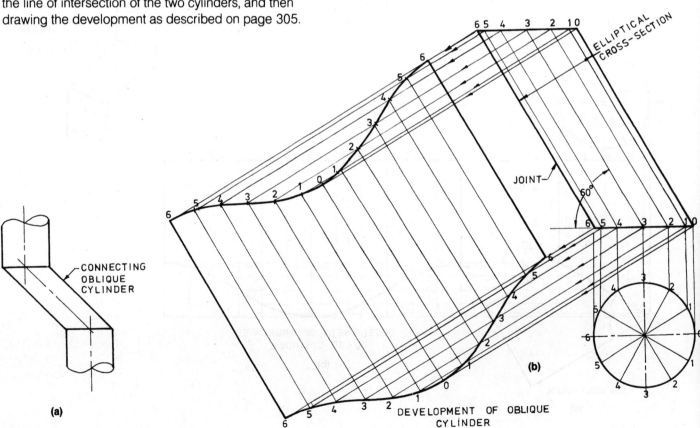

(a)

Fig. 12.21 *Oblique cylinder*

(b)

DEVELOPMENT OF OBLIQUE CYLINDER

7. The development of branch A (see pp. 304–5, development of a truncated right cylinder) is found by developing the overall rectangle for the cylinder, drawing on it the surface element lines, and projecting points a, b, c, d, e, f and g onto the relevant element lines to give the required points for the intersecting end of branch A.

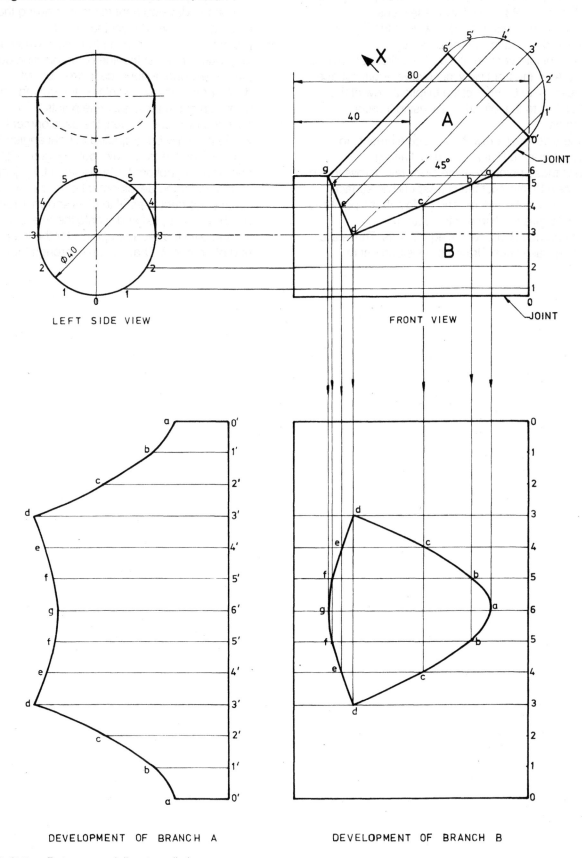

LEFT SIDE VIEW

FRONT VIEW

DEVELOPMENT OF BRANCH A

DEVELOPMENT OF BRANCH B

Fig. 12.22 *Oblique T piece—equal diameter cylinders*

Offset oblique T piece—unequal diameter cylinders

To develop both branches of the offset oblique T piece, refer to Figure 12.23 and follow these steps:

1. Draw both the front and side views of the T piece as an aid in drawing the developments.
2. The line of intersection of A and B is first determined. Divide A on the front view into twelve equal surface elements 0' to 6'. Project the elliptical view of the end of A onto the left side view using transfer ordinates.
3. On the side view, draw the surface elements to intersect branch B at a, b, c, d, e, f and g.
4. Project these points from the side view across to the front view to intersect the corresponding surface elements at a, b, c, d, e, f and g. Join these points as shown to give the line of intersection.
5. Divide branch B into twelve equal parts. These are the long-dash lines. Draw the development of branch B below the front view, showing the long-dash dividing lines. An extra line is drawn tangential to the line of intersection on the front view at g and intersecting the end of X between 4 and 5. Plot this line on the development to aid in drawing the looping curve, which just touches it.
6. Project the points from the front view where the long-dash lines cut the line of intersection down to the corresponding long-dash lines on the development to give points on the curved hole. Draw a smooth curve through these points.
7. As in the previous section, the development of A would normally be projected at right angles to branch A on the front view, but in Figure 12.23 it is placed at the bottom for better layout. Project the rectangle for the development of A, marking on it the surface element lines 0' to 6' as shown. Project the points a, b, c, d, e, f and g onto the relevant element lines to give the required points for the intersecting end of branch A. Draw a curve through these points.

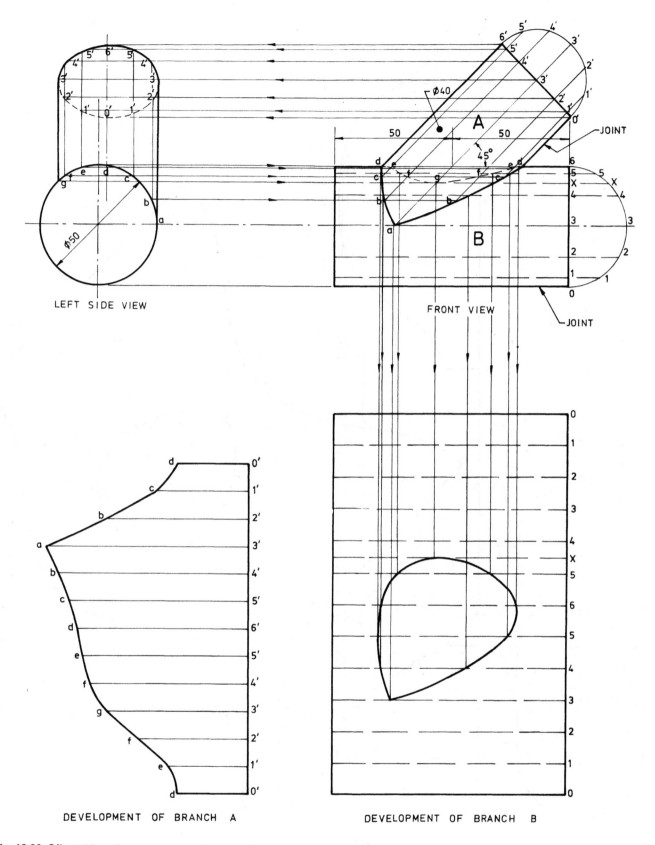

LEFT SIDE VIEW

FRONT VIEW

DEVELOPMENT OF BRANCH A

DEVELOPMENT OF BRANCH B

Fig. 12.23 *Offset oblique T piece—unequal diameter cylinders*

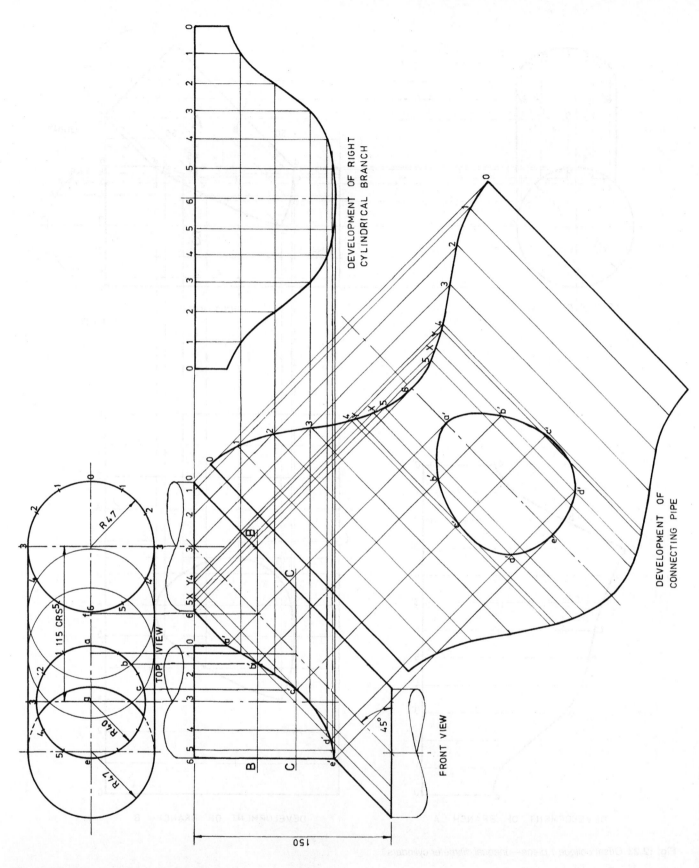

DEVELOPMENT OF RIGHT CYLINDRICAL BRANCH

DEVELOPMENT OF CONNECTING PIPE

TOP VIEW

FRONT VIEW

Fig. 12.24 *Oblique cylindrical connecting pipe*

Oblique cylindrical connecting pipe

Figure 12.24 illustrates the development of an oblique cylindrical connecting pipe with a cylindrical pipe insert. The development of the connecting pipe without the hole for the insert is described on page 304. The line of intersection between the insert and the connecting pipe for this problem must be determined before the development can be completed.

1. Draw the front and top views as shown.
2. To determine the line of intersection, use the method of sections described on page 302. Consider a horizontal section B-B cutting both pipes. Project this section onto the top view where it is represented by two circles, centres f and g, intersecting at b. Project b down to B-B to give b′, a point on the required line of intersection.
3. Draw another section C-C, project it onto the top view to give c, and project c back to the section C-C to give c′, another point on the line of intersection. In the same way indicate d′ and as many more points as are required to plot a satisfactory curve.
4. Draw the development of the oblique cylinder as described on page 306.
5. Points a′ and e′ are projected directly across to the centre line 6 on the development.
6. Draw surface element lines on the front view parallel to the axis and passing through b′, c′ and d′ to meet the end of the pipe at X and Y. Note b′ and d′ are on the one line.
7. Project X and Y onto the end of the development (two positions each), and draw surface element lines from these points across the development.
8. Project points b′ and d′ from the front view onto the two surface element lines from X. Project c′ onto the

two lines from Y. Join up the points plotted on the development to give the egg-shaped hole shown.

As this is a symmetrical development, each point found on the front view represents two points on the development. If the cylindrical insert were offset (e.g. if its axis were not in the same vertical plane as the axis of the connecting pipe), two curves would be required on the front view to give a non-symmetrical hole on the development similar to the problem for the offset oblique T piece development of Figure 12.23.

The development of the right cylindrical insert is shown on the right-hand side of Figure 12.24.

Figure 12.25 illustrates a second method of obtaining the line of intersection between the insert and the connecting pipe.

1. Draw the front and half top views.
2. Instead of using horizontal sections as in Figure 12.24, a series of vertical sections are taken through the insert and the connecting pipe. The small diagram illustrates how one such section A-A through points 4 and 2 on the front side of the insert pipe is used to determine points c and e (circled) on the line of intersection. (The shaded area represents the side view of the vertical section taken along plane A-A.) Points c and e can also be found by using points 4 and 2 on the back side, as is done on the large diagram when only a half plan is used.
3. Other points on the line of intersection are determined in a similar manner to c and e.

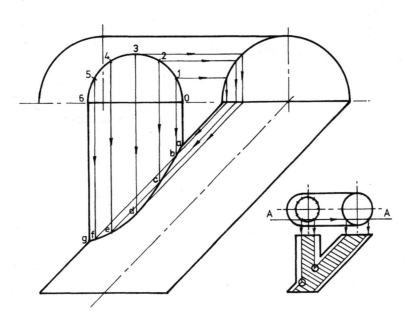

Fig. 12.25 *Alternative method of determining the line of intersection*

Exercises on development of cylinders and T pieces

The following constructions are given for reference. Practise them and become familiar with the method used.

12.5

Given the top view and front view of a cylindrical pipe, diameter of ends 50 mm, axis 75 mm, cut by two inclined planes at 30° to the horizontal plane, draw the development of the curved surface.

Solution

1. Obtain the length of the stretchout line for a cylinder. Draw a horizontal line tangential to point 3 in the top view to intersect lines drawn at 60° from points 0 and 6. This distance is equal to the semicircumference of the cylinder.
2. Produce stretchout lines from the front view and set off the semicircumference from 0 to 6 and from 6 to 0.
3. Divide the stretchout line into twelve equal parts and draw vertical lines to intersect the stretchout lines at a, b, c, d, e and f and a', b', c', d', e' and f'.
4. Draw a freehand curve through each point to complete the development of the curved surface of the cylinder.

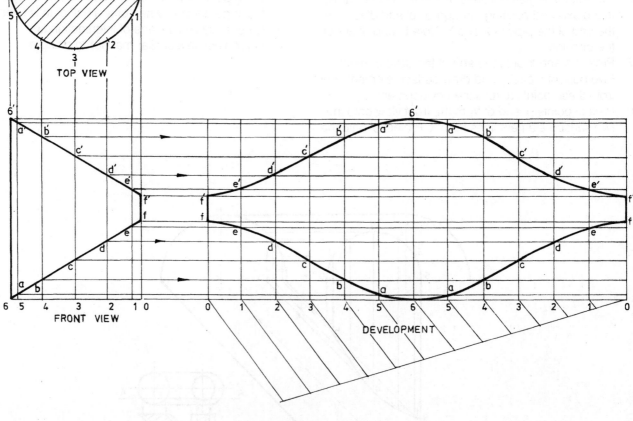

12.6

You are given the top view and the front view of a cylinder, diameter of end 76 mm, axis 88 mm, cut by two planes CB and AB. CB is a vertical plane and AB an inclined plane at 45° to the horizontal plane. The plane AB cuts the cylinder 20 mm above the base and CB is positioned at the mid-point between 3 and 4. Draw the development of the cylindrical part, and draw in one piece the two planes of section together with the remainder of the top.

Solution
1. Produce stretchout lines from the front view.
2. Obtain the length of the stretchout line and divide it into twelve equal parts. Number all points (see Exercise 12.5).
3. Draw vertical lines from points on the stretchout lines to intersect horizontal lines drawn from the front view to give points A, e, f and g on the development.
4. Mark the position of point B on the stretchout line and draw vertical lines from each point to intersect horizontal lines from the front view.
5. Join points A, e, f, g and B with a freehand curve and line in the development as shown.
6. To draw the true shape of the surface ABCD, project lines horizontally from the top view.
7. Set off on the centre line AD the distances Ae, ef, fg, gB and BC.
8. Draw vertical lines from e, f, g, B and C to intersect the lines drawn from the top view.
9. Join all points as shown.
10. Mark the centre point of the cylinder on the true shape and describe the arc CDC.
11. Line in and cross-hatch to complete the true shape.

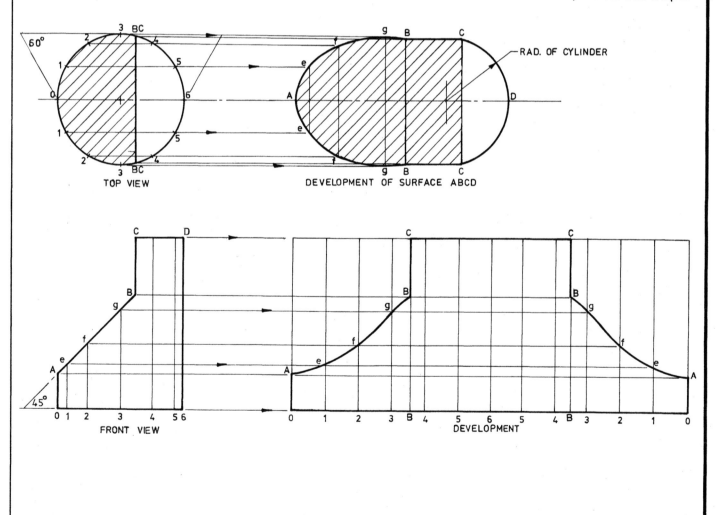

313

Development of pyramids

Right pyramid

A right pyramid may be defined as a surface with a number of identical triangular sides that have a common apex situated vertically above the centre of the base. An important fact to remember about all right pyramids is that the sloping edges may be totally contained within the surface of an enveloping cone. This is illustrated in Figure 12.26(a); Figure 12.26(b) is a pictorial view of a hexagon-based right pyramid—its development is described as follows.

1. Draw the front and half top view of the pyramid, either looking across the points (Fig. 12.26(c)) or across the flats (Fig. 12.26(d)). Figure 12.26(c) gives the true length of the edges (called the slant height) directly. Figure 12.26(d) requires the top view of the edge to be rotated into the base line, and this point joined to the apex A to give the slant height.

2. To develop the pyramid, an arc of radius equal to the slant height is described, and one of the base edges (taken from the half top view) is marked around the arc six times (Fig. 12.26(e)). These points are joined to the centre of the arc A and represent the fold lines along which the development is bent when 'forming up' the pyramid (Fig. 12.26(b)).

3. A pyramid may be truncated by a plane either parallel to the base (X–X) or at an angle to the base (Y–Y) (Fig. 12.26(c)). In the first case, the portion of the slant height (AX) from the apex to the truncating plane is used to describe an arc on the development, cutting the edge lines at points that are joined to give the line of truncation (XXX).

4. The angular truncation Y–Y which intersects the sloping edges at various distances from the apex needs to be plotted on the development. Project the points a and b (where Y–Y cuts the sloping edges) horizontally onto the slant height at points a' and b' (Fig. 12.26(c)). Then Aa' and Ab' are the true lengths of that part of A2 and A1 cut off by the plane Y–Y.

5. Transfer these lengths (Aa' and Ab') from the front view onto the lines A2 and A1 respectively on the development.

6. The two lengths of AY on the front view are transferred to the edges A0 and A3 on the development to complete the points that when joined give the line of truncation (YYY).

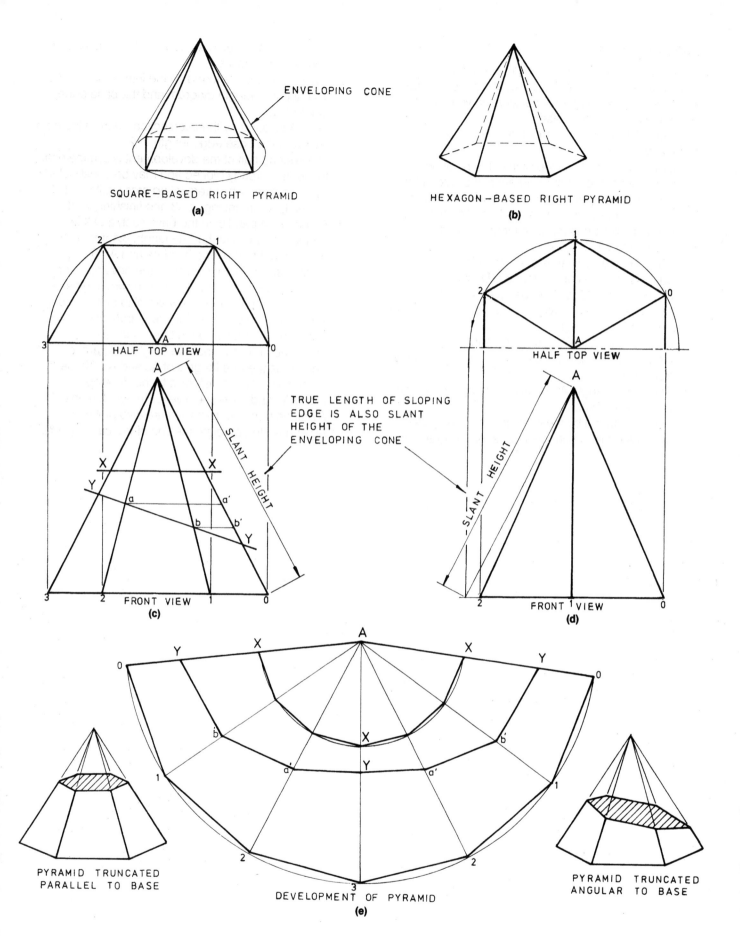

ENVELOPING CONE

SQUARE–BASED RIGHT PYRAMID
(a)

HEXAGON–BASED RIGHT PYRAMID
(b)

HALF TOP VIEW

TRUE LENGTH OF SLOPING
EDGE IS ALSO SLANT
HEIGHT OF THE
ENVELOPING CONE

SLANT HEIGHT

FRONT VIEW
(c)

HALF TOP VIEW

SLANT HEIGHT

FRONT VIEW
(d)

PYRAMID TRUNCATED
PARALLEL TO BASE

DEVELOPMENT OF PYRAMID
(e)

PYRAMID TRUNCATED
ANGULAR TO BASE

Fig. 12.26 *Right pyramid*

315

Oblique pyramid

The oblique pyramid (Fig. 12.27) may be defined as a surface with a number of flat unequal triangular sides which have a common apex not situated vertically above the centre of the base.

Figure 12.27 demonstrates the development of the oblique pyramid:

1. Draw a front and half top view of the pyramid.
2. Construct a true length diagram at the side of the front view in order to obtain the true length of the sloping edges. The diagram is based on the true length triangle described in Figure 12.8, and is constructed as follows.
3. Draw AP, the vertical height of the triangle, also equal to the vertical difference (VD) of the sides. From P plot P1 and P2 equal to A1 and A2 respectively on the half top view. Join A1 and A2 on the true length diagram, and these are the true lengths of the sloping edges A1 and A2. The true lengths of A0 and A3 may be taken directly from the front view.
4. Set down a length A0 on either the right or left side of the development.
5. Point 1 is found by the intersection of two arcs of radius A1 (taken from the true length diagram) and 1-2 which is equal to an edge of the base taken from the half top view.
6. Similarly, point 2 is found by the intersection of A2 from the true length diagram and the base edge length 1-2.
7. Point 3 is found by the intersection A3 from the front view and the base edge length 1-2.
8. The second half of the development is symmetrical to the half already plotted and may be constructed by projection or by intersecting arcs similar to the first half, commencing at A3 and finishing at A0.
9. A truncation parallel to the base such as XX is projected across to the true length diagram to determine the true lengths of Aa and Ab. These lengths are then transferred to the development along A2 and A1 respectively. The two lengths of AX taken from the front view are also plotted along A0 and A3 to complete the line of truncation (XXX).
10. A truncation angular to the base, such as YY, is projected across to the true length diagram to determine the true lengths of Ac and Ad. These lengths are then transferred to the development along A2 and A1 respectively. The two lengths of AY taken from the front view are also plotted along A0 and A3 to complete the line of truncation (YYY).

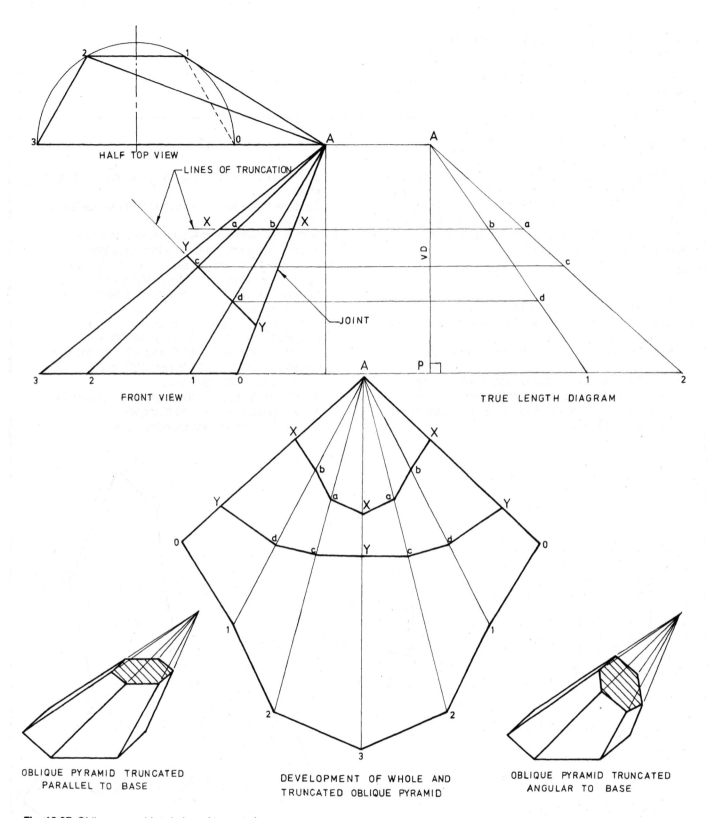

HALF TOP VIEW

LINES OF TRUNCATION

FRONT VIEW

TRUE LENGTH DIAGRAM

VD

JOINT

DEVELOPMENT OF WHOLE AND
TRUNCATED OBLIQUE PYRAMID

OBLIQUE PYRAMID TRUNCATED
PARALLEL TO BASE

OBLIQUE PYRAMID TRUNCATED
ANGULAR TO BASE

Fig. 12.27 *Oblique pyramid—whole and truncated*

Exercises on development of pyramids

Practise the methods outlined in these exercises to become familiar with them.

12.7

You are given the top view and front view of a square pyramid, edges of base 60 mm, axis 70 mm, cut by an inclined plane at 30° to the horizontal plane. The cutting plane intersects the axis of the pyramid 40 mm below the apex. Draw the true shape made by the cut and the development of the pyramid.

Solution

1. Mark points 1/2 and 3/4 on the cutting plane. Project these points to the top view to intersect the slant edges at 1, 2, 3 and 4.
2. Join all points and cross-hatch the top view of the cut.
3. To draw the true shape, rebate points 1/2 and 3/4 to ab and cd produced from the front view. Project vertically to intersect horizontal lines drawn from the top view of the cut.
4. Join points 1, 2, 3 and 4 and cross-hatch the true shape.
5. To draw the development, first find the true length of AC.
6. With A as centre and Ac as radius, describe an arc to intersect a horizontal line from A at point c′.
7. Project c′ to ab and cd produced and join to the apex. The line Ac′ is the true length of the pyramid.
8. With Ac′ as radius describe an arc. Step off one side of the square base four times around the arc and join all points to A. Join c, d, a, b and c.
9. Project 1/2 and 3/4 to the true length line.
10. Transfer the distance A1/2 and A3/4 on the true length line to points A1′2′ and A3′4′ on the development. Join all points.
11. Add the true shape of the truncated top and the square base to one of the triangles.
12. Line in to complete the development.

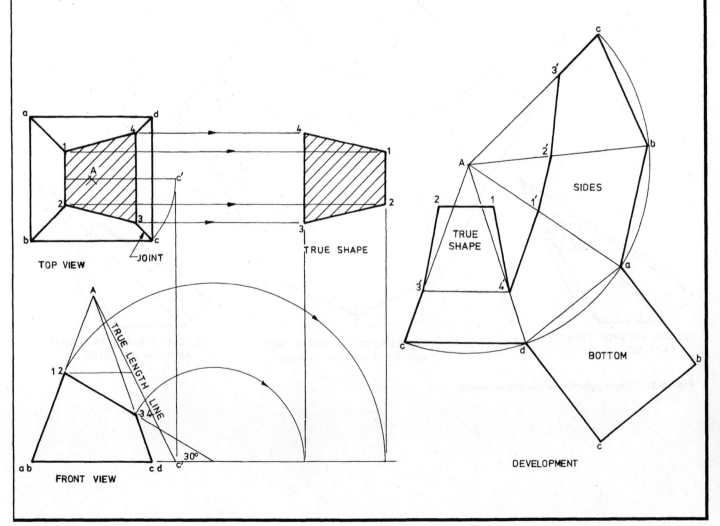

TOP VIEW

JOINT

TRUE SHAPE

FRONT VIEW

TRUE LENGTH LINE

30°

A

TRUE SHAPE

SIDES

BOTTOM

DEVELOPMENT

12.8

You are given the top view and front view of a square pyramid, edges of base 50 mm, axis 65 mm, truncated by a vertical plane at 45° to the vertical plane. The HT passes through the mid-point of the side ab in the top view. Draw the true shape made by the cut and the development of the pyramid.

Solution

1. Mark points 1, 2, 3 and 4 on the HT. Project these points to the corresponding slant edges in the front view.
2. Join all points and cross-hatch the front view of the cut.
3. To draw the true shape, rebate points 1, 2, 3 and 4 to the horizontal line as shown. Project these points down to intersect horizontal lines drawn from the front view of the cut.
4. Join points 1, 2, 3 and 4 and cross-hatch the true shape.
5. To draw the development, first find the true length of Ac.
6. With A as centre, and Ac as radius, describe an arc to intersect a horizontal line from A at point c.
7. Project c' to ac produced in the front view and join to the apex. The line Ac' is the true length of the pyramid.
8. With Ac' as radius describe an arc, step off one side of the square base four times around the arc and join all points to A. Join b, c, d, a and b.
9. Project 3 and 2 to the true length line.
10. Transfer the distance of A3 and A2 on the true length line to points A3 and A2 on the development.
11. From the top view, transfer the distance of a4 and d1 to the development. Join all points as shown.
12. Add the base and true shape as shown and line in to complete the development.

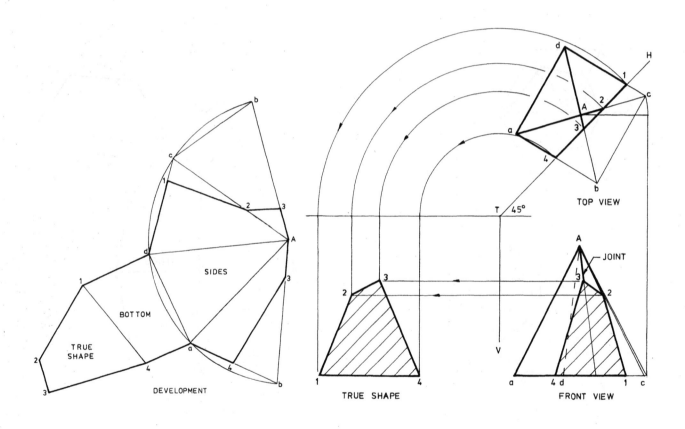

319

12.9

You are given the top view and the front view of a hexagonal pyramid, side of base 35 mm, axis 70 mm, truncated by an inclined plane at 30° to the horizontal plane. The cutting plane intersects the axis of the pyramid 40 mm below the apex. Draw the true shape made by the cut and the development of the pyramid.

Solution

1. Mark points 1, 2/6, 3/5 and 4 on the VT. Project these points to the top view to intersect the slant edges of 1, 2, 3, 4, 5 and 6.
2. Join all points and cross-hatch the top view of the cut.
3. To draw the true shape, draw a centre line parallel to the VT and project points 1, 2/6, 3/5 and 4 at right angles to the cutting plane to intersect the centre line.
4. Mark points 1 and 4.
5. Transfer the distance of 6 to the centre line in the top view either side of centre line on the true shape to obtain points 2 and 6. Points 3 and 5 are obtained in the same way.
6. Join all points and cross-hatch the true shape.
7. To draw the development, take Od, the true length of the pyramid, as radius and scribe an arc.
8. Step off one side of the base six times around the arc and join all points to o. Join d, e, f, a, b, c and d.
9. Project 1, 2/6, 3/5 and 4 to the true length line.
10. Transfer the distance of 0-1, 0-2/6, 0-3/5 and 0-4 on the true length line to their respective lines on the development. Join all points.
11. Add the true shape of the truncated top and the base as shown, and line in to complete the development.

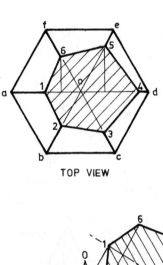

TOP VIEW

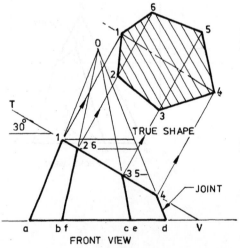

FRONT VIEW

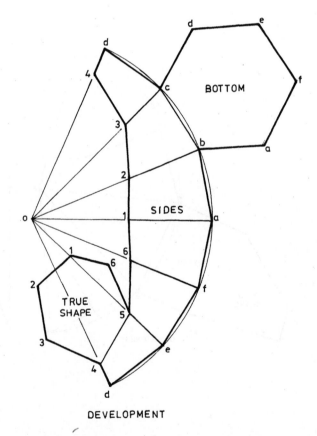

DEVELOPMENT

12.10

You are given the top and front views of an oblique square-based pyramid, side of base 50 mm, axis 50 mm. The axis has an inclination of 50°. Draw the development of the pyramid.

Solution

1. Draw the given views and letter all points.
2. Construct the true length diagram at the side of the front view.
3. With centre o and radius oba, the true length, describe an arc.
4. With centre o and radius ocd, the true length, describe an arc.
5. At a convenient position mark point d and join to the centre o.
6. Commencing at point d, mark off the length of the side of base. Letter all points.
7. Attach the base to complete the development.

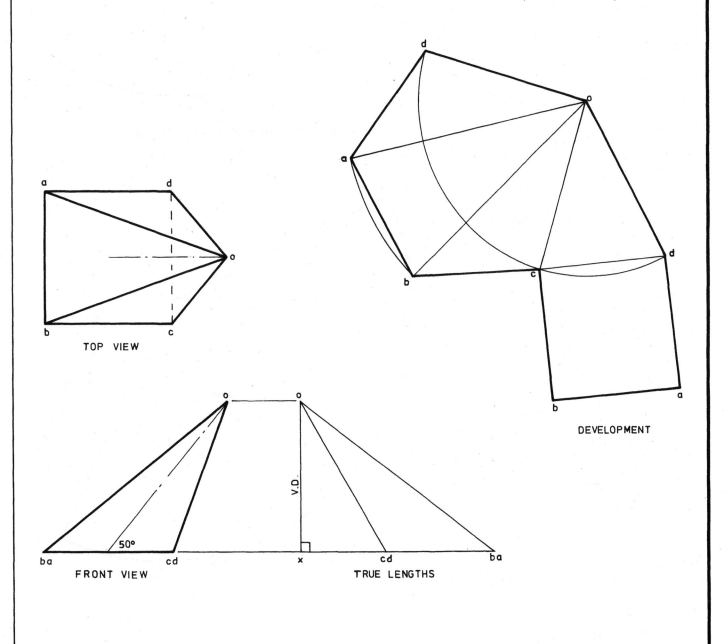

TOP VIEW

FRONT VIEW

TRUE LENGTHS

DEVELOPMENT

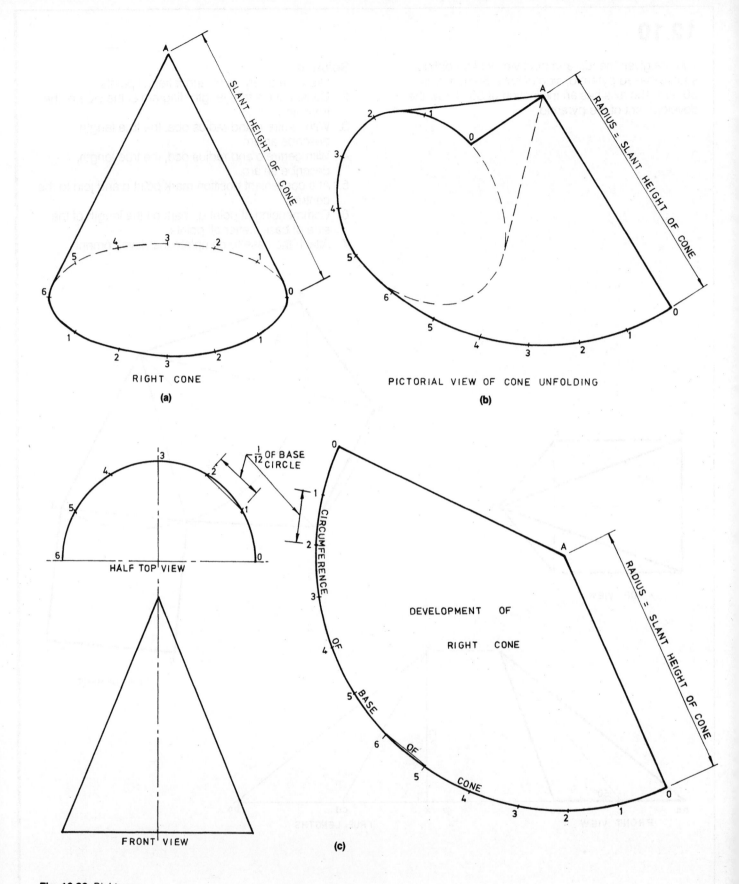

RIGHT CONE

(a)

PICTORIAL VIEW OF CONE UNFOLDING

(b)

HALF TOP VIEW

FRONT VIEW

$\frac{1}{12}$ OF BASE CIRCLE

CIRCUMFERENCE OF BASE OF CONE

DEVELOPMENT OF RIGHT CONE

RADIUS = SLANT HEIGHT OF CONE

(c)

Fig. 12.28 *Right cone*

Development of cones

Right cone

A right cone can be defined as a surface that has a circular base and a curved sloping side which radiates from a point situated vertically above the centre of the base. This point is called the *apex* of the cone. The length of any straight line drawn down the sloping side from the apex to the base is constant and is called the *slant height* of the cone. Figure 12.28(a) shows a pictorial view of the right cone.

The shape of a plane surface required for the development of a right cone is shown in Figure 12.28(b), and is part of a circle, the radius of which is equal to the slant height of the cone.

Figure 12.28(c) shows the development, which is obtained as follows:

1. Draw the front and half top view of the cone, dividing the base circumference into twelfths.
2. With radius A0, the slant height of the cone, describe an arc and mark off one-twelfth of the base circumference around this arc twelve times. Join the ends of the arc 0 to A to complete the development.

Right cone truncated parallel to the base

Figure 12.29 shows the development.

1. First draw the development of the whole cone as described in the previous section.
2. Describe a second arc on the development (R2) equal in length to the slant distance of the line of truncation from the apex A. The portion of the development between R1 and R2 is the truncated portion.

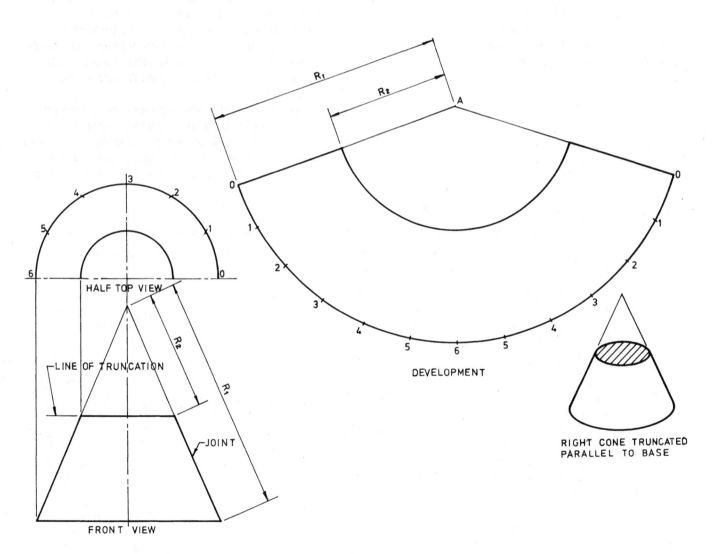

Fig. 12.29 *Right cone truncated parallel to the base*

323

Right cone truncated at an angle to the base

Figure 12.30 shows the development.
1. Draw the front and top views, showing the line of truncation on each view.
2. Divide the base into twelve equal parts, and draw surface element lines connecting these points to the apex, A1, A2 and so on.
3. The true lengths of the surface element lines between the apex A and the line of truncation 0'6' are found by projecting horizontally onto the slant height. That is, A5', A4', A3' and so on are the true lengths of these elements.
4. With centre A on the development and radius the slant height of the cone, describe an arc. Mark one-twelfth the base circumference around the arc twelve times and join these points to A. These are the surface element lines on the development corresponding to those on the front view.
5. Taking the true lengths A0', A1', A2', ... A6' from the slant height, mark them off successively along the corresponding surface element line on the development. Join the points with a smooth curve to complete the line of truncation.

Right cone-vertical cylinder intersection

Figure 12.31 shows the development.
1. Draw the front and half top views.
2. The line of intersection is determined by the method of horizontal sections. Draw X–X, a horizontal section cutting both surfaces.
3. Project this section onto the half top view, where it is represented by the two semicircles intersecting at x. Project x back to section X–X to point x', a point on the required line of intersection.
4. Continue taking a series of horizontal sections, all of which cut both surfaces, until sufficient points on the line of intersection have been found to draw a satisfactory curve (say four points). The line of intersection could also have been determined by the surface element method (see Fig. 12.16).
5. The development of the right cone is drawn, including surface element lines radiating from equal positions around the cone base. Corresponding surface element lines intersect the line of intersection at b, c, d, e and f on the front view.
6. Project these points horizontally onto the slant height of the cone to find the true lengths of the surface element lines between the apex A and the line of intersection.
7. To gain the line of intersection for joint 1, mark off the true lengths of Aa, Ab, Ac and so on along the corresponding surface element lines, A6, A5, A4 and so on, on the development. Draw a smooth curve through the points to give the line of intersection for joint 1.

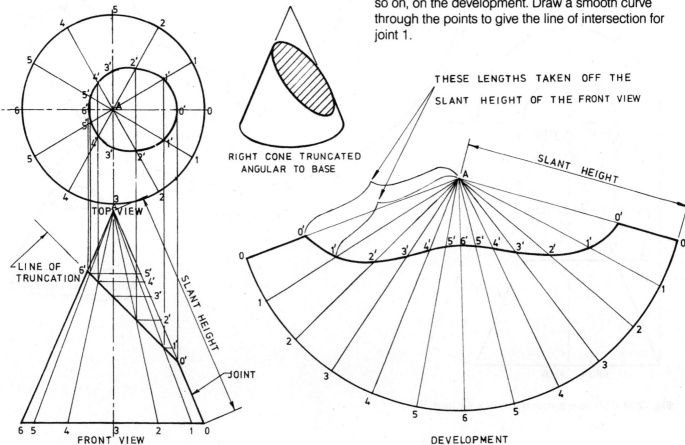

Fig. 12.30 *Right cone truncated at an angle to the base*

8. To gain the line of intersection for joint 2, mark off the true lengths of Aa, Ab, Ac and so on along A0, A1, A2 and so on. Draw a smooth curve through the points to give the line of intersection for joint 2.

The effect on the development of using one or the other of the two joints is shown. For joint 1 the line of intersection narrows the development at the ends and widens it at the centre. For joint 2 the reverse applies, and the development is wider at the ends and narrow in the centre. Joint 1 is more desirable as the joint is shorter and the development, with narrow ends and a wide centre, easier to handle.

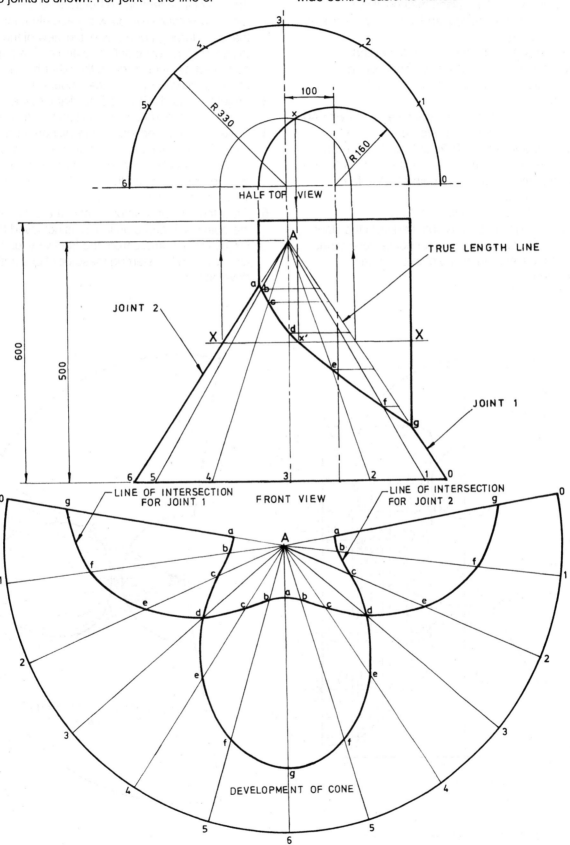

Fig. 12.31 *Right cone-vertical cylinder intersection*

Truncated right cone-right cylinder intersection

Figure 12.32 shows the development.

1. Draw the line of intersection of the cylinder and cone as described on page 302 and shown in Figure 12.17, using the cutting plane method. If the developments only are required, a half top view is all that is necessary.
2. Join the apex A on the top view to the cylinder intersection points b, c, d, e and f and extend onto the base circle at points 1', 2', 4', 5' and 3', respectively.
3. Draw the development of the truncated cone with A0 as the centre line of the development.
4. Step the distances 0' 1', 0' 2', 0' 3' and so on, taken from the base on the top view to either side of 0' on the development. Join these points to A to form surface elements.
5. Step along these elements from A the corresponding true lengths obtained from the slant height on the front view. Join the points determined to give the developed shape of the cylinder intersection.

Right cone-right cylinder, oblique intersection

Figure 12.33 shows the development.

1. It is necessary to draw an auxiliary view of the cone and cylinder showing the true shape of the cross-section of the intersecting cylinder in order to plot the line of intersection on the front view.
2. On the auxiliary view, draw surface element lines that pass through one-half of the view of the cylinder to intersect the base at 0, a, b and c. There is no need to draw lines through the other half as the line of intersection is symmetrical about A0.
3. Project a, b and c from the auxiliary view across to the base of the front view and join to the apex.
4. Project the intersections of the cylinder and the surface element lines on the auxiliary view across to the corresponding surface element lines on the front view to give points on the line of intersection. Join with a smooth curve.
5. Draw the overall development of the cone and plot the points a, b and c on it. This is achieved by projecting a, b and c from the front view up to the top view and transferring these positions onto the development.

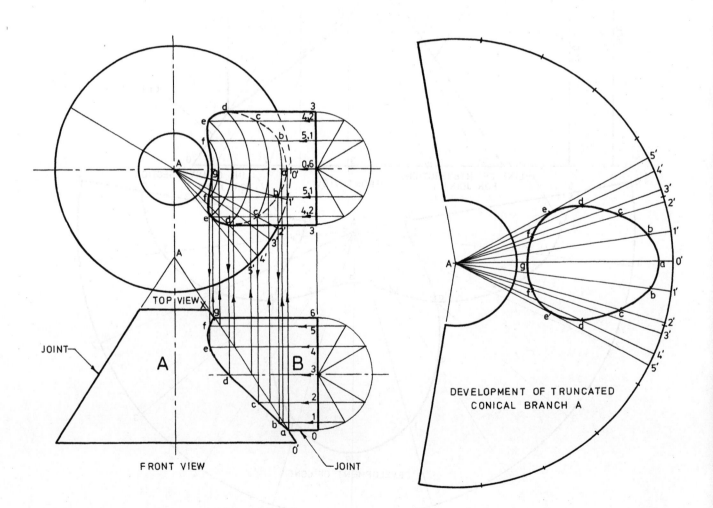

DEVELOPMENT OF TRUNCATED
CONICAL BRANCH A

Fig. 12.32 *Truncated right cone–right cylinder intersection*

326

6. Determine points on the line of intersection on the top view by projecting points on the line of intersection on the front view to the corresponding lines A0, Aa, Ab and Ac on the top view. Join these points on the top view with a smooth curve.
7. Project points from the line of intersection on the front view across to the slant height to give the true lengths of these elements, which are in turn transferred to the corresponding elements on the development to give points on the line of intersection. Draw a smooth curve through the points.

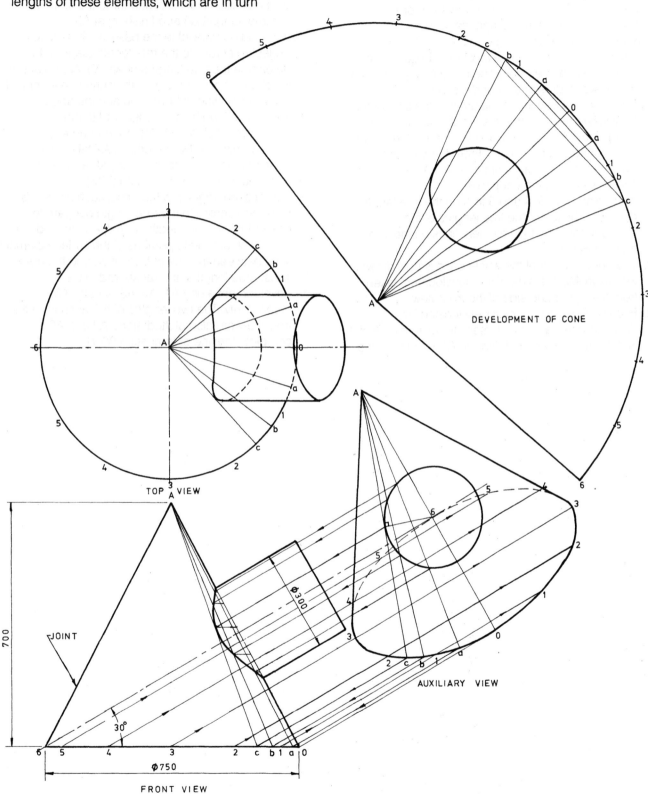

DEVELOPMENT OF CONE

TOP VIEW

AUXILIARY VIEW

JOINT

30°

Ø300

Ø750

FRONT VIEW

Fig. 12.33 *Right cone–right cylinder, oblique intersection*

327

Oblique cone

An oblique cone can be defined as a surface that has a circular base and a curved sloping side which radiates from a point not situated vertically above the centre of the base. The length of any straight line drawn down the sloping side from the apex to the base is not constant, hence the oblique cone does not have a constant slant height, and its development is somewhat more complicated than that of the right cone. The development is shown in Figure 12.34.

1. Draw the front and top views of the oblique cone, showing surface element lines connecting the apex to the twelve divisions around the base.
2. Construct the true length diagram to the side of the front view based on the true length triangle (Fig. 12.8). Draw AP, the vertical difference in the heights of all surface element lines. From P, mark off P1, P2, P3, P4 and P5 equal in length to A1, A2, A3, A4 and A5 respectively on the top view.
3. Join points 1, 2, 3, 4 and 5 to A on the true length diagram. These are the true lengths of the corresponding surface element lines on the front and top views.
4. The development of the whole cone is now drawn. Set down A0, one side of the development taken from the right-hand side of the front view.
5. Each point is now successively located by describing two arcs to intersect, for example point 1 is determined by describing arc A1 taken from the true length diagram to intersect with arc 01 equal to one-twelfth of the base circumference taken from the top view. Points, 2, 3, 4, 5 and 6 are plotted similarly, although A6 is taken from the left-hand side of the front view.
6. The second half of the development, which is symmetrical to the first, is determined by projection or is plotted in a similar manner to the first half commencing at A6 and finishing at A0.
7. A truncation parallel to the base, such as X–X, is projected across to the true length diagram to determine the true lengths of Aa, Ab, Ac, Ad and Ae, which are those portions of the surface element lines between the line of truncation and the apex.
8. These true lengths are transferred to the development along A1, A2, A3, A4 and A5, respectively. The two lengths of AX taken from the front view are also plotted along A0 and A6 to complete the line of truncation (XXX).
9. A truncation angular to the base, such as Y–Y, is projected across to the true length diagram to determine the true lengths of Aa′, Ab′, Ac′, Ad′ and Ae′, which are those portions of the surface element lines between the line of truncation and the apex.
10. These true lengths are transferred to the development along A1, A2, A3, A4 and A5, respectively. The two lengths of AY taken from the front view are also plotted along A0 and A6 to complete the line of truncation (YYY).

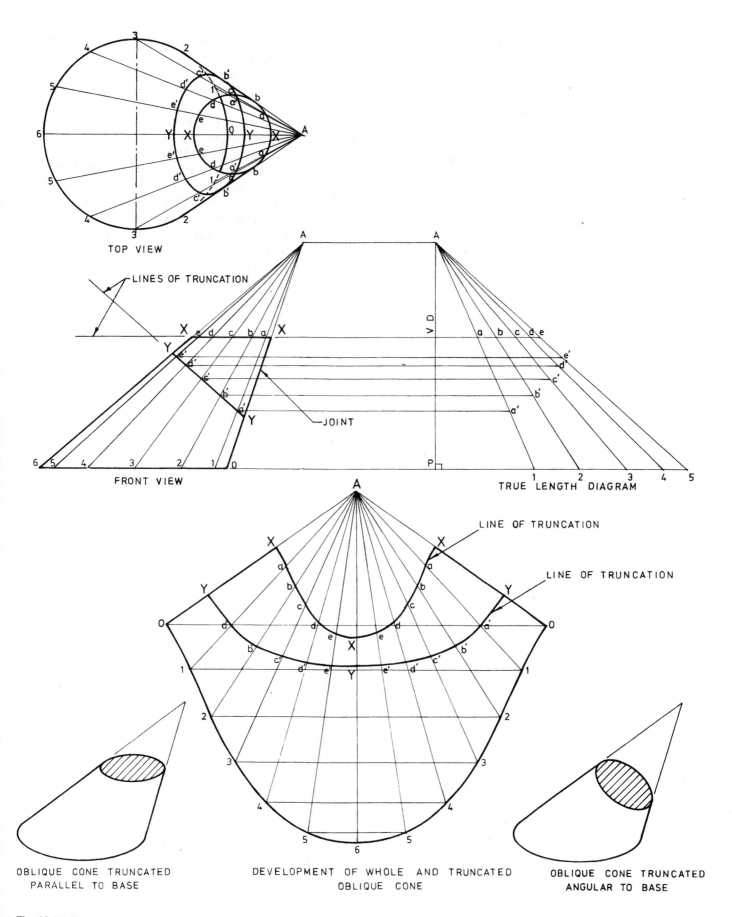

TOP VIEW

LINES OF TRUNCATION

FRONT VIEW

JOINT

TRUE LENGTH DIAGRAM

A

LINE OF TRUNCATION

LINE OF TRUNCATION

DEVELOPMENT OF WHOLE AND TRUNCATED
OBLIQUE CONE

OBLIQUE CONE TRUNCATED
PARALLEL TO BASE

OBLIQUE CONE TRUNCATED
ANGULAR TO BASE

Fig. 12.34 *Oblique cone—whole and truncated*

Oblique cone–oblique cylinder intersection

This is shown in Figure 12.35.

1. Draw the front and top views. These are required to draw the line of intersection between the two surfaces and hence plot it on the development.

2. The line of intersection is determined by the method of sections. Draw any horizontal section A–A cutting both surfaces of the cone and cylinder. Where the section cuts the axis of the cone, construct a semicircle on a diameter equal to the cross-section of the cone at this level.

3. Similarly construct another semicircle on a diameter equal to the cross-section of the cylinder at this level to cut the first semicircle at point a.

4. Project point a downwards onto the section line at point a′ to give a point on the line of intersection.

5. Project point a upwards to the centre line of the top view, and mark off a distance on either side of this line equal to aa′ taken from the front view. This locates two points on the line of intersection on the top view.

6. A sufficient number of horizontal sections are taken to provide enough points to enable smooth curves to be drawn on the front and top views to give the lines of intersection of the cone and cylinder. Usually four or five sections are required.

7. The development of the oblique cone can now proceed. First develop the whole cone, then plot on it the line of truncation as described on page 328. Note that the cone is inverted in this case, as is the true length diagram.

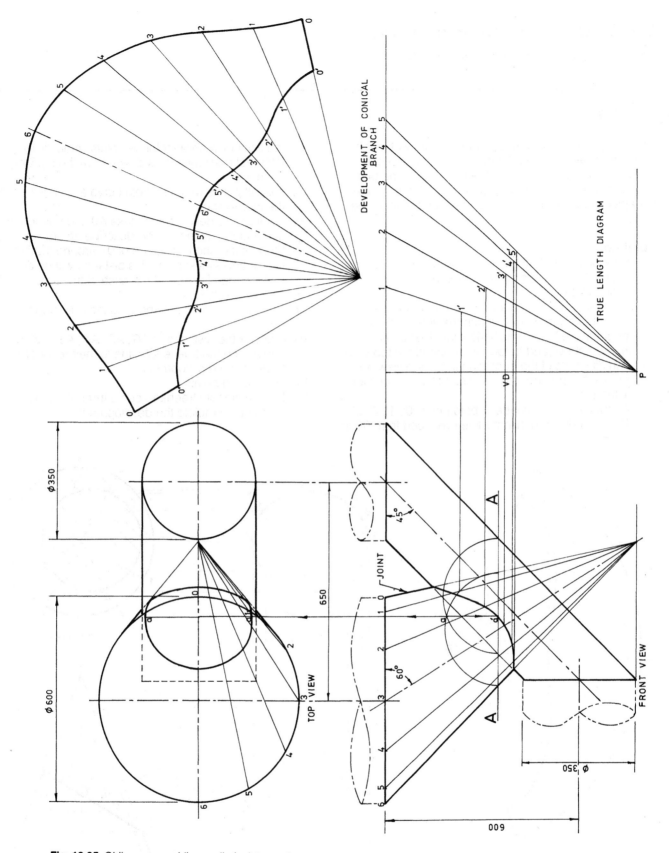

Fig. 12.35 *Oblique cone–oblique cylinder intersection*

Exercises on development of cones

The following exercises demonstrate the methods used in these constructions.

12.11

You are given the top view and the front view of a right cone, diameter of base 60 mm, axis 70 mm, truncated by an inclined plane at 30° to the HP. The cutting plane intersects the axis of the cone at a point 40 mm below the apex. Draw the true shape and development of the cone.

Solution

1. Mark points 0', 1', 2', 3', 4', 5' and 6' on the VT. Project these points to the corresponding generators in plan.
2. To obtain point 3' in plan, 3' in the front view is projected horizontally to touch the line A0, the true length of the cone. This horizontal distance is marked above and below the centre of the top view.
3. Join all points of the truncation in the top view with a smooth freehand curve. Cross-hatch the top view of the cut.
4. To obtain the true shape, rebate points 0', 1', 2', 3', 4', 5' and 6' to a horizontal line produced from 06 in the front view. Project these points vertically to intersect horizontal lines drawn from the top view of the cut.
5. Join all points with a smooth curve and cross-hatch the true shape.
6. To draw the development, take A0, the true length of the cone, as radius and describe an arc.
7. Step off the chord distance of 01 around the arc twelve times as shown. This distance will be equal to the circumference of the base of the cone.
8. Join all points to A.
9. Project 0', 1', 2', 3', 4' and 5' to the true length line A6.
10. Transfer the distance of A6', A5', A4', A3', A2', A1' and A0' on the true length line to their respective lines on the development.
11. Number all points.
12. Draw a smooth freehand curve through all points. Line in to complete the development.

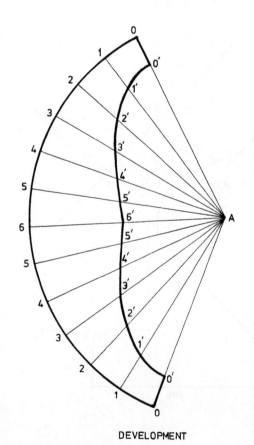

DEVELOPMENT

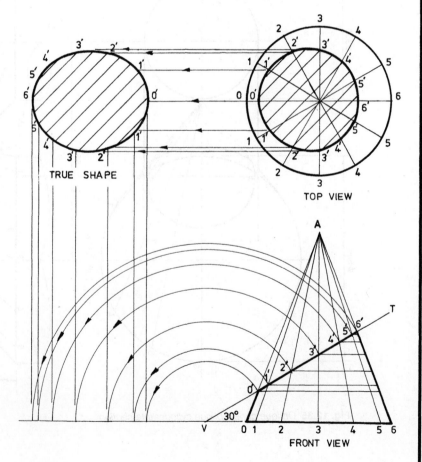

TRUE SHAPE

TOP VIEW

FRONT VIEW

12.12

You are given the top view and front view of a right cone, diameter of base 60 mm, axis 66 mm, truncated by two inclined planes at 30° and 45° respectively. Both planes intersect at the mid-point of the axis. Draw the true shapes of the truncated cone and the development of the remaining portion of the cone.

Solution

1. Mark points 0', 1', 2', 3', 4', 5' and 6' on the cutting planes. Project these points to the corresponding generators in the top view.
2. To obtain point 3' in the top view, 3' in the front view is projected horizontally to meet the the line A0. This horizontal distance is marked above and below the centre of the top view.
3. Join all points of the truncation in the top view with a smooth freehand curve. Cross-hatch the top view of the truncations.
4. To draw the true shape of the 45° truncation, draw a centre line parallel to the cutting plane and project points 0', 1', 2' and 3' at right angles to intersect the centre line. Mark point 0'.
5. Transfer the distance of 1' to the centre line in the top view either side of the centre line on the true shape to obtain points 1' and 1'. Points 2' and 2' and 3' and 3' are obtained in the same way.
6. Join all points with a freehand curve and cross-hatch the true shape.
7. Obtain the true shape of the 30° truncation using points 3', 4', 5' and 6'.
8. To draw the development, take A0, the true length of the cone, as radius and describe an arc.
9. Step off the chord distance of 01 around the arc twelve times as shown. Join all points to A.
10. Project points 5', 4', 3', 2' and 1' to their true length lines.
11. Transfer the distance of A6', A5', A4', A3', A2', A1' and A0' on the true length lines to their respective lines on the development. Number all points.
12. Draw a smooth freehand curve joining all points. Line in to complete the development.

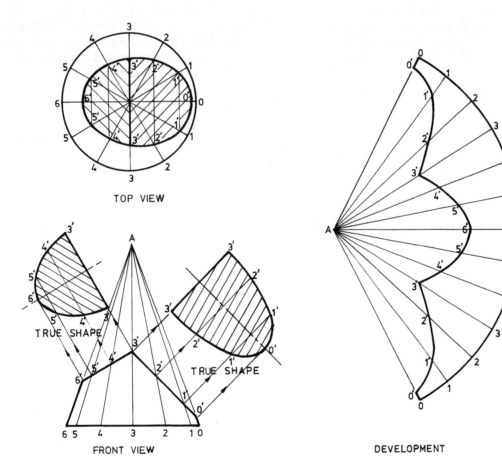

TOP VIEW

TRUE SHAPE

TRUE SHAPE

FRONT VIEW

DEVELOPMENT

Development of breeches or Y pieces

The breeches piece is a three-way junction between cylindrical pipes or between cylindrical pipes and conical sections. The angles between the various branches can be equal or unequal. The main requirement when drawing the front view and determining the line of intersection of the branches is that each branch should envelop a common sphere represented on the front view by a circle. This is shown in the three examples in Figures 12.36, 12.37 and 12.38. On breeches pieces involving equal cylinders (Figs 12.36 and 12.37), the common point of intersection of the three cylinders on the front view is also coincident with the intersection of the axes and the centre of the common sphere.

Once the front view has been drawn and the line of truncation determined, the development of the branches is merely that of truncated cylinders and cones.

Breeches piece—equal angle, equal diameters; unequal angle, equal diameters

This method applies to both Figures 12.36 and 12.37.

1. Draw the front view of the breeches piece, determining the line of intersection of the three branches A, B and C by the common sphere method outlined on page 304 and illustrated in Figure 12.18(b).

2. The development of the branches is determined by the surface element method described on page 323.

Note: The development of branches B and C would have been more conveniently projected at right angles to the side of B or C, as was the development of A. However, the positions have been chosen for the sake of page layout. The top view has been included for clarity.

Breeches piece—cylinder and two cones, equal angle

This method is shown in Figure 12.38.

1. Draw the front view of the breeches piece determining the line of intersection of the three branches A, B and C by the common sphere method outlined on page 304 and illustrated in Figure 12.18(b).

2. The developments of the conical branches B and C are identical. The method used is the surface element method described on page 323 and illustrated in Figures 12.29 and 12.30.

Note: The junction of the lines of intersection, point d on the front view, does not fall on a surface element line of the cone, and an extra line is drawn through d to intersect the base at D. Point D is then plotted on the development between points 2 and 3, and the surface element line AD is drawn in.

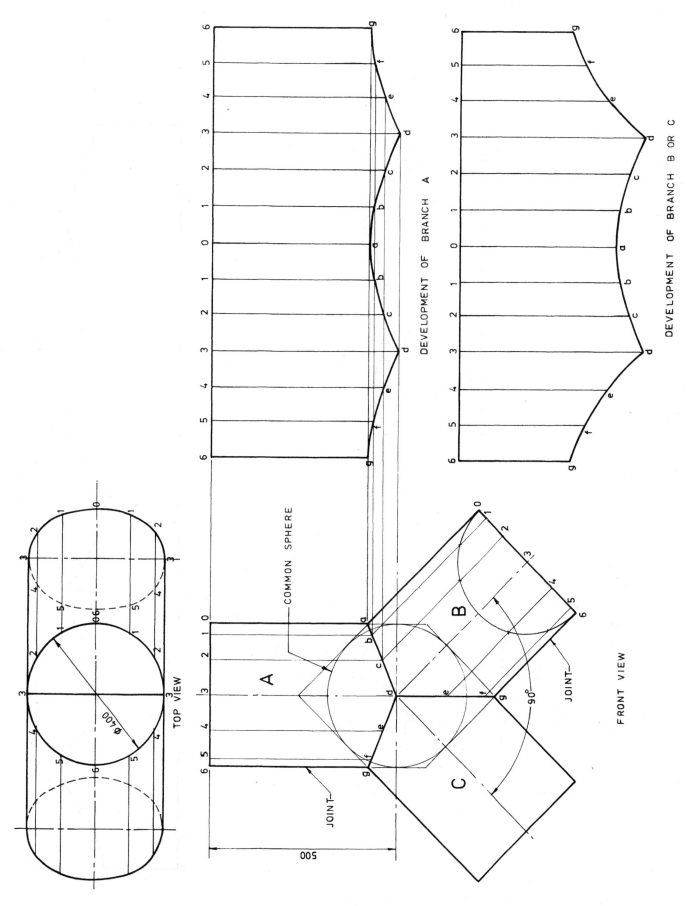

Fig. 12.36 *Breeches piece—equal angle, equal diameter*

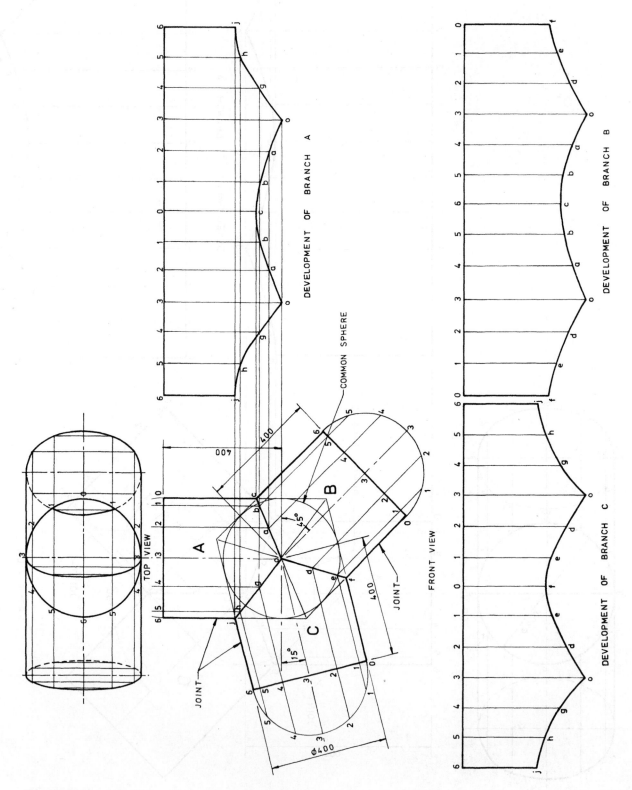

Fig. 12.37 *Breeches piece—unequal angle, equal diameter*

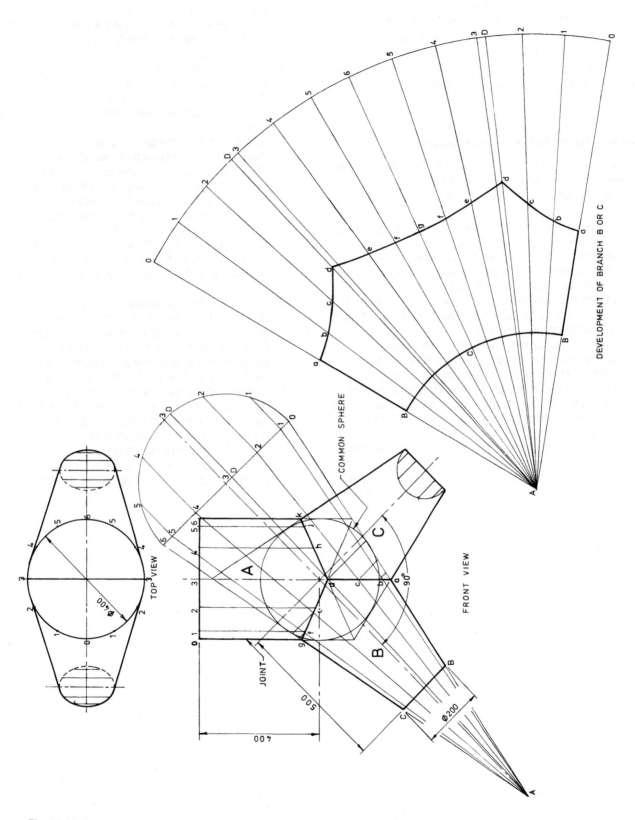

DEVELOPMENT OF BRANCH B OR C

COMMON SPHERE

TOP VIEW

FRONT VIEW

JOINT

Fig. 12.38 *Breeches piece—cylinder and two cones, equal angle*

337

Development of transition pieces

Transition pieces are used to connect pipes of different cross-sectional shapes and areas. Their development is generally achieved by a technique called *triangulation*. The method involves dividing the curved surface into a number of segments resembling triangles, finding the true shape of each, and laying these down side by side to form the true surface development.

Round-to-round transition piece

A pictorial view of the transition piece that connects two circular sections not in parallel planes is shown in Figure 12.39(a). Its development is obtained as follows (refer to Fig. 12.39(b)):

1. Draw the front and top views of the transition piece to be used as an aid to triangulate the curved surface, that is, consider the surface as consisting of a number of flat triangles lying side by side and having their bases at one end or the other.
2. Divide the two openings into twelve equal parts 0, 1, 2, 3 and so on, and a, b, c and so on, with 0 and a located on the joint line.
3. These divisions are now joined with surface element lines in a zigzag pattern which has the effect of dividing the surface into the triangular pattern. It is convenient to distinguish the surface element lines by making them alternately dash and full lines in order to avoid confusion on the true length diagram.
4. Commencing at 0 on both the front and top views, draw a full line up to b, dash line down to 1 and so on until the final line is a full line up to g. A line is considered to connect g6 although it is not shown.

5. Construct the true length diagram to the side of the front view. A common vertical difference (VD) line is used, and the heights of the points b, c, d, e, f and g are projected across to it.
6. The top view lengths of the full lines are now taken off the top view, set off to the right of the VD line and numbered 0, 1, 2, 3, 4 and 5.
7. These numbers are now joined to the corresponding top points b, c, d, e, f and g to give the true length of the full element lines. For example 0 joins b, 1 joins c (notice these cross over) and 2 joins d.
8. Similarly, a true length diagram is constructed for the dash lines to the left of the VD line.
9. The development can now be drawn. Set down a0 taken from the right-hand side of the front view (it is a true length). Draw the true shape of the first triangle a0b as follows: from a describe an arc ab equal to one-twelfth of the top opening circumference; from 0 describe another arc equal to 0b taken from the full line side of the true length diagram to cut the first arc at b; join 0b with a full line.
10. Now draw the true shape of the second triangle 0b1 as follows: from 0 describe an arc 01 equal to one-twelfth of the bottom opening circumference; from b describe another arc equal to b1 taken from the dash line side of the true length diagram to cut the first arc at 1; join b1 with a dash line.
11. Continue constructing all the triangles until line g6 is set down. Its true length is taken from the left-hand side of the front view. Line g6 represents the dividing line of the development.
12. The second half of the development is best obtained by projecting points across at right angles to g6 and using transfer ordinates to plot the second half.

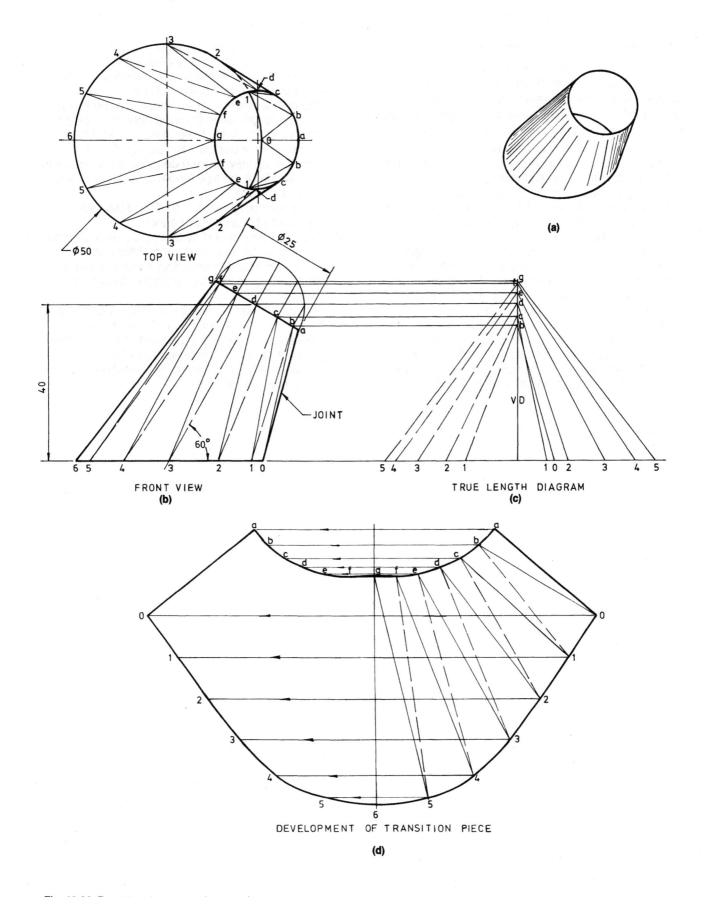

TOP VIEW

Ø50

Ø25

40

60°

JOINT

FRONT VIEW
(b)

TRUE LENGTH DIAGRAM
(c)

V D

(a)

DEVELOPMENT OF TRANSITION PIECE

(d)

Fig. 12.39 *Transition piece—round-to-round*

Square-to-round transition piece

The development is carried out in much the same way as the previous exercise. The triangulation of the surface is somewhat different because, as can be seen from Figure 12.40, there are four flat triangles, the bases of which correspond to the sides of the square end. Only the curved surfaces are triangulated.

1. Draw the front and top views to be used as an aid to triangulate the curved surfaces.
2. Divide the circular end into twelve equal divisions such that 0, the bottom of the joint, is at the apex of the shortest triangular side and is also one of the circular end divisions.
3. Draw two sets of three triangles on the front view, each set having a common apex at c and b and bases coinciding with a division of the circular end as shown. The pictorial view in Figure 12.40(a) identifies the triangles more clearly.
4. Draw these triangles on the top view as well.
5. Construct the true length diagram to the side of the front view. Care must be taken to ensure that the top view lengths set out to the right of the VD line are joined to the correct height point. If difficulty is experienced, make one set of triangles full lines and the other set dash lines to distinguish between them more easily. Alternatively, put one set on the left and one set on the right of the VD line.
6. The development is now set out, commencing at line a0, with length taken from the right-hand side of the front view. The true shape of triangle ab0 is found as follows: from a describe an arc equal to ab taken from the top view; from 0 describe an arc equal to 0b taken from the true length diagram to intersect the first arc at b.
7. The three triangles b0/1, b1/2 and b2/3 are constructed in a similar manner.
8. Triangle b3c is found by describing an arc from b equal to bc taken from the top of the front view (not the top view) and intersecting it with another arc from 3 equal to 3c taken from the true length diagram.
9. The second set of triangles is now constructed as before, then finally triangle cd6, which is an isosceles triangle with cd equal to twice ab taken from the top view. The remainder of the development is completed by projection and the use of transfer ordinates.

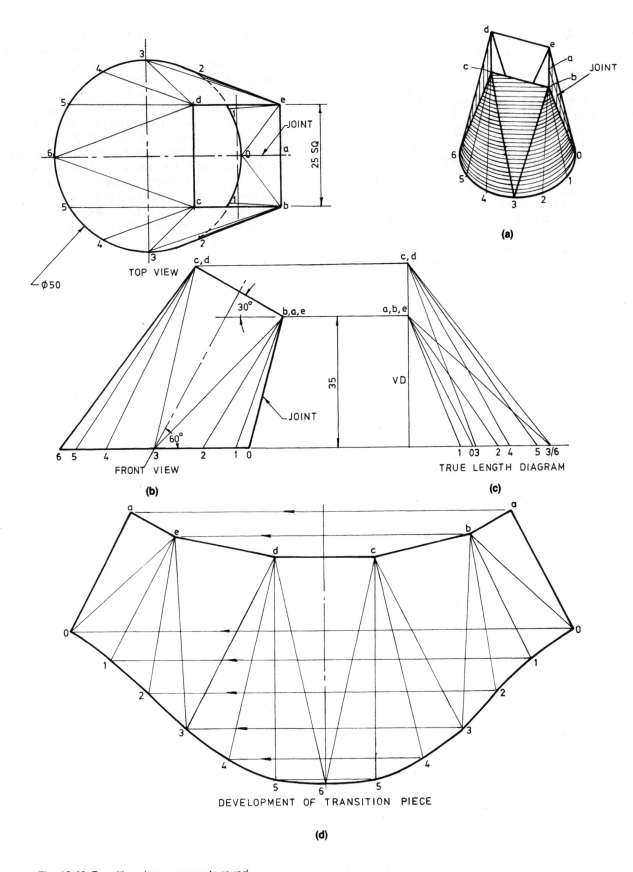

TOP VIEW

JOINT

25 SQ

Ø50

(a)

JOINT

30°

b, a, e

a, b, e

35

VD

60°

JOINT

6 5 4 3 2 1 0

FRONT VIEW

1 0 3 2 4 5 3/6

TRUE LENGTH DIAGRAM

(b)

(c)

DEVELOPMENT OF TRANSITION PIECE

(d)

Fig. 12.40 *Transition piece—square-to-round*

Oblique hood

The method used to construct an oblique hood is shown in Figure 12.41.

1. Draw the front and top views, showing the lines of triangulation joining equal divisions of the top and base. Note that the true shape of the base is elliptical and the top view is circular.
2. Construct the true length diagram using a common base along the top. Top view lengths of the element lines taken from the top view are set off along the common base line on either side of the VD line to give points a, b, c, d, e and f. The dash lines are set off to the left of the VD line and full lines to the right to lessen confusion.
3. Project points 1, 2, 3, 4, 5 and 6 across to the VD line, and join them to the appropriate base line point to give the true length of the surface element lines.
4. Set down vertically the centre line of the development, a0, equal to a0 on the front view. Now construct triangle a0/1 by describing an arc equal to a division of the elliptical base from 0 and intersecting it at 1 with another arc equal to a1 (full line) from the true length diagram described from a.
5. Continue constructing triangles until line g6 is laid down. Draw smooth curves through the apex points of the triangles to complete half the development.
6. The second half of the development is plotted quickly by projecting each point horizontally across from the first half and marking ordinates on the left of the centre line equivalent to those on the right. Draw a smooth curve through these points to complete the development.

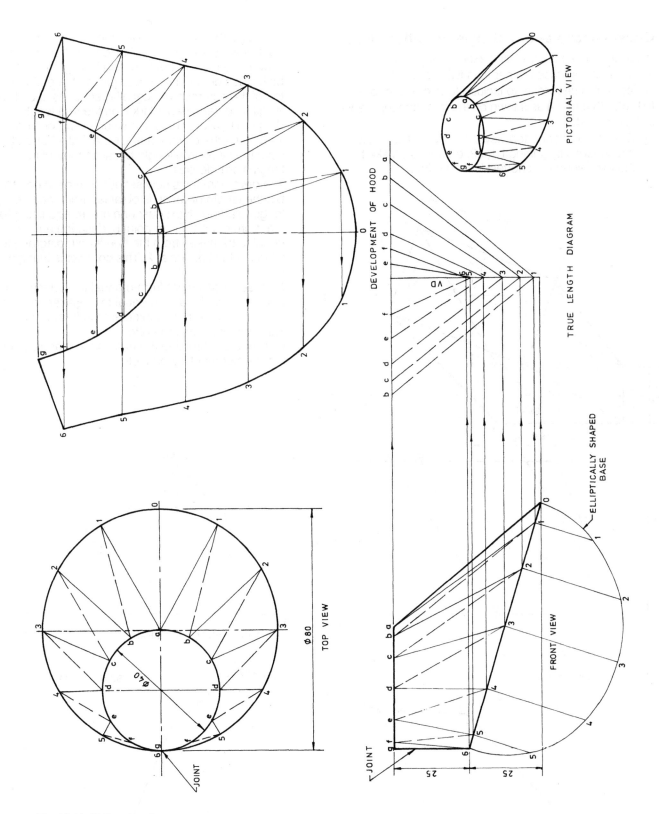

Fig. 12.41 *Oblique hood*

DEVELOPMENT OF HOOD

TRUE LENGTH DIAGRAM

PICTORIAL VIEW

TOP VIEW

FRONT VIEW

ELLIPTICALLY SHAPED BASE

JOINT

Ø80

Ø40

Offset rectangle-to-rectangle transition piece

In order to make the transition piece using a series of flat triangular surfaces rather than twisted quadrangular surfaces, it is necessary to include four kinked edges (b1, c2, d3 and a4) as shown on the front and top views in Figure 12.42.

1. Draw the front and top views in order to triangulate the surfaces by joining b1, c2, d3 and a4. Make b1 the joint along the shortest kinked edge.

2. Construct the true length diagram to the side of the front view by transferring plan lengths from the top view to the base of the true length diagram using both sides of the VD line and joining the ends to the top of the common vertical difference. As each true length is determined, mark it on the true length diagram to avoid confusion when taking off true lengths for the development. Note that a1 and c2 have the same true length because their plan lengths are identical.

3. The development is now set out commencing at line b1, the length of which is obtained from the true length diagram. Next describe an arc equal to 340 mm from point 1. From point b, describe an arc equal to b2 (taken from the true length diagram) to intersect the first arc at 2. This completes triangle b1/2.

4. Next describe an arc from b equal to 150 mm. From point 2, describe an arc equal to c2 (taken from the true length diagram) to intersect the first arc at c. This completes triangle b2c.

5. Continue constructing the true shape triangles until the development is complete.

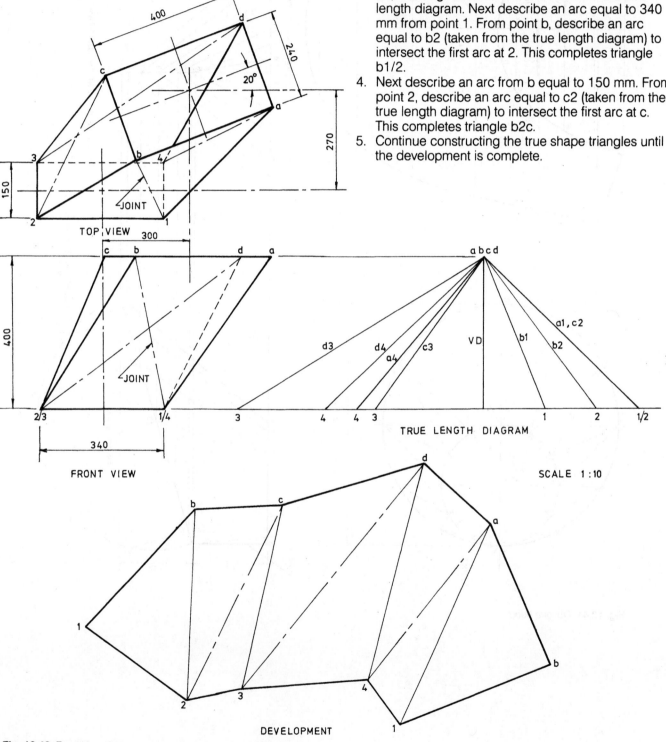

Fig. 12.42 *Transition piece—rectangle-to-rectangle, angular offset*

344

Practical metalwork developments

The method used to draw metalwork development is shown in the following figures. These are the basic methods used in all metalwork developments.

Figure 12.43 shows the layout method and illustrates allowances for lap seams, notching and folded edges. Figure 12.44 illustrates the allowances for folded seams, notching and wired edges and Figure 12.45 illustrates the allowances for a container having separate ends and wired edges. Finally, Figure 12.46 shows the method used to obtain the layout development for a container with sloping slides.

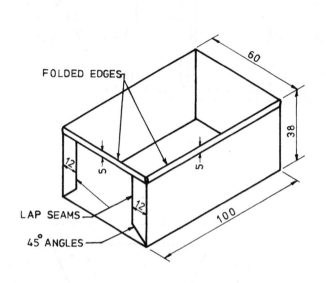

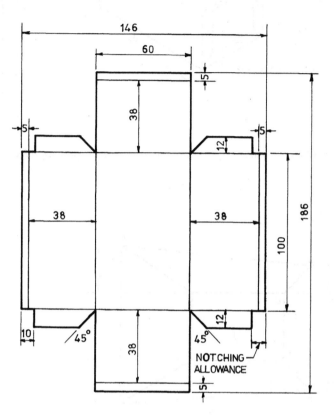

Fig. 12.43 *Allowances for lap seams, notching and folded edges*

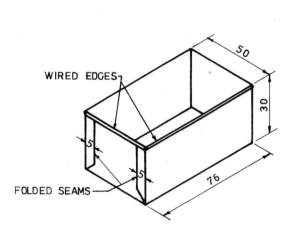

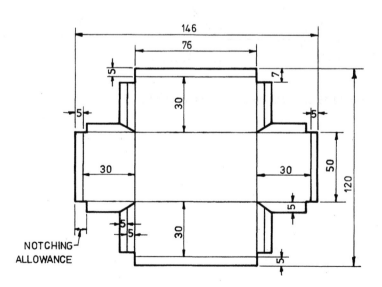

Fig. 12.44 *Allowances for folded seams, notching and wired edges*

345

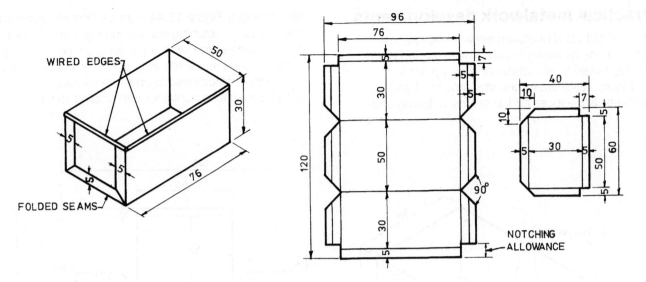

Fig. 12.45 *Allowances for container with separate ends and wired edges*

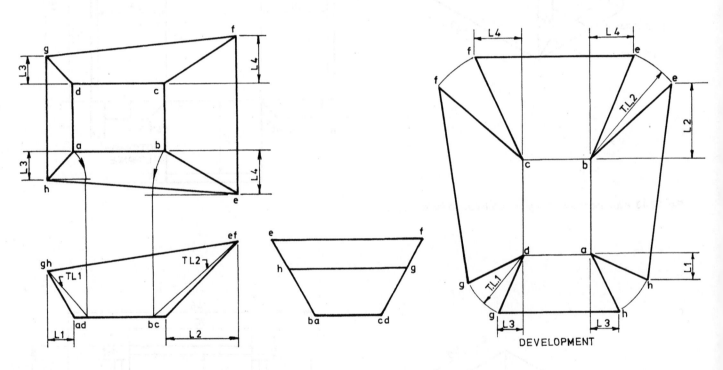

Fig. 12.46 *Layout for container with sloping sides*

Exercises on practical metalwork

These exercises have been compiled to give further practice in the development of practical metalwork constructions by parallel and radial line methods. The primary objective is to relate true length of line exercises to the practical developments of metalwork projects.

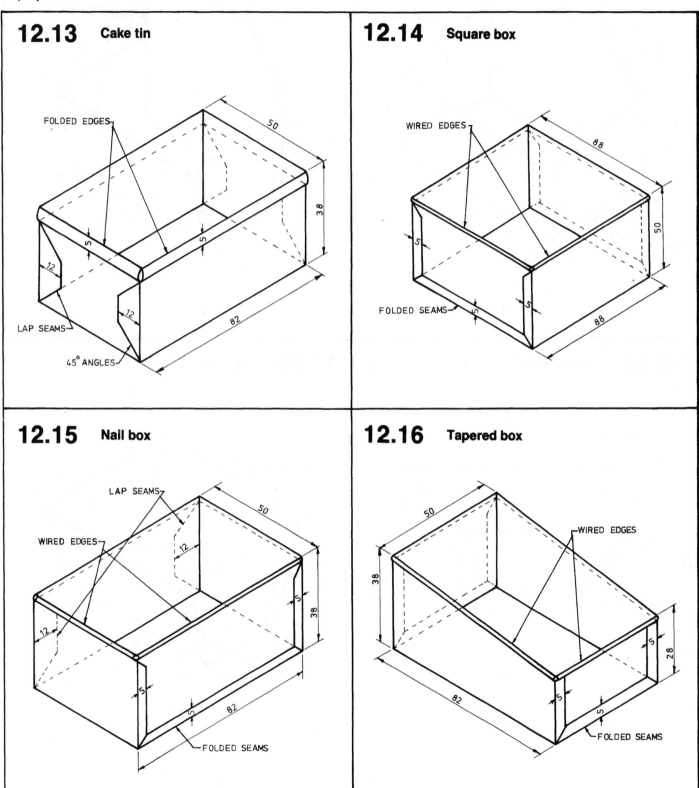

12.13   Cake tin

FOLDED EDGES
50
38
5
5
12
12
LAP SEAMS
45° ANGLES
82

12.14   Square box

WIRED EDGES
88
50
5
FOLDED SEAMS
5
5
88

12.15   Nail box

LAP SEAMS
50
WIRED EDGES
12
5
38
12
5
5
82
FOLDED SEAMS

12.16   Tapered box

50
WIRED EDGES
38
5
28
82
5
5
FOLDED SEAMS

12.17 Screw tin

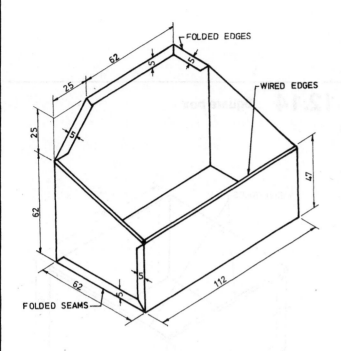

12.18 Feed tin

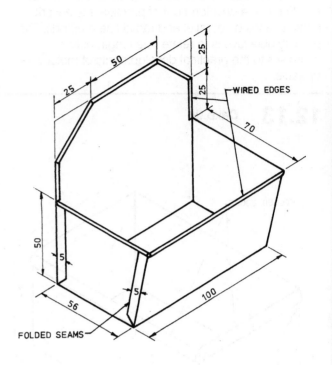

12.19 Tapered dish

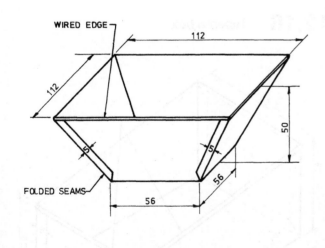

12.20 Parts box

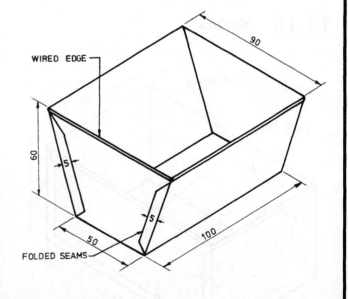

12.21 Scoop

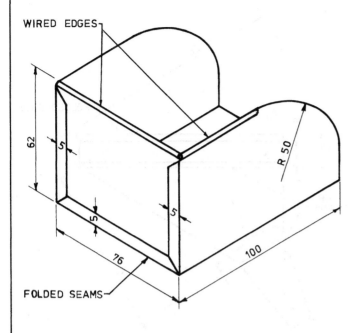

WIRED EDGES

R 50

62

5

5

5

76

100

FOLDED SEAMS

12.22 Tapered scoop

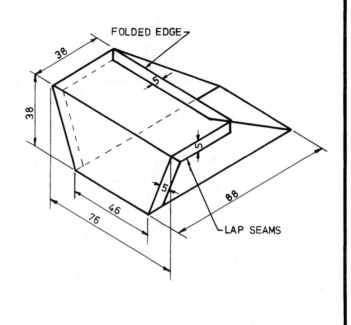

FOLDED EDGE

38

38

5

5

5

46

76

88

LAP SEAMS

12.23 Feed hopper

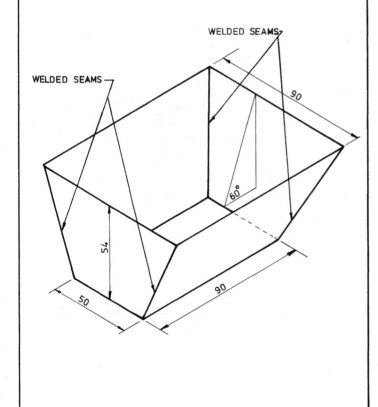

WELDED SEAMS

WELDED SEAMS

90

60°

54

50

90

12.24 Wheelbarrow tray

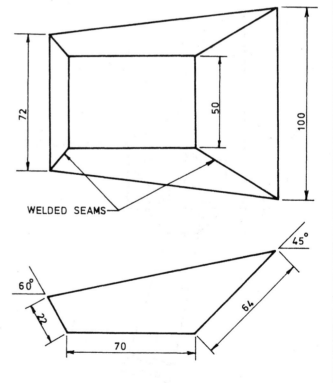

72

50

100

WELDED SEAMS

45°

60°

22

64

70

12.25 Sugar tin

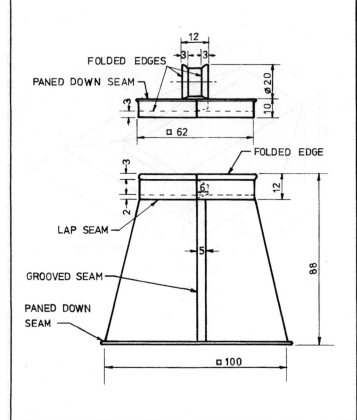

FOLDED EDGES

PANED DOWN SEAM

12

3 3

3

□ 20

10

□ 62

3

6

2

12

FOLDED EDGE

LAP SEAM

5

GROOVED SEAM

88

PANED DOWN
SEAM

□ 100

12.26 Square funnel

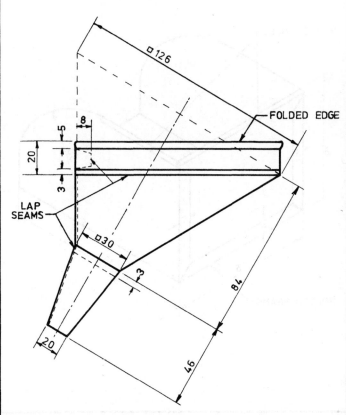

□ 126

FOLDED EDGE

5

8

20

3

LAP
SEAMS

□ 30

3

20

84

46

12.27 Grain scoop

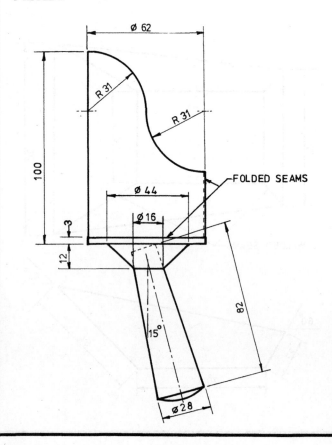

Ø 62

R 31

R 31

100

FOLDED SEAMS

Ø 44

Ø 16

3

12

82

15°

ø 28

12.28 Fuel measure

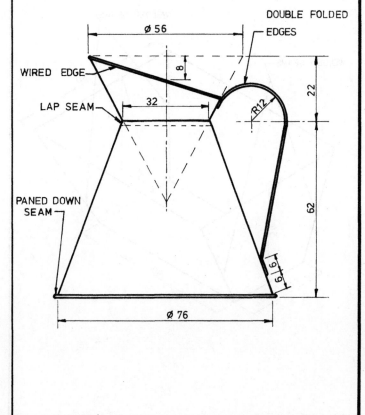

DOUBLE FOLDED
EDGES

Ø 56

WIRED EDGE

8

LAP SEAM

32

R12

22

PANED DOWN
SEAM

62

6
6

Ø 76

12.29

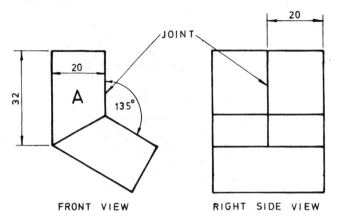

JOINT

20

20

32

A

135°

FRONT VIEW

RIGHT SIDE VIEW

Draw the development of part A of the rectangular pipe bend illustrated.
Scale 2:1

12.30

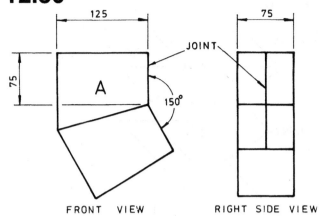

125

75

JOINT

75

A

150°

FRONT VIEW

RIGHT SIDE VIEW

Draw the development of part A of the rectangular pipe bend illustrated.
Scale 1:2

12.31

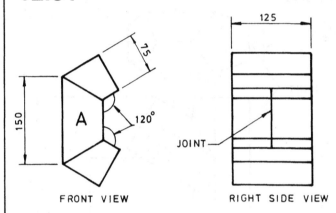

75

125

150

A

120°

JOINT

FRONT VIEW

RIGHT SIDE VIEW

Draw the development of part A of the rectangular pipe bend illustrated.
Scale 1:2

12.32

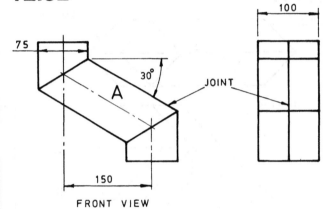

75

100

30°

A

JOINT

150

FRONT VIEW

Draw the development of part A of the rectangular pipe offset illustrated.
Scale 1:2

12.33

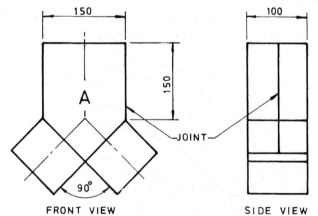

150

100

150

A

JOINT

90°

FRONT VIEW

SIDE VIEW

Draw the development of part A of the rectangular pipe junction illustrated.
Scale 1:2

12.34

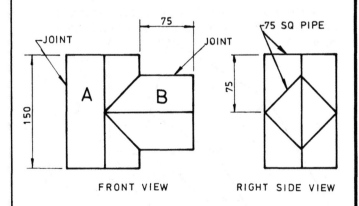

JOINT

75

JOINT

75 SQ PIPE

150

A

B

75

FRONT VIEW

RIGHT SIDE VIEW

Draw the development of parts A and B of the square pipe T piece illustrated.
Scale 1:2

12.35

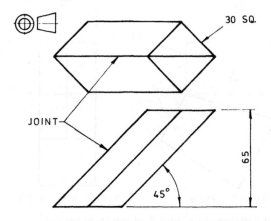

30 SQ.

JOINT

65

45°

Draw the development of the open, oblique square-based prism shown above.
Scale 1:1

12.36

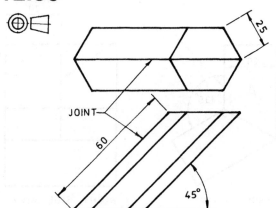

25

JOINT

60

45°

Draw the development of the open, oblique, hexagon-based prism shown above.
Scale 1:1

12.37

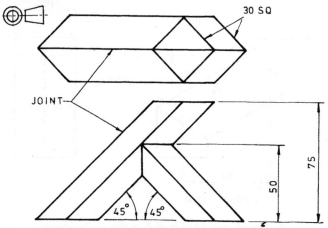

30 SQ

JOINT

75

50

45° 45°

Draw the development of both branches of the oblique junction piece shown above.
Scale 1:1

12.38

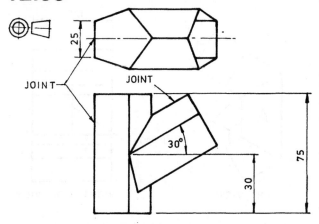

25

JOINT   JOINT

30°

75

30

Draw the development of both parts of the pentagonal pipe junction shown above.
Scale 1:1

12.39

Draw the development of the full and half segments required to make the 90° lobster-back bend shown.
Scale 1:1

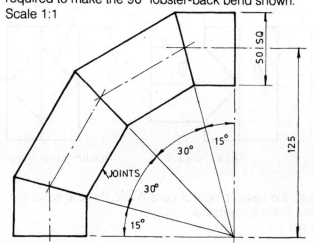

50 SQ

15°
30°
30°
JOINTS
15°

125

12.40

Draw the development of the segment required to make the 90° lobster-back bend shown.
Scale 1:1

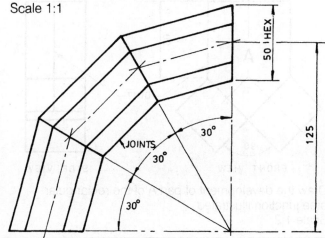

50 HEX

30°
JOINTS
30°
30°

125

12.41

Complete the front view, showing the line of intersection of the branches of the T piece, and draw the development of both branches.
Scale 1:1

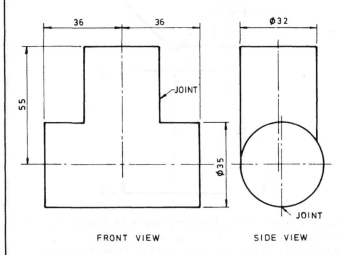

FRONT VIEW   SIDE VIEW

12.42

Develop a pattern for the segments of the lobster-back quarter bend of round pipe shown.
Scale 1:5

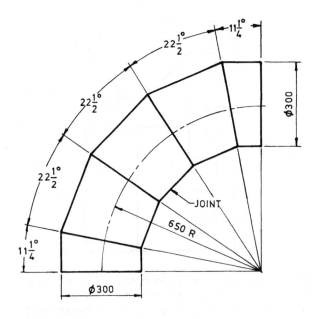

12.43

The front view of a conical sheet-metal hood is given. Draw the development of each part of the hood.
Scale 1:5

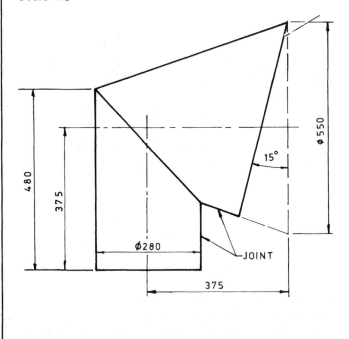

12.44

The front view of a conical fume extraction hood is shown. Draw its development.
Scale 1:5

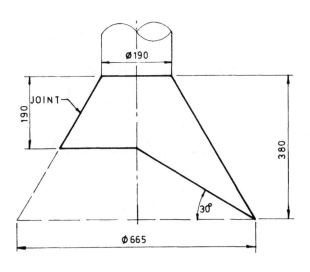

12.45

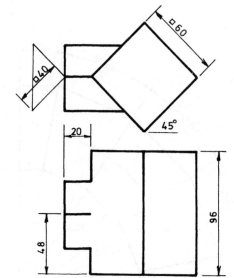

Given the top view and incomplete front view of two intersecting square prisms, draw the lines of intersection in the front view and the development of the vertical prism showing the hole.
Scale 1:1

12.46

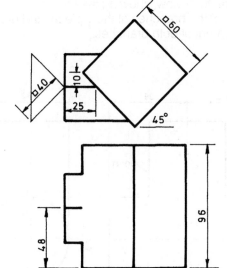

Given the top view and incomplete front view of two intersecting square prisms, draw the lines of intersection in the front view and the development of both prisms.
Scale 1:1

12.47

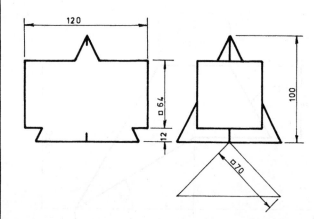

Given the side view and incomplete front view of a square pyramid penetrating a square prism, draw the lines of intersection in the front view and project a top view.
Scale 1:1

12.48

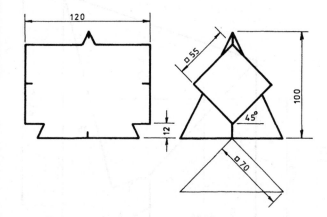

Given the side view and incomplete front view of a square pyramid penetrating a square prism, draw the lines of intersection in the front view and project a top view. Also draw the development of the square prism.
Scale 1:1

12.49

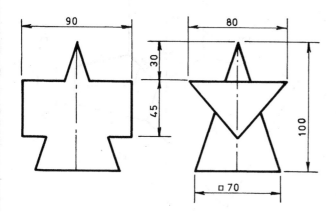

Given the side view and incomplete front view of a square pyramid penetrating a triangular prism, draw the lines of intersection in the front view and project a top view.
Scale 1:1

12.50

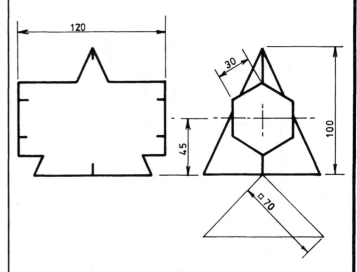

Given the side view and incomplete front view of a square pyramid penetrating a hexagonal prism, draw the lines of intersection in the front view and project a top view.
Scale 1:1

12.51

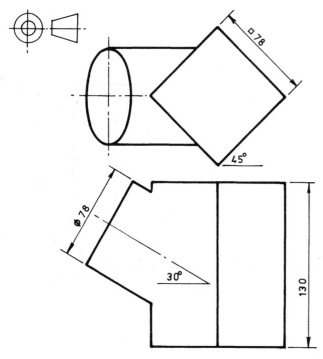

Complete the front view showing the lines of intersection and project a left side view. Also draw the development of the cylinder.
Scale 1:1

12.52

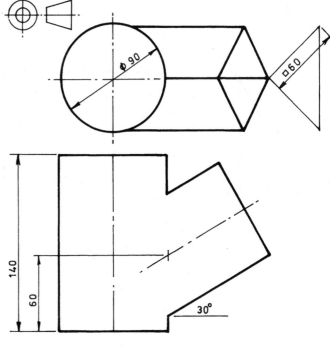

Complete the front view showing the lines of intersection and project a right side view.
Scale 1:1

12.53

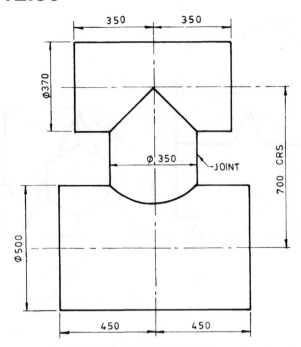

The view above is a front view of two parallel cylindrical pipes connected by another cylindrical pipe at right angles. Draw the front view, and complete the development of the connecting pipe.
Scale 1:10

12.54

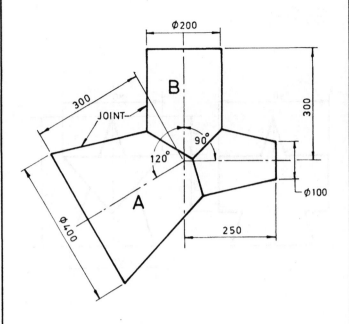

The front view of a breeches piece is given above. Draw this view, and complete the development of branches A and B.
Scale 1:5

12.55

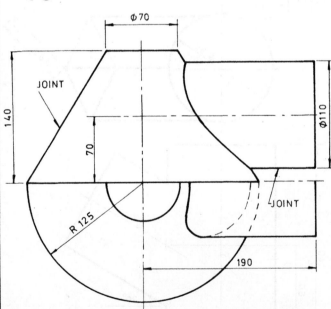

A combined front and half top view of a conical hood with a cylindrical branch is shown. Draw the development of both.
Scale 1:2

12.56

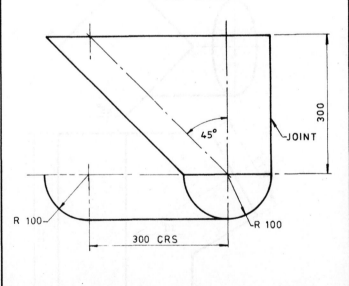

A combined front and half top view of a shoe-shaped funnel is shown above. Develop the pattern for the funnel.
Scale 1:5

12.57

A view of a right-angled cylindrical bend with a cylindrical branch is given. Complete this view, showing the line of intersection, and draw the development of the cylindrical branch.
Scale 1:1

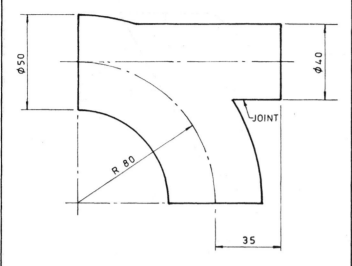

12.58

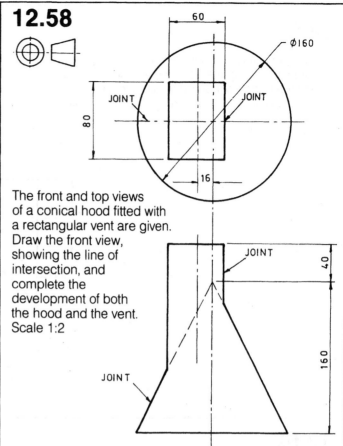

The front and top views of a conical hood fitted with a rectangular vent are given. Draw the front view, showing the line of intersection, and complete the development of both the hood and the vent.
Scale 1:2

12.59

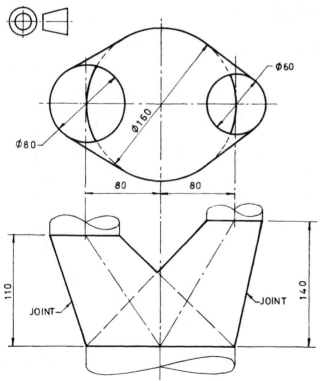

The front and the top views of a junction piece for three unequal cylindrical pipes are shown.
1. Draw the given front and top views, showing the line of intersection.
2. Develop the patterns for the two arms of the junction piece.
Scale 1:2

12.60

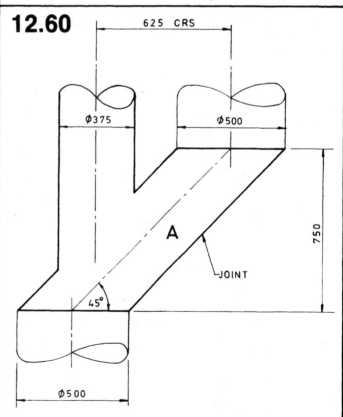

An oblique cylindrical connecting piece fitted with a cylindrical branch is shown. Draw this view, showing the line of intersection of the two pipes, and draw the development of pipe A.
Scale 1:10

12.61

An oblique cylindrical connecting piece is given.
1. Draw the complete front and top views of the cylinder and oblique cone intersection.
2. Develop both the oblique cone and the cylindrical branch.

Scale 1:5

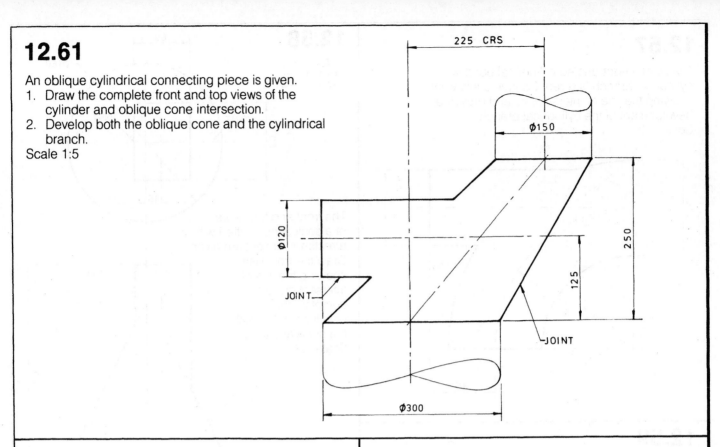

225 CRS

Ø150

Ø120

JOINT

250

125

JOINT

Ø300

12.62

Two views of a transition piece are given. Draw the development.
Scale 1:10

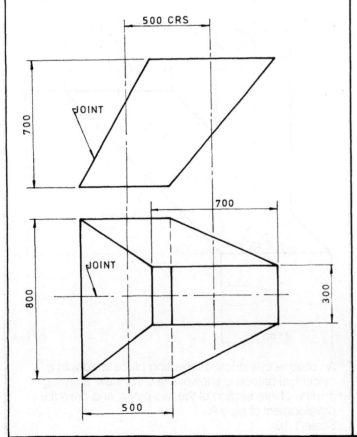

500 CRS

700

JOINT

700

800

JOINT

300

500

12.63

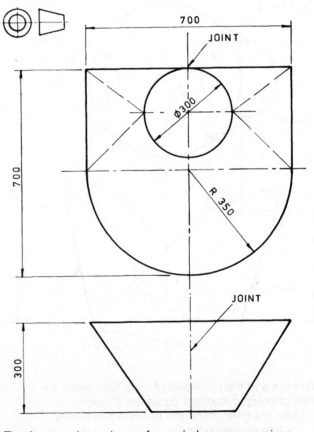

700

JOINT

Ø300

R 350

700

JOINT

300

JOINT

The front and top views of a grain hopper are given. Develop the pattern necessary to make the hopper.
Scale 1:10

12.64

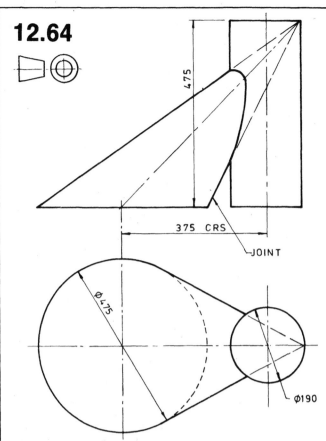

The front and top views of an oblique conical hood fitting into a cylindrical pipe are given. Draw the front view, showing the line of intersection, and complete the development of the hood.
Scale 1:5

12.65

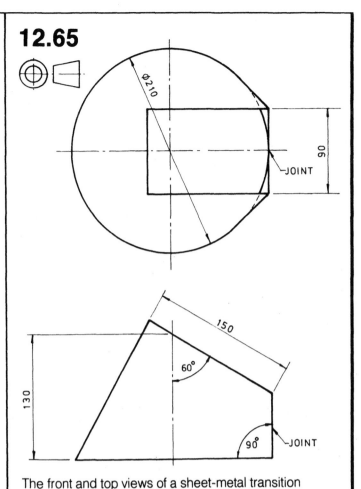

The front and top views of a sheet-metal transition piece are given. Draw its development.
Scale 1:2

12.66

The front and top views of an oblique conical connecting pipe with an oblique cylindrical branch are given.
1. Complete the front and top views, showing the line of intersection.
2. Develop the connecting pipe and the branch.
Scale 1:2

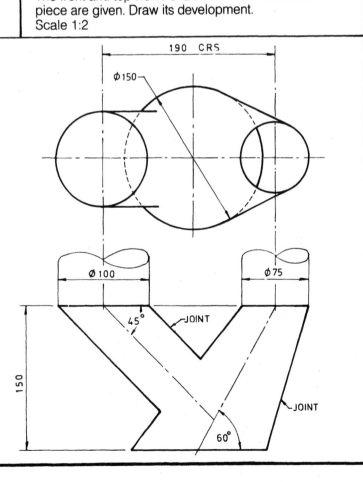

359